Lecture Notes in Artificial Intelligence   12256

Subseries of Lecture Notes in Computer Science

Series Editors

Randy Goebel
  University of Alberta, Edmonton, Canada
Yuzuru Tanaka
  Hokkaido University, Sapporo, Japan
Wolfgang Wahlster
  DFKI and Saarland University, Saarbrücken, Germany

Founding Editor

Jörg Siekmann
  DFKI and Saarland University, Saarbrücken, Germany

More information about this series at http://www.springer.com/series/1244

Vicenç Torra · Yasuo Narukawa ·
Jordi Nin · Núria Agell (Eds.)

Modeling Decisions
for Artificial Intelligence

17th International Conference, MDAI 2020
Sant Cugat, Spain, September 2–4, 2020
Proceedings

Springer

Editors
Vicenç Torra 🆔
Department of Computing Science
Umeå University
Umeå, Sweden

Yasuo Narukawa 🆔
Department of Management Science
Tamagawa University
Tokyo, Japan

Jordi Nin 🆔
Department of Operations,
Innovation and Data Sciences
ESADE
Sant Cugat, Spain

Núria Agell 🆔
Department of Operations,
Innovation and Data Sciences
ESADE
Sant Cugat, Spain

ISSN 0302-9743   ISSN 1611-3349   (electronic)
Lecture Notes in Artificial Intelligence
ISBN 978-3-030-57523-6   ISBN 978-3-030-57524-3   (eBook)
https://doi.org/10.1007/978-3-030-57524-3

LNCS Sublibrary: SL7 – Artificial Intelligence

This Springer imprint is published by the registered company Springer Nature Switzerland AG
The registered company address is: Gewerbestrasse 11, 6330 Cham, Switzerland

Preface

This volume contains papers that were presented at the 17th International Conference on Modeling Decisions for Artificial Intelligence (MDAI 2020), in Sant Cugat del Vallès, Spain, September 2–4, 2020. Due to the COVID-19 pandemic the conference was canceled. Nevertheless, the submission process was already opened and we remained committed to the publication of the conference proceedings.

This conference followed MDAI 2004 (Barcelona), MDAI 2005 (Tsukuba), MDAI 2006 (Tarragona), MDAI 2007 (Kitakyushu), MDAI 2008 (Sabadell), MDAI 2009 (Awaji Island), MDAI 2010 (Perpinyà), MDAI 2011 (Changsha), MDAI 2012 (Girona), MDAI 2013 (Barcelona), MDAI 2014 (Tokyo), MDAI 2015 (Skövde), MDAI 2016 (Sant Julià de Lòria), MDAI 2017 (Kitakyushu), MDAI 2018 (Mallorca), and MDAI 2019 (Milan).

The aim of MDAI is to provide a forum for researchers to discuss different facets of decision processes in a broad sense. This includes model building and all kinds of mathematical tools for data aggregation, information fusion, and decision-making; tools to help make decisions related to data science problems (including, e.g., statistical and machine learning algorithms as well as data visualization tools); and algorithms for data privacy and transparency-aware methods so that data processing procedures and the decisions made from them are fair, transparent, and avoid unnecessary disclosure of sensitive information.

The MDAI conference included tracks on the topics of (a) data science, (b) machine learning, (c) data privacy, (d) aggregation functions, (e) human decision-making, (f) graphs and (social) networks, and (g) recommendation and search.

The organizers received 46 papers from 17 different countries, 24 of which are published in this volume. Each submission received at least three reviews from the Program Committee and a few external reviewers. We would like to express our gratitude to them for their work.

The conference was supported by ESADE-Institute for Data-Driven Decisions (esadeD3), the European Society for Fuzzy Logic and Technology (EUSFLAT), the Catalan Association for Artificial Intelligence (ACIA), the Japan Society for Fuzzy Theory and Intelligent Informatics (SOFT), and the UNESCO Chair in Data Privacy.

July 2020

Vicenç Torra
Yasuo Narukawa
Jordi Nin
Núria Agell

Organization

General Chairs

Jordi Nin	ESADE, Universitat Ramon Llull, Spain
Núria Agell	ESADE, Universitat Ramon Llull, Spain

Program Chairs

Vicenç Torra	Umeå University, Sweden
Yasuo Narukawa	Tamagawa University, Japan

Advisory Board

Didier Dubois	Institut de Recherche en Informatique de Toulouse, CNRS, France
Jozo Dujmović	San Francisco State University, USA
Lluis Godo	IIIA-CSIC, Spain
Kaoru Hirota	Beijing Institute of Technology, JSPS Beijing Office, China
Janusz Kacprzyk	Systems Research Institute, Polish Academy of Sciences, Poland
Sadaaki Miyamoto	University of Tsukuba, Japan
Sandra Sandri	Instituto Nacional de Pesquisas Espaciais, Brazil
Michio Sugeno	Tokyo Institute of Technology, Japan
Ronald R. Yager	Machine Intelligence Institute, Iona Collegue, USA

Program Committee

Laya Aliahmadipour	Shahid Bahonar University, Iran
Esteve Almirall	ESADE, Spain
Plamen Angelov	Lancaster University, UK
Eva Armengol	IIIA-CSIC, Spain
Edurne Barrenechea	Universidad Pública de Navarra, Spain
Gloria Bordogna	Consiglio Nazionale delle Ricerche, Italy
Humberto Bustince	Universidad Pública de Navarra, Spain
Alina Campan	North Kentucky University, USA
Francisco Chiclana	De Montfort University, UK
Susana Díaz	Universidad de Oviedo, Spain
Josep Domingo-Ferrer	Universitat Rovira i Virgili, Spain
Yasunori Endo	University of Tsukuba, Japan
Vladimir Estivill-Castro	Griffith University, Australia
Zoe Falomir	Universität Bremen, Germany

Additional Referees

Jordi Casas
Julián Salas
Najeeb Jebreel
Sergio Martinez Lluis
Ashneet Khandpur Singh
Rami Haffar

Supporting Institutions

ESADE-Institute for Data-Driven Decisions
The European Society for Fuzzy Logic and Technology (EUSFLAT)
The Catalan Association for Artificial Intelligence (ACIA)
The Japan Society for Fuzzy Theory and Intelligent Informatics (SOFT)
The UNESCO Chair in Data Privacy

Contents

Aggregation Operators and Decision Making

A Characterization of Belief Merging Operators in the Regular Horn Fragment of Signed Logic

Pilar Dellunde[1,2,3(⊠)]

[1] Universitat Autònoma de Barcelona, Barcelona, Spain
pilar.dellunde@uab.cat
[2] Barcelona Graduate School of Mathematics, Barcelona, Spain
[3] Artificial Intelligence Research Institute IIIA-CSIC, Barcelona, Spain

Abstract. In this paper, we present a set of logical postulates for belief merging in the set of all regular signed Horn formulas. Proving a representation result, we show that regular Horn merging in signed logic can be seen as an aggregation problem on rankings of possible interpretations in this many-valued setting.

Keywords: Belief merging · Belief change · Signed logic · Horn clause

1 Introduction

Belief change is an active area of artificial intelligence, offering a common methodology for the understanding of different processes such as revision [1], update [13] or contraction [1]. Belief merging [3,18] shares also this methodology and studies the combination of consistent knowledge bases (possibly mutually inconsistent) in order to obtain a single consistent knowledge base. In general, the knowledge bases considered in all these works are expressed in full classical propositional logic. However, recent important contributions to the field have restricted the study of belief changes to some fragments of propositional logic with good computational properties. For instance, to the Horn fragment, that affords very efficient algorithms, being the Horn clauses a natural way to represent facts and rules. For a reference of these works see [7,9,20] in revision, [6,10,22] in contraction, and [8,11] in merging.

Following recent trends in the field, our present work is a contribution to the study of belief merging in the Horn fragment of many-valued logics. In [5,21] an abstract framework was introduced for the revision of fuzzy belief bases. Our research differs from [21], because we take a semantic approach, not a syntactic one, and also because we focus on the Horn fragment of signed logic, instead of giving a general account of merging for the whole logic. In order to study merging in many-valued logics, we have chosen to start with signed logic. Signed logic was introduced as a generic treatment of many-valued logics (see [12]), being the basic underlying idea to generalise the classical notion of a literal, attaching

© Springer Nature Switzerland AG 2020
V. Torra et al. (Eds.): MDAI 2020, LNAI 12256, pp. 3–15, 2020.
https://doi.org/10.1007/978-3-030-57524-3_1

a sign or label (which in general consists of a finite set of values) to an atomic formula. Signed formulas are a logical language for knowledge representation that lies in the intersection of the areas of constraint programming, many-valued logic, and annotated logic programming, this last formalism can be embedded into signed logic (see [16]). A *knowledge base* can be expressed by a finite set of signed Horn propositional formulas, defining a *profile* as a nonempty finite multiset of consistent, but not necessarily mutually consistent knowledge bases. It is useful to think of a profile as a multiset of agents, represented by their sets of beliefs. Each agent's belief induces a ranking on possible states of affairs, which can be thought of as the way in which the agent ranks possible worlds in terms of their plausibility. Merging is then the task of finding a common ranking that approximates, as best as possible, the individual rankings.

The motivation for our research on many-valued Horn merging comes from a previous work. In [2] a database with information about the quality of life of people with intellectual disabilities was analyzed. The database included the answers to the GENCAT questionnaire [17]. The questionnaire had 69 questions divided into 8 blocks, one for each dimension of quality of life, according to the Shalock and Verdugo model in [17]: emotional well-being (EW), interpersonal relations (IR), material well-being (MW), personal development (PD), physical well-being (PW), self-determination (SD), social inclusion (SI) and rights (RI). In the study presented in [2], using machine learning techniques, we obtained a set of rules regarding the contribution of the different dimensions to the global quality of life of a person. All these rules were naturally expressed by many-valued Horn formulas. Discussion with the social practitioners gave rise to sets of rules different from the ones obtained by the machine, sometimes, mutually inconsistent. Our aim to find a method to merge these databases, brought us to define new merging operators. The results we present in this paper are the first steps of a general program that involves different stages. In the following list, steps 1–3 describe the original contributions of the present article (we put in parenthesis the number of the sections where the results can be found) steps 4–6 are ongoing research and future work.

1. Characterization of the class of models of a Horn signed formula in terms of the closure of this class under certain operations (Proposition 1 Sect. 3).
2. Prove a sufficient condition for a signed Horn merging operator to satisfy the logical IC-postulates, showing that signed Horn merging can be seen as an aggregation problem on rankings of signed interpretations (Proposition 2 Sect. 4).
3. Characterization of a signed Horn merging operator in terms of syncretic assignments with total preorders on the class of all possible interpretations (Theorem 1 Sect. 4).
4. Definition of different signed Horn merging operators, by giving a notion of distance between signed interpretations (to induce preorders for each knowledge base), and using aggregation functions (to combine the individual rankings into a final preorder for the profile).

5. Apply the signed Horn operators obtained to merge knowledge-bases in real-world aggregation problems.
6. Generalization of the theoretical results to other many-valued logics.

2   Preliminaries

In this first section we introduce the syntax and semantics of signed logic. Since in this paper we focus on the study of belief merging in regular Horn signed formulas, we introduce the definition and basics facts related to the regular Horn fragment of signed logic

Syntax. A set of values N is a linearly ordered nonempty finite set $N = \{i_0, \ldots, i_n\}$, we will denote the smallest and largest elements in N by 0 and 1, respectively. A sign S is a subset of N. Let $\mathcal{P}$ be a nonempty finite set of propositional variables $\mathcal{P} = \{p_0, \ldots, p_k\}$. A *signed literal* is an expression of the form $S : p$, where S is a sign and p a propositional variable. The informal meaning is p *takes one of the values in* S. The *complement of a signed literal* $S : p$ is $N - S : p$, sometimes denoted by $\overline{S} : p$. A *signed clause* is a finite set of signed literals (understood as a classical disjunction). A signed formula in *conjunctive normal form* (CNF) is a finite conjunction of signed clauses. When N is considered to be a truth value set, signed CNF formulas turn out to be a generic representation for finite-valued logics. The problem of deciding the satisfiability of formulas (SAT problem) of any finite-valued logic is in a natural way polynomially reducible to the problem of deciding satisfiability of formulas in signed CNF (signed SAT).

Semantics. A signed interpretation I is a function $I : \mathcal{P} \to N$. We denote by $\mathcal{W}$ the set of all signed interpretations. A signed interpretation I of the propositional variables satisfies a signed literal $S : p$, if $I(p) \in S$; *satisfies a signed clause*, if I satisfies at least one of its signed literals; and *satisfies a signed CNF*, if I satisfies all its signed clauses. If a signed interpretation I satisfies a signed formula φ, we say that I is a *model* of φ, and we denote by $[\varphi]$ the set of models of φ.

Signed Regular Horn Formulas. Let $\uparrow i$ denote the sign $\{j \in N : j \geq i\}$, for some $i \in N$; and let $\downarrow i$ denote the sign $\{j \in N : j \leq i\}$, for some $i \in N$. It is said that a sign S is *positive regular*, if it is of the form $\uparrow i$, for some $i \in N$; and it is said to be *negative regular*, if it is of the form $\downarrow i$, for some $i \in N$. A *signed regular clause* is a signed clause with all the signs regular (positive or negative). A *signed regular Horn clause* is a regular clause containing at most one signed regular positive literal. A *signed regular Horn formula* is a signed CNF formula consisting solely of signed regular Horn clauses. The set of all signed regular Horn clauses is called the *regular Horn fragment of signed logic*.

Example: Let $\mathcal{P} = \{p_0, \ldots, p_7, q_v\}$ be the set of variables corresponding to the 8 dimensions, and to the global index of quality of life of the database of [2]. The set of values $N = \{0, 0.25, 0.5.0.75, 1\}$ with the usual linear order represents different grades of the quality of life dimensions. For instance, the fact that the interpersonal relations of a person are *very high* is represented by assigning to the corresponding variable the number 1 (*very-low* by 0, *low* by 0.25...). The clause $\downarrow 0.75 : p_5 \vee \uparrow 0.5 : p_1 \vee \downarrow 0.75 : q_v$ symbolizes *If interpersonal relations are very high, and self-determination is low or very-low, then the quality of life is not very high*, equivalent to a disjunctive form without negations, because the negation of $\uparrow 1 : p_5$ is equivalent to $\downarrow 0.75 : p_5$, the negation of $\downarrow 0.25 : p_1$ is equivalent to $\uparrow 0.5 : p_1$, and the negation of $\uparrow 1 : q_v$ is equivalent to $\downarrow 0.75 : q_v$.

3   Models of Signed Regular Horn Formulas

In this section we give a characterization of the class of models of a signed Horn formula in terms of the closure of the set of its models under the minimum operation. In order to do that, first we present a transformation, first introduced in [4], of signed regular Horn formulas into classical Horn formulas. In [4], it was proved that the transformation is linear in the size of the transformed formula and quadratic in the size of the truth-value lattice. For the sake of clarity, from now on, we will refer to signed regular Horn formulas simply as *signed Horn formulas*.

Transformation of Regular Signed Horn formulas into Classical Horn Formulas. Given a set of propositional variables $\mathcal{P}$, we define $\mathcal{P}^*$ to be the set of propositional variables $r_{\uparrow i:p}$ such that $i \in N$, $i > 0$ and $p \in \mathcal{P}$. Let $\mathcal{SH}$ be the set of signed Horn $\mathcal{P}$-formulas that do not contain literals either of the form $\downarrow 1 : p$ or of the form $\uparrow 0 : p$ in their clauses. Let $\mathcal{CH}$ be the set of classical Horn $\mathcal{P}^*$-formulas. Observe that every signed Horn formula is equivalent to one in $\mathcal{SH}$.

Definition 1. *The transformation* $*: \mathcal{SH} \to \mathcal{CH}$ *is defined as follows: for every signed Horn formula* $\phi = C_1 \wedge \cdots \wedge C_k$, *where each* C_i *is a signed Horn clause of the form* $\downarrow j_1 : p_1 \vee \cdots \vee \downarrow j_s : p_s \vee \uparrow l : q$, *we take* $C_i^* = \neg r_{\uparrow j_1+1:p_1} \vee \cdots \vee \neg r_{\uparrow j_s+1:p_s} \vee r_{\uparrow l:q}$ *and* $\phi^* = C_1^* \wedge \cdots \wedge C_k^*$.

Notice that the transformation * is well-defined because the formulas in $\mathcal{SH}$ do not contain literals either of the form $\downarrow 1 : p$, or of the form $\uparrow 0 : p$. It is possible to define also an inverse transformation:

Definition 2. *The transformation from* $\# : \mathcal{CH} \to \mathcal{SH}$ *is defined as follows: for every classical Horn formula* $\psi = T_1 \wedge \cdots \wedge T_k$, *where each* T_i *is a classical Horn clause of the form* $T_i = \neg r_{\uparrow j_1:p_1} \vee \cdots \vee \neg r_{\uparrow j_s:p_s} \vee r_{\uparrow l:q}$, *we take* $T_i^{\#} = \downarrow j_1 - 1 : p_1 \vee \cdots \vee \downarrow j_s - 1 : p_s \vee \uparrow l : q$ *and* $\phi^* = T_1^{\#} \wedge \cdots \wedge T_k^{\#}$.

The inverse transformation $\#$ is also well-defined because literals of the form $r_{\uparrow 0:p}$ are not contained in the classical Horn formulas in $\mathcal{CH}$. Notice that both transformations are injective. Observe also that to every signed interpretation $I : \mathcal{P} \to N$ we can associate a classical interpretation $I^* : \mathcal{P}^* \to \{0,1\}$ in the following way: $I^*(r_{\uparrow i:p}) = 1$, if $I(p) \geq i$, and $= 0$, otherwise. However, not every classical interpretation is associated to a signed interpretation, only those that satisfy the classical Horn formula D defined as follows: for every $i, j \in N$ with $i < j$, and every propositional variable $p \in \mathcal{P}$, let $D_{i,j}^p$ be the classical Horn clause $\neg r_{\uparrow j:p} \vee r_{\uparrow i:p}$. We denote by D the conjunction of all the $D_{i,j}^p$ clauses. In Theorem 12 of [4], it is proved that, for every signed Horn formula in $\mathcal{SH}$, ϕ is satisfiable iff $\phi^* \wedge D$ is satisfiable.

Characterization of the Class of Models of a Signed Horn Formula. Let I_1 and I_2 be interpretations (either classical or signed), we denote by $\min(I_1, I_2)$ the pairwise minimum interpretation, that is, for every propositional variable p, $\min(I_1, I_2)(p) = \min\{I_1(p), I_2(p)\}$. Given a set of interpretations $\mathcal{M}$, we denote by $Cl_{\min}(\mathcal{M})$ the closure of $\mathcal{M}$ under the minimum operation. Let us prove a preliminary lemma, before the introduction of the characterization result.

Lemma 1. *Let $\mathcal{P}$ be a set of propositional variables, and $\mathcal{M}$ be a set of signed interpretations over $\mathcal{P}$. If $Cl_{\min}(\mathcal{M}) = \mathcal{M}$, then $Cl_{\min}(\mathcal{M}^*) = \mathcal{M}^*$, where $\mathcal{M}^*$ is the set of classical interpretations $\{I^* : I \in \mathcal{M}\}$.*

Proof. It is enough to show that, given I_1 and I_2, two signed interpretations over $\mathcal{P}$, $(\min(I_1, I_2))^* = \min(I_1^*, I_2^*)$. For every propositional variable p we have $(\min(I_1, I_2))^*(r_{\uparrow i:p}) = 1$ iff $\min(I_1, I_2)(p) \geq i$, iff $\min\{I_1(p), I_2(p)\} \geq i$, iff $I_1^*(r_{\uparrow i:p}) = 1$ and $I_2^*(r_{\uparrow i:p}) = 1$. Finally, this happens iff $\min(I_1^*, I_2^*)(r_{\uparrow i:p}) = 1$.

Proposition 1. *Let $\mathcal{P}$ be a set of propositional variables. A set of signed interpretations $\mathcal{M}$ over $\mathcal{P}$ is the set of models of a signed Horn formula over $\mathcal{P}$ if and only if $Cl_{\min}(\mathcal{M}) = \mathcal{M}$.*

Proof. First assume that $Cl_{\min}(\mathcal{M}) = \mathcal{M}$. Then, by Lemma 1, $Cl_{\min}(\mathcal{M}^*) = \mathcal{M}^*$. By the known result of Schaefer [19] on the closure of the class of models of a classical Horn formula, $\mathcal{M}^*$ is the set of models of a Horn formula α over $\mathcal{P}^*$. We have that α is a conjunction of classical Horn clauses $\{C_1, \ldots, C_k\}$. Remark that all the models in $\mathcal{M}^*$ are models of the classical Horn formula D, mentioned before when we were speaking of the transformation. Consider now the signed Horn clause $C_i^\#$ of Definition 2. Take φ to be the conjunction of the clauses $\{C_1^\#, \ldots, C_k^\#\}$, it is easy to check that $\mathcal{M}$ is the set of models of φ. For the other direction, let $\mathcal{M}$ be the set of models of a signed Horn formula φ. We have that φ is a conjunction of Horn clauses of the form $\{D_1, \ldots, D_t\}$. For every $1 \leq i \leq t$, the clause D_i is of the form $\downarrow j_1 : p_1 \vee \cdots \vee \downarrow j_s : p_s \vee \uparrow l : q$. Given two signed interpretations, I_1 and I_2, models of φ, I_1 and I_2 are also models of D_i. If $I_1(q) \geq l$ and $I_2(q) \geq l$, then $\min(I_1, I_2)(q) \geq l$, and therefore $\min(I_1, I_2)$ is a model of D_i. Otherwise, without loss of generality, there is $1 \leq m \leq s$ such that $I_1(p_m) \leq j_m$. In this case, $\min(I_1, I_2)(p_m) \leq j_m$ and thus, $\min(I_1, I_2)$ is a model of D_i. Consequently, applying the same argument to all the clauses, we have that $\min(I_1, I_2)$ is a model of φ, and thus $\min(I_1, I_2) \in \mathcal{M}$.

For the sake of clarity, sometimes we present the interpretations as tuples of the following form $I = (j_0, \ldots, j_k)$, where for every $0 \leq l \leq k$, $I(p_l) = j_l$. When a class of signed interpretations $\mathcal{M}$ is the set of models of a formula

φ, it is said that $\mathcal{M}$ is *representable* by φ. For example, consider the set of truth-values $N = \{0, 0.5, 1\}$, and the set of propositional variables $\mathcal{P} = \{p, q\}$. Let $\mathcal{M} = \{(0, 0.5), (1, 0)\}$, then $Cl_{\min}(\mathcal{M}) = \{(0, 0.5), (1, 0), (0, 0)\}$, and the set $Cl_{\min}(\mathcal{M})$ is representable by the following signed Horn formula

$$\downarrow 0.5 : q \wedge (\downarrow 0 : p \vee \uparrow 1 : p) \wedge (\downarrow 0.5 : p \vee \downarrow 0 : q)$$

Remark that every signed interpretation is representable by a signed Horn formula. For instance the singleton $\{(0, 0.5)\}$ is representable by $\downarrow 0 : p \wedge \downarrow 0.5 : q \wedge \uparrow 0.5 : q$

4   Logical Postulates for Signed Horn Merging

In this section we present a set of logical postulates for belief merging under constraints in signed logic. The logical postulates for classical belief merging we take as a reference were introduced in [14, 15] but, as explained in [8], if we restrict to the Horn fragment, it is not possible in general to use the operators in [14, 15] as they stand. It could be the case that the databases we want to merge are expressed by Horn formulas, but the result of merging these databases is not expressible by a Horn formula. The same happens in many-valued logic, for this reason some notions have to be adapted, and new postulates have to be introduced. A *knowledge base* is a finite set of signed Horn formulas. A *profile* is a nonempty finite multiset $\{K_1, \ldots K_n\}$ of consistent, but not necessarily mutually consistent knowledge bases. Given a knowledge base K we denote by $[K]$ the set of all models of K, by $[E]$ the set of all models of E (that is, the intersection of all sets of models $[K_i]$ of the knowledge bases K_i of the profile) and by $D \sqcup E$ the multiset union of D and E. Given two profiles D, E, it is said that D and E are *equivalent* (in symbols $D \equiv E$), if there is a bijection $f : D \to E$ such that for every $K \in D$, $[K] = [f(K)]$. A signed Horn merging operator Δ is a function that maps every profile E, and a signed Horn formula μ called the *constraint*, to a knowledge base $\Delta_\mu(E)$. Let $\mathcal{K}$ be the set of all knowledge bases, and $\mathcal{E}$ the set of all profiles. Let us now introduce the logical postulates of [14, 15], and prove a sufficient condition for a signed Horn merging operator to satisfy them, later we will introduce some new postulates. Let $D, E \in \mathcal{E}$, $K, M \in \mathcal{K}$, $[K, M] = [K] \cap [M]$, and μ, μ' signed Horn formulas:

(**IC$_0$**) $\Delta_\mu(E) \models \mu$
(**IC$_1$**) If μ is consistent, then $\Delta_\mu(E)$ is consistent
(**IC$_2$**) If $\bigwedge E$ is consistent with μ, then $\Delta_\mu(E) \equiv \bigwedge E \wedge \mu$
(**IC$_3$**) If $D \equiv E$ and $\mu \equiv \mu'$, then $\Delta_\mu(D) \equiv \Delta_{\mu'}(E)$
(**IC$_4$**) If $K \models \mu$ and $M \models \mu$, then $\Delta_\mu([K, M]) \wedge K$ is consistent if and only if $\Delta_\mu([K, M]) \wedge M$ is consistent.
(**IC$_5$**) $\Delta_\mu(D) \wedge \Delta_\mu(E) \models \Delta_\mu(D \sqcup E)$
(**IC$_6$**) If $\Delta_\mu(D) \wedge \Delta_\mu(E)$ is consistent, then $\Delta_\mu(D \sqcup E) \models \Delta_\mu(D) \wedge \Delta_\mu(E)$
(**IC$_7$**) $\Delta_\mu(E) \wedge \mu' \models \Delta_{\mu \wedge \mu'}(E)$
(**IC$_8$**) If $\Delta_\mu(E) \wedge \mu'$ is consistent, then $\Delta_{\mu \wedge \mu'}(E) \models \Delta_\mu(E) \wedge \mu'$

The set of postulates represent the conditions that any rational merging operator should satisfy: postulate $(\mathbf{IC_0})$ guarantees that the merging result satisfies the constraint μ; $(\mathbf{IC_1})$ ensures that the result is consistent if μ is consistent; $(\mathbf{IC_2})$ states that the result is exactly the conjunction of the given knowledge bases whenever this is possible; $(\mathbf{IC_3})$ is the principle of irrelevance of syntax; $(\mathbf{IC_4})$ expresses fairness: if exactly two knowledge bases are merged, none should be preferred over the other; $(\mathbf{IC_5})$ and $(\mathbf{IC_6})$ combined state that if two profiles agree on at least one alternative, then merging the combination of these two profiles results in exactly those alternatives; $(\mathbf{IC_7})$ and $(\mathbf{IC_8})$ express conditions on handling conjunctions of constraints.

Now we show that Horn merging can be seen as an aggregation problem on rankings of the interpretations. Given two signed interpretations I_1 and I_2, it is said that $I_1 \subseteq I_2$, if for every $p \in \mathcal{P}$, $I_1(p) \leq I_2(p)$. A *preorder* $\leq$ on the set of interpretations $\mathcal{W}$ is a binary, reflexive and transitive relation on $\mathcal{W}$. A preorder is *total*, if for any interpretations I_1, I_2, either $I_1 \leq I_2$ or $I_2 \leq I_1$. We will use in the proof of our results the notion of *Horn compliance* introduced in [9], allowing only those rankings that are coherent with the semantics of Horn formulas

Definition 3. *A preorder $\leq$ on the set of all interpretations $\mathcal{W}$ is Horn compliant, if for any signed Horn formula μ, $\min_\leq[\mu]$ is representable by a signed Horn formula, where $\min_\leq[\mu]$ is the set of minimal models of $[\mu]$ wrt. the preorder $\leq$.*

Given a Horn compliant preorder $\leq$, we denote by $\phi_{min_\leq[\mu]}$ the signed Horn formula that represents the set $\min_\leq[\mu]$. Moreover, giving two interpretations I, J, we denote by $I \approx_E J$ the fact that $I \leq_E J$ and $J \leq_E I$; and by $I <_E J$ the fact that $I \leq_E J$ but it is not the case that $I \leq_E J$. Let us now introduce the definition of syncretic assignment of [14,15], in order to prove our first proposition:

Definition 4. *Let s be a function that maps every profile in $\mathcal{E}$ to a preorder $\leq_E$ on the set of all interpretations $\mathcal{W}$. It is said that s is a* syncretic assignment *if and only if for any $D, E \in \mathcal{E}$, $K, M \in \mathcal{K}$, and $I, J \in \mathcal{W}$, the following conditions hold:*

$(\mathbf{s_1})$ *If $I, J \in [E]$, then $I \approx_E J$*
$(\mathbf{s_2})$ *If $I \in [E]$ and $J \notin [E]$, then $I <_E J$*
$(\mathbf{s_3})$ *If $D \equiv E$, then $\leq_D = \leq_E$*
$(\mathbf{s_4})$ *If $I \in [K]$, then there is $J \in [M]$ such that $J \leq_{[K,M]} I$*
$(\mathbf{s_5})$ *If $I \leq_D J$, and $I \leq_E J$, then $I \leq_{D \sqcup E} J$*
$(\mathbf{s_6})$ *If $I \leq_D J$ and $I <_E J$, then $I <_{D \sqcup E} J$*

Intuitively speaking, these conditions state that all models of the profile are equally preferred $(\mathbf{s_1})$; that the models of the profile are preferred compared to other interpretations $(\mathbf{s_2})$; the principle of irrelevance of syntax $(\mathbf{s_3})$; fairness (that is, if exactly two knowledge bases are involved, no model of one of these knowledge bases should be preferred over all the models of the other) $(\mathbf{s_4})$; and the fact that if two profiles agree on the ordering of two interpretations, then the combination of these profiles, orders these interpretations in the same way $(\mathbf{s_5})$ and $(\mathbf{s_6})$.

Proposition 2. *Assume that there is a syncretic assignment mapping each profile E to a Horn compliant total preorder $\leq_E$. Then the operator Δ defined as: for every profile E, and every signed Horn formula μ, $\Delta_\mu(E) = \phi_{min_{\leq_E}[\mu]}$ is a signed Horn merging operator that satisfies postulates $(\mathbf{IC_0}) - (\mathbf{IC_8})$.*

Proof. Assume that there is a syncretic assignment mapping each profile E to a Horn compliant total preorder $\leq_E$. Let Δ be defined as: for every profile E, and every signed Horn formula μ, $\Delta_\mu(E) = \phi_{min_{\leq_E}[\mu]}$. Observe that $\Delta_\mu(E)$ is a signed Horn formula, because $\leq_E$ is Horn compliant, and thus Δ is a signed Horn merging operator. Let us see now that Δ satisfies the logical postulates: $(\mathbf{IC_0})$: By definition, $\Delta_\mu(E) \models \mu$. $(\mathbf{IC_1})$: If μ is consistent, since $[\mu]$ is finite, $\phi_{min_{\leq_E}[\mu]}$ is also consistent. $(\mathbf{IC_2})$: Assume that $\bigwedge E$ is consistent with μ. By $(\mathbf{s_1})$ and $(\mathbf{s_1})$, it is easy to check that $\Delta_\mu(E) \equiv \bigwedge E \wedge \mu$.

$(\mathbf{IC_3})$: Clear by $(\mathbf{s_3})$. $(\mathbf{IC_4})$: Assume that $K \models \mu$, $M \models \mu$, and $\Delta_\mu([K,M]) \wedge K$ is consistent. If $I \in [\Delta_\mu([K,M])] \cap [K]$, by $(\mathbf{s_4})$, there is $J \in [M]$ such that $J \leq_{[K,M]} I$. Since $M \models \mu$, we have that $J \in [\Delta_\mu([K,M])]$, and thus $\Delta_\mu([K,M]) \wedge M$ is consistent.

$(\mathbf{IC_5})$: If $I \in [\Delta_\mu(D) \wedge \Delta_\mu(E)]$, then for every $J \in [\mu]$, $I \leq_D J$ and $I \leq_E J$. Consequently, by $(\mathbf{s_5})$, $I \leq_{D \sqcup E} J$, and thus $I \in [\Delta_\mu(D \sqcup E)]$.

$(\mathbf{IC_6})$: Assume that $\Delta_\mu(D) \wedge \Delta_\mu(E)$ is consistent. Let $I \in [\Delta_\mu(D \sqcup E)]$ and, searching for a contradiction, assume that $I \notin [\Delta_\mu(D) \wedge \Delta_\mu(E)]$. Without loss of generality, suppose that $I \notin [\Delta_\mu(D)]$. Let $J \in [\Delta_\mu(D) \wedge \Delta_\mu(E)]$, by $(\mathbf{IC_5})$, $J \in [\Delta_\mu(D \sqcup E)]$. Then we have $J <_D I$, $J \leq_E I$ and, by $(\mathbf{s_6})$, $J <_{D \sqcup E} I$, contradicting the fact that $I \approx_{D \sqcup E} J$ because $I, J \in [\Delta_\mu(D \sqcup E)]$.

$(\mathbf{IC_7})$: Clear by definition of Δ. $(\mathbf{IC_8})$: Assume $\Delta_\mu(E) \wedge \mu'$ is consistent. Let $I \in [\Delta_{\mu \wedge \mu'}(E)]$ and, searching for a contradiction, assume that $I \notin [\Delta_\mu(E)]$. Let $J \in [\Delta_\mu(E)] \cap [\mu']$. Then we have both $J \leq_E I$, because $I \in [\mu]$, and $I \leq_E J$, because $J \in [\mu \wedge \mu']$. However, by $(\mathbf{s_2})$, this contradicts the fact that $I \notin [\Delta_\mu(E)]$. $\square$

We would like to go beyond Proposition 2, and prove a sufficient and necessary condition for a merging operator to satisfy postulates $(\mathbf{IC_0}) - (\mathbf{IC_8})$, but we are subject to the same restrictions explained in [11]. We need to introduce one new postulate $(\mathbf{T})$, and one new condition $(\mathbf{s_T})$ for the assignments. The new postulate and condition essentially ensure us that the preorders will be total. First we present some additional notation. Given a sequence of sets of signed interpretations $L_1, \ldots L_n$, we denote by $\langle L_1, \ldots L_n \rangle$ the closure under the minimum operation of their union, that is, $\langle L_1, \ldots L_n \rangle = Cl_{min}(L_1 \cup \cdots \cup L_n)$. We denote by $\phi_{\langle L_1, \ldots L_n \rangle}$ the signed Horn formula that represents the set $\langle L_1, \ldots L_n \rangle$. If for every $0 < i \leq n$, $L_i = \{I_i\}$ is a singleton, we write $\langle I_1, \ldots I_n \rangle$ instead of $\langle L_1, \ldots L_n \rangle$.

Theorem 1. *Let $(\mathbf{T})$ be the postulate: for every profile E, and signed interpretations, I, J, either $I \in [\Delta_{\phi_{\langle I,J \rangle}}(E)]$ or $J \in [\Delta_{\phi_{\langle I,J \rangle}}(E)]$. And let $(\mathbf{s_T})$ be the condition: for every profile E, and every signed interpretations, I, J, either*

$I \in min_{\leq_E}\langle I, J\rangle$ or $J \in min_{\leq_E}\langle I, J\rangle$. Then, a signed Horn merging operator Δ satisfies the postulates $(\mathbf{IC_0}) - (\mathbf{IC_8}) + (\mathbf{T})$ if and only if there is a syncretic assignment satisfying $(\mathbf{s_T})$ such that to each profile E assigns a Horn compliant total preorder $\leq_E$ with the property that for every signed Horn formula μ, $[\Delta_\mu(E)] = min_{\leq_E}[\mu]$.

Proof. $(\Rightarrow)$ Assume that Δ is a signed Horn merging operator that satisfies postulates $(\mathbf{IC_0}) - (\mathbf{IC_8}) + (\mathbf{T})$. First we define for every profile E, the relation $\leq_E$ as follows: given two signed interpretations I, J, $I \leq_E J$ if and only if $I \in [\Delta_{\phi_{\langle I,J\rangle}}(E)]$. We show that $\leq_E$ is a total preorder: since every signed interpretation I is representable by a signed Horn formula, we have $\Delta_{\phi_{\langle I,I\rangle}}(E) = \Delta_{\phi_I}(E)$, where ϕ_I is the formula representing I. By $(\mathbf{IC_1})$, $\Delta_{\phi_I}(E)$ is consistent, and by $(\mathbf{IC_0})$, I is its unique model. Therefore $I \in [\Delta_{\phi_{\langle I,I\rangle}}(E)]$ and thus, by definition of the relation $\leq_E$, $I \leq_E I$, showing that the relation is reflexive. Moreover, by postulate $(\mathbf{T})$, $\leq_E$ is total.

Let us see that $\leq_E$ is transitive. Assume $I \leq_E J$, $J \leq_E L$ and, searching for a contradiction, that $I \not\leq_E L$. For the proof of the transitivity property, we need to prove first these five statements:

(1) $I \notin [\Delta_{\phi_{\langle I,J,L\rangle}}(E)]$: otherwise, by $(\mathbf{IC_7})$ $[\Delta_{\phi_{\langle I,J,L\rangle}}(E)] \cap \langle I, L\rangle \subseteq [\Delta_{\phi_{\langle I,L\rangle}}(E)]$, and then $I \in [\Delta_{\phi_{\langle I,L\rangle}}(E)]$ contradicting the fact that $I \not\leq_E L$.

(2) $[\Delta_{\phi_{\langle I,J,L\rangle}}(E)] \cap \langle I, J\rangle = \emptyset$: otherwise, by $(\mathbf{IC_7})$ and $(\mathbf{IC_8})$, $[\Delta_{\phi_{\langle I,J,L\rangle}}(E)] \cap \langle I, J\rangle = [\Delta_{\phi_{\langle I,J\rangle}}(E)]$. Then, since $I \leq_E J$, we have $I \in [\Delta_{\phi_{\langle I,J\rangle}}(E)]$, and consequently, $I \in [\Delta_{\phi_{\langle I,J,L\rangle}}(E)]$, contradicting (1).

(3) $[\Delta_{\phi_{\langle I,J,L\rangle}}(E)] \cap \langle J, L\rangle = \emptyset$: otherwise, by $(\mathbf{IC_7})$ and $(\mathbf{IC_8})$, $[\Delta_{\phi_{\langle I,J,L\rangle}}(E)] \cap \langle J, L\rangle = [\Delta_{\phi_{\langle J,L\rangle}}(E)]$. Then, since $J \leq_E L$, we have $J \in [\Delta_{\phi_{\langle J,L\rangle}}(E)]$, and consequently, $J \in [\Delta_{\phi_{\langle I,J,L\rangle}}(E)] \cap \langle I, J\rangle$, contradicting (2).

(4) $[\Delta_{\phi_{\langle I,J,L\rangle}}(E)] \cap \langle I, L\rangle = \emptyset$: otherwise, by $(\mathbf{IC_7})$ and $(\mathbf{IC_8})$, $[\Delta_{\phi_{\langle I,J,L\rangle}}(E)] \cap \langle I, L\rangle = [\Delta_{\phi_{\langle I,L\rangle}}(E)]$. Then, since $L \leq_E I$ (because we have proved that the relation $\leq_E$ is total, and by assumption $I \not\leq_E L$), we have $L \in [\Delta_{\phi_{\langle I,L\rangle}}(E)]$. Consequently, $L \in [\Delta_{\phi_{\langle I,J,L\rangle}}(E)] \cap \langle J, L\rangle$ contradicting (3).

(5) $[\Delta_{\phi_{\langle I,J,L\rangle}}(E)] \cap \langle min(I,J), min(J,L)\rangle = \emptyset$: otherwise, by $(\mathbf{IC_7})$ and $(\mathbf{IC_8})$, $[\Delta_{\phi_{\langle I,J,L\rangle}}(E)] \cap \langle min(I,J), min(J,L)\rangle = [\Delta_{\phi_{\langle min(I,J),min(J,L)\rangle}}(E)]$. Since the relation $\leq_E$ is total, either $min(I,J) \leq_E min(J,L)$ or $min(J,L) \leq_E min(I,J)$. Assume without loss of generality that $min(I,J) \leq_E min(J,L)$. Then, $min(I,J) \in [\Delta_{\phi_{\langle min(I,J),min(J,L)\rangle}}(E)]$, contradicting (2).

Now by (1)–(5) we can conclude that $[\Delta_{\phi_{\langle I,J,L\rangle}}(E)] = \emptyset$, contradicting $(\mathbf{IC_1})$ (because $\langle I, J, L\rangle \neq \emptyset$). Therefore we have shown that the relation $\leq_E$ is also transitive, and thus a preorder. Now let us see that the assignment is syncretic, that is, that satisfies conditions $(\mathbf{s_1})$–$(\mathbf{s_6})$.

$(\mathbf{s_1})$: assume $I, J \in [E]$. Then $[E] \cap \langle I, J\rangle \neq \emptyset$, and by $(\mathbf{IC_2})$, $[\Delta_{\phi_{\langle I,J\rangle}}(E)] = [E] \cap \langle I, J\rangle$. Therefore $I, J \in [\Delta_{\phi_{\langle I,J\rangle}}(E)]$, and thus $I \approx_E J$.

$(\mathbf{s_2})$: assume $I \in [E]$ and $J \notin [E]$. Then $[E] \cap \langle I, J\rangle \neq \emptyset$, and by $(\mathbf{IC_2})$, $[\Delta_{\phi_{\langle I,J\rangle}}(E)] = [E] \cap \langle I, J\rangle$. Therefore $I \in [\Delta_{\phi_{\langle I,J\rangle}}(E)]$ but $J \notin [\Delta_{\phi_{\langle I,J\rangle}}(E)]$, and thus $I <_E J$.

(s_3): if $D \equiv E$, we have that $[D] = [E]$. Thus, by (**IC$_3$**), $[\Delta_{\phi_{\langle I,J \rangle}}(D)] = [\Delta_{\phi_{\langle I,J \rangle}}(E)]$. Then, if $I \leq_D J$, $I \in [\Delta_{\phi_{\langle I,J \rangle}}(D)] = [\Delta_{\phi_{\langle I,J \rangle}}(E)]$, and thus $I \leq_E J$. We can conclude that $\leq_D = \leq_E$.

(s_4): Let K, M be knowledge bases. Assume that $I \in [K]$ and, searching for a contradiction, that there is no $J \in [M]$ such that $J \leq_{\{K,M\}} I$. To get a contradiction from this assumption, let us prove first these four facts:

(**A**) $[\Delta_{\phi_{\langle [K],[M] \rangle}}(K,M)] \neq \emptyset$: by (**IC$_1$**), because $I \in \langle [K],[M] \rangle$.

(**B**) $[\Delta_{\phi_{\langle [K],[M] \rangle}}(K,M)] \cap [M] = \emptyset$: otherwise there will be $J \in [\Delta_{\phi_{\langle [K],[M] \rangle}}(K,M)] \cap [M]$. Then $J \in [\Delta_{\phi_{\langle [K],[M] \rangle}}(K,M)] \cap \langle I, J \rangle$, and by (**IC$_7$**) and (**IC$_8$**),

$$[\Delta_{\phi_{\langle [K],[M] \rangle}}(K,M)] \cap \langle I,J \rangle = [\Delta_{\phi_{\langle I,J \rangle}}(K,M)]$$

consequently, $J \in [\Delta_{\phi_{\langle I,J \rangle}}(K,M)]$. But the fact that $J \leq_{\{K,M\}} I$ contradicts our original assumption that there was not such J.

(**C**) $[\Delta_{\phi_{\langle [K],[M] \rangle}}(K,M)] \cap [K] = \emptyset$: by (**IC$_8$**) and (**B**).

(**D**) For every signed interpretation S, if $S \in \langle [K],[M] \rangle$, then

$$S \notin [\Delta_{\phi_{\langle [K],[M] \rangle}}(K,M)]$$

The closure by the minimum operation of a set of interpretations is defined inductively, allowing us to provide a proof by induction. For the initial step the property holds by (**B**) and (**C**). Now assume inductively that, for two signed interpretations $S, T \in \langle [K],[M] \rangle$ the property holds, and searching for a contradiction, that $min(S,T) \in [\Delta_{\phi_{\langle [K],[M] \rangle}}(K,M)]$. Then, by (**IC$_7$**) and (**IC$_8$**),

$$[\Delta_{\phi_{\langle [K],[M] \rangle}}(K,M)] \cap \langle S,T \rangle = [\Delta_{\phi_{\langle S,T \rangle}}(K,M)]$$

Since $\leq_{\{K,M\}}$ is a total preorder, assume without loss of generality, that $S \leq_{\{K,M\}} T$. Then $S \in [\Delta_{\phi_{\langle S,T \rangle}}(K,M)]$, and consequently, $S \in [\Delta_{\phi_{\langle [K],[M] \rangle}}(K,M)]$ contradicting the inductive hypothesis. Now taking into account the facts (**A**) − (**D**) we have just proved, by (**IC$_0$**) $[\Delta_{\phi_{\langle [K],[M] \rangle}}(K,M)] \subseteq \langle [K],[M] \rangle$. But then the fact (**A**) contradicts (**D**). Therefore we can conclude that exists $J \in [M]$ such that $J \leq_{\{K,M\}} I$.

(s_5): assume that $I \leq_D J$ and $I \leq_E J$. By definition, $I \in [\Delta_{\phi_{\langle I,J \rangle}}(D)]$ and $I \in [\Delta_{\phi_{\langle I,J \rangle}}(E)]$. Then, by (**IC$_5$**) $I \in [\Delta_{\phi_{\langle I,J \rangle}}(D \sqcup E)]$, and thus $I \leq_{D \sqcup E} J$.

(s_6): assume that $I \leq_D J$, $I <_E J$ and, searching for a contradiction, that $I \not\leq_{D \sqcup E} J$. Then, by (**IC$_5$**) and (**IC$_6$**), since $I \in [\Delta_{\phi_{\langle I,J \rangle}}(D)] \cap [\Delta_{\phi_{\langle I,J \rangle}}(E)]$, we have $[\Delta_{\phi_{\langle I,J \rangle}}(D \sqcup E)] = [\Delta_{\phi_{\langle I,J \rangle}}(ED)] \cap [\Delta_{\phi_{\langle I,J \rangle}}(E)]$. Now, since $\leq_{D \sqcup E}$ is a total preorder, and we have assumed that $I \not\leq_{D \sqcup E} J$, we have $J \leq_{D \sqcup E} I$. But then $J \in [\Delta_{\phi_{\langle I,J \rangle}}(D \sqcup E)] = [\Delta_{\phi_{\langle I,J \rangle}}(D)] \cap [\Delta_{\phi_{\langle I,J \rangle}}(E)]$, contradicting our original assumptions.

In order to finish the proof of the ($\Rightarrow$) direction of the theorem, we have to prove three more facts, the first is that for every profile E, and signed Horn formula μ, $[\Delta_\mu(E)] = min_{\leq_E}[\mu]$. Without loss of generality, let μ be consistent, and take $I \in [\Delta_\mu(E)]$. Then for every $J \in [\mu]$, $I \in [\Delta_\mu(E)] \cap \langle I, J \rangle$, and by $(\mathbf{IC_7})$ and $(\mathbf{IC_8})$, $[\Delta_\mu(E)] \cap \langle I, J \rangle = [\Delta_{\langle I,J \rangle}(E)]$. Thus $I \in [\Delta_{\langle I,J \rangle}(E)]$, and consequently, for every $J \in [\mu]$, $I \leq_E J$, and therefore $I \in min_{\leq_E}[\mu]$.

For the other inclusion, assume that $I \in min_{\leq_E}[\mu]$. Since μ is consistent, by $(\mathbf{IC_0})$ and $(\mathbf{IC_1})$, $[\Delta_\mu(E)] \neq \emptyset$ and $[\Delta_\mu(E)] \subseteq [\mu]$. Let $J \in [\Delta_\mu(E)]$. Then, by $(\mathbf{IC_7})$ and $(\mathbf{IC_8})$, $[\Delta_\mu(E)] \cap \langle I, J \rangle = [\Delta_{\langle I,J \rangle}(E)]$. Since $I \in min_{\leq_E}[\mu]$, $I \leq_E J$, and thus, $I \in [\Delta_{\langle I,J \rangle}(E)]$. We can conclude that $I \in [\Delta_\mu(E)]$.

Now we can prove $(\mathbf{s_T})$: by $(\mathbf{T})$, given two signed interpretations, I, J, either $I \in [\Delta_{\phi_{\langle I,J \rangle}}(E)]$ or $J \in [\Delta_{\phi_{\langle I,J \rangle}}(E)]$. Since we have just proved in general that $[\Delta_\mu(E)] = min_{\leq_E}[\mu]$, for any signed Horn formula μ, we have that either $I \in min_{\leq_E}\langle I, J \rangle$ or $J \in min_{\leq_E}\langle I, J \rangle$.

Finally, we conclude this part of the proof showing that the preorder $\leq_E$ is Horn compliant. For every signed Horn formula μ, since $[\Delta_\mu(E)] = min_{\leq_E}[\mu]$ and Δ is a Horn operator, we have that $min_{\leq_E}[\mu]$ is representable by a signed Horn formula.

$(\Leftarrow)$ Assume that there is a syncretic assignment satisfying $(\mathbf{s_T})$ and such that to each profile E assigns a Horn compliant total preorder $\leq_E$. Let us check that the operator Δ defined as: for every profile E, and signed Horn formula μ, $\Delta_\mu(E) = \phi_{min_{\leq_E}[\mu]}$ satisfies postulates $(\mathbf{IC_0}) - (\mathbf{IC_8}) + (\mathbf{T})$. By Proposition 2, Δ is a signed Horn merging operator satisfying postulates $(\mathbf{IC_0}) - (\mathbf{IC_8})$. Now, by $(\mathbf{s_T})$, for every two signed interpretations, I, J, either $I \in min_{\leq_E}\langle I, J \rangle$ or $J \in min_{\leq_E}\langle I, J \rangle$. Consequently either $I \in [\Delta_{\phi_{\langle I,J \rangle}}(E)]$ or $J \in [\Delta_{\phi_{\langle I,J \rangle}}(E)]$ and thus, postulate $(\mathbf{T})$ is also satisfied.

5 Future Work

There are different lines of research that can continue the work presented in this paper. Firstly, we would like to define different signed Horn merging operators, and evaluate which are more suitable to work in a many-valued setting. Due to the model-based approach we adhere to, new notions of distance have to be introduced to induce preorders for each knowledge base. We will start by studying a generalization of the Hamming distance. We will also compare different fuzzy aggregation functions in order to combine the individual rankings into a final preorder for a profile. Secondly, we would like to apply the signed Horn operators obtained, to merge knowledge-bases we have obtained in previous works. We consider that this application will shed light to the relevance of the different distances and aggregation functions proposed in the theoretical study. Finally, we would like to generalize the theoretical results to other many-valued formalisms, starting with Gödel logic, that seems to us to be a good candidate, due to the possibility to define, in a natural way, distances between its interpretations. We plan to open the study to other fragments of many-valued logics with good computational properties, and to other graded logics.

Acknowledgements. The author would like to thank S. Rümmele and E. Lee for their contributions to a preliminary draft of this paper, and the referees for their useful comments. This article has received funding from the project 2018-Recercaixa AppPhil.

References

1. Alchourrnón, C.E., Gärdenfors, P., Makinson, M.: On the logic of theory change: partial meet contraction and revision functions. J. Symb. Log. **50**, 2 (1985)
2. Armengol, E., García-Cerdaña, À., Dellunde, P.: Experiences using decision trees for knowledge discovery. In: Torra, V., Dahlbom, A., Narukawa, Y. (eds.) Fuzzy Sets, Rough Sets, Multisets and Clustering. SCI, vol. 671, pp. 169–191. Springer, Cham (2017). https://doi.org/10.1007/978-3-319-47557-8_11
3. Baral, Ch., Kraus, S., Minker, J.: Combining multiple knowledge bases. IEEE Trans. Knowl. Data Eng. **3**, 2 (1991)
4. Beckert, B., Hähnle, R., Manyà, F.: Transformations between signed and classical clause logic. In: Proceedings of the 29th ISMVL, pp. 248–255 (1999)
5. Booth, R., Richter, E.: On revising fuzzy belief bases. In: Proceedings of the UAI, pp. 81–88 (2003)
6. Booth, R., Meyer, T.A., Varzinczak, I.J., Wassermann, R.: On the link between partial meet, kernel, and infra contraction and its application to Horn logic. J. Artif. Intell. Res. (JAIR) **42**, 31–53 (2011)
7. Creignou, N., Papini, O., Pichler, R., Woltran, S.: Belief revision within fragments of propositional logic. J. Comput. Syst. Sci. **80**(2), 427–449 (2014)
8. Creignou, N., Papini, O., Pichler, R., Rümmele, S., Woltran, S.: Belief merging within fragments of propositional logic. ACM Trans. Comput. Log. **17**(3), 20 (2016)
9. Delgrande, J.P., Peppas, P.: Belief revision in Horn theories. Artif. Intell. **218**, 1–22 (2015)
10. Delgrande, J.P., Wassermann, R.: Horn clause contraction functions. J. Artif. Intell. Res. **48**, 475–511 (2013)
11. Haret, A., Rümmele, S., Woltran, S.: Merging in the Horn fragment. ACM Trans. Comput. Log. **18**(1), 6 (2017)
12. Hähnle, R.: Automated Deduction in Multiple-Valued Logics. Oxford University Press, Oxford (1994)
13. Katsuno, H., Mendelzon, A.O.: On the difference between updating a knowledge base and revising it. In: Proceedings of KR, pp. 387–394 (1991)
14. Konieczny, S., Pino, R.: Merging information under constraints: a logical framework. J. Log. Comput. **12**(5), 773–808 (2002)
15. Konieczny, S., Pino, R.: Logic based merging. J. Philos. Log. **40**(2), 239–270 (2011)
16. Lu, J.J., Murray, N.V., Rosenthal, E.: Signed formulas and annotated logics. In: Proceedings of the 20th ISMVL, pp. 48–53 (1993)
17. Verdugo, M.V., Arias, B., Gómez, L.E., Schalock, R.L.: Formulari de l'Escala GEN-CAT de Qualitat de Vida. Departament d'Acció Social i Ciutadania, Generalitat de Catalunya (2008)
18. Revesz, P.Z.: On the semantics of theory change: arbitration between old and new information. In: Proceedings of the 12th ACM SIGACT-SIGMOD-SIGART Symposium on Principles of Database Systems, pp. 71–82 (1993)
19. Schaefer, T.J.: The complexity of satisfiability problems. In: Proceedings of STOC, pp. 216–226 (1978)
20. Van De Putte, F.: Prime implicates and relevant belief revision. J. Log. Comput. **23**(1), 109–119 (2013)

21. Witte, R.: Fuzzy belief revision. In: Proceedings of NMR, pp. 311–320 (2002)
22. Zhuang, Z.Q., Pagnucco, P.: Model based Horn contraction. In: Proceedings of KR, pp. 169–178 (2012)

Bivariate Risk Measures
and Stochastic Orders

Yuji Yoshida[✉]

Faculty of Economics and Business Administration, University of Kitakyushu,
4-2-1 Kitagata, Kokuraminami, Kitakyushu 802-8577, Japan
yoshida@kitakyu-u.ac.jp

Abstract. Bivariate value-at-risks and bivariate average value-at-risks
are defined with copula functions, and stochastic orders are induced
from these bivariate value-at-risks and bivariate average value-at-risks.
Value-at-risk order is almost equivalent to the first-order stochastic dominance, however it is made clear that average value-at-risk order is weaker
than the second-order stochastic dominance and it has similar properties.
We find that average value-at-risk order is an important criterion as a
stochastic order in multi-object case-based reasoning with risk aversion.

1 Introduction

The first-order stochastic dominance (FSD) is used commonly to compare
stochastic events, and the second-order stochastic dominance (SSD) is also an
important stochastic order which is well-known in relation to risk aversity in economics and which is applicable to pairwise comparison in case-based reasoning
with stochastic risks [1,6]. This paper deals with these stochastic dominances
for random vectors. Value-at-risks (VaR) and average value-at-risks (AVaR) are
known as risk measures in finance [2]. This paper defines stochastic orders $\preceq_{\mathrm{VaR}}$
and $\preceq_{\mathrm{AVaR}}$ which are induced from value-at-risks and average value-at-risks,
and we investigate their properties. We make clear relations among the induced
value-at-risk order $\preceq_{\mathrm{VaR}}$ and average value-at-risk order $\preceq_{\mathrm{AVaR}}$, the first-order
stochastic dominance $\preceq_{\mathrm{FSD}}$, the second-order stochastic dominance $\preceq_{\mathrm{SSD}}$ and
the characterization by weighted quasi-arithmetic means [8,10,11].

For multi-object decision making, using copula functions [5], we define bivariate value-at-risks and bivariate average value-at-risks and we induce stochas-
tic orders $\preceq_{\mathrm{VaR}}$ and $\preceq_{\mathrm{AVaR}}$ from bivariate value-at-risks and bivariate average
value-at-risks. We also adopt the second-order stochastic dominance for ran-
dom vectors from the view point of risk estimation of random events in finance
[7]. Value-at-risk order $\preceq_{\mathrm{VaR}}$ is almost equivalent to the first-order stochastic
dominance $\preceq_{\mathrm{FSD}}$, however we will find that average value-at-risk order $\preceq_{\mathrm{AVaR}}$
is weaker than the second-order stochastic dominance $\preceq_{\mathrm{SSD}}$ and it has similar
properties. We find that average value-at-risk order $\preceq_{\mathrm{AVaR}}$ is an important criterion, which is useful for various multi-object case-based reasoning as a stochastic
order with risk aversion. The proofs of this paper is also given in Appendix.

© Springer Nature Switzerland AG 2020
V. Torra et al. (Eds.): MDAI 2020, LNAI 12256, pp. 16–27, 2020.
https://doi.org/10.1007/978-3-030-57524-3_2

2 Bivariate Value-at-Risks and Bivariate Average Value-at-Risks

In this section we introduce bivariate value-at-risks and bivariate average value-at-risks and we investigate their properties. Let $\mathbb{R} = (-\infty, \infty)$ and let I be a non-empty open interval in $\mathbb{R}$. Let (Ω, P) be a probability space, where P is a non-atomic probability measure on a sample space Ω. Let $\mathcal{X}$ be a family of random variables $X : \Omega \mapsto \mathbb{R}$ which have C^1-class density functions on I and which also satisfy the following tail condition:

$$\lim_{z \to \inf I} zP(X \leq z) = 0.$$

Value-at-risks and average value-at-risks are used to estimate risks of random variables (Artzner [2]). For a positive probability $p \in (0, 1]$, a *value-at-risk* $\mathrm{VaR}_p(X)$ of a random variable $X (\in \mathcal{X})$ is defined by

$$\mathrm{VaR}_p(X) = \sup\{x \in \mathbb{R} \mid P(X \leq x) \leq p\}, \tag{2.1}$$

and then an *average value-at-risk* $\mathrm{AVaR}_p(X)$ is also given by

$$\mathrm{AVaR}_p(X) = \frac{1}{p} \int_0^p \mathrm{VaR}_q(X)\, dq = E(X \mid X \leq \mathrm{VaR}_p(X)). \tag{2.2}$$

Let $D = I_1 \times I_2$ be a fixed domain in $\mathbb{R}^2$, where I_1 and I_2 are non-empty open intervals in $\mathbb{R}$. Denote by $\mathcal{X}^2$ a family of *random vectors* $(X_1, X_2) : \Omega \mapsto \mathbb{R}^2$ which have C^1-class density functions $w : D \mapsto [0, \infty)$ and which also satisfy the following tail condition:

$$\lim_{z \to \inf I_i} zP(X_i \leq z) = 0$$

for all $i = 1, 2$. For simplicity we put $w = 0$ on $\mathbb{R}^2 \setminus D$. For a random vector $(X_1, X_2) \in \mathcal{X}^2$, C^1-class marginal density functions for X_1 and X_2 are given respectively by

$$w_1(x_1) = \int_{-\infty}^{\infty} w(x_1, z)\, dz \quad \text{and} \quad w_2(x_2) = \int_{-\infty}^{\infty} w(z, x_2)\, dz \tag{2.3}$$

for $(x_1, x_2) \in \mathbb{R}^2$. In this paper, we assume for $\mathcal{X}^2$ there exists a C^2-class function $C : [0, 1]^2 \mapsto [0, 1]$ such that

$$C(P(X_1 \leq x_1), P(X_2 \leq x_2)) = P(X_1 \leq x_1, X_2 \leq x_2) \tag{2.4}$$

holds for all random vectors $(X_1, X_2) \in \mathcal{X}^2$ and $(x_1, x_2) \in \mathbb{R}^2$. Hence C is called a *copula function* (Nelson [5]). For a probability $p \in (0, 1]$, we also define a set of pairs of probabilities

$$\Gamma(p) = \{(p_1, p_2) \in (0, 1]^2 \mid C(p_1, p_2) = p\}. \tag{2.5}$$

Hence we use the following partial order $\preceq$ on $\mathbb{R}^2$, and then we introduce coherent risk measures on bivariate space $\mathcal{X}^2$ (Jouini et al. [3]).

Definition 2.1

(i) For two points $(x_1, x_2), (y_1, y_2) \in \mathbb{R}^2$, an order $(x_1, x_2) \preceq (y_1, y_2)$ implies $x_1 \leq y_1$ and $x_2 \leq y_2$.

(ii) For two sets $A, B(\subset \mathbb{R}^2)$, an order $A \preceq B$ implies the following (a) and (b):

    (a) For any $(x_1, x_2) \in A$ there exists $(y_1, y_2) \in B$ with $(x_1, x_2) \preceq (y_1, y_2)$.

    (b) For any $(y_1, y_2) \in B$ there exists $(x_1, x_2) \in A$ with $(x_1, x_2) \preceq (y_1, y_2)$.

Definition 2.2. A map $\rho : \mathcal{X}^2 \mapsto \mathbb{R}^2$ is called a *coherent risk measure* on $\mathcal{X}^2$ if it satisfies the following (i)–(iv):

(i) If $(X_1, X_2) \preceq (Y_1, Y_2)$, then $\rho(X_1, X_2) \succeq \rho(Y_1, Y_2)$ for $(X_1, X_2), (Y_1, Y_2) \in \mathcal{X}^2$. (*monotonicity*)

(ii) $\rho(\gamma X_1, \gamma X_2) = \gamma\rho(X_1, X_2)$ for $(X_1, X_2) \in \mathcal{X}^2$ and $\gamma > 0$. (*positively homogeneity*)

(iii) $\rho(X_1 + \gamma_1, X_2 + \gamma_2) = \rho(X_1, X_2) - (\gamma_1, \gamma_2)$ for $(X_1, X_2) \in \mathcal{X}^2$ and $(\gamma_1, \gamma_2) \in \mathbb{R}^2$. (*translation invariance*)

(iv) $\rho(X_1 + Y_1, X_2 + Y_2) \preceq \rho(X_1, X_2) + \rho(Y_1, Y_2)$ for $(X_1, X_2), (Y_1, Y_2) \in \mathcal{X}^2$. (*sub-additivity*)

From (2.1) and (2.2), for a random vector $(X_1, X_2) \in \mathcal{X}^2$ and a probability $p \in (0, 1]$ we define a *bivariate value-at-risk* $\mathrm{VaR}_p(X_1, X_2)$ by

$$\begin{aligned} \mathrm{VaR}_p(X_1, X_2) &= \{(x_1, x_2) \in \mathbb{R}^2 \mid P((X_1, X_2) \preceq (x_1, x_2)) = p\} \\ &= \{(x_1, x_2) \in \mathbb{R}^2 \mid P(X_1 \leq x_1, X_2 \leq x_2) = p\} \end{aligned} \tag{2.6}$$

and we also give a *bivariate average value-at-risk* $\mathrm{AVaR}_p(X_1, X_2)$ by

$$\mathrm{AVaR}_p(X_1, X_2) = E((X_1, X_2) \mid (X_1, X_2) \preceq \mathrm{VaR}_p(X_1, X_2)). \tag{2.7}$$

Now for copula function C we define a *copula density function* c on $(0, 1)^2$ as follows:

$$c(q_1, q_2) = \frac{\partial^2}{\partial q_1 \partial q_2} C(q_1, q_2) \tag{2.8}$$

for $(q_1, q_2) \in (0, 1)^2$. Then we have

$$C(p_1, p_2) = \iint_{[0, p_1] \times [0, p_2]} c(q_1, q_2)\, dq_1 dq_2 \tag{2.9}$$

for $(p_1, p_2) \in [0, 1]^2$ because $C(0, p_2) = C(p_1, 0) = 0$ from (2.4). Hence we have the following lemma for copula density function c.

Lemma 2.1. *Let a random vector* $(X_1, X_2) \in \mathcal{X}^2$. *If* $(x_1, x_2) \in \mathbb{R}^2$ *and* $(q_1, q_2) \in (0, 1)^2$ *satisfy relations* $x_1 = \mathrm{VaR}_{q_1}(X_1)$ *and* $x_2 = \mathrm{VaR}_{q_2}(X_2)$, *then the following (i)–(iii) hold:*

(i) $w(x_1, x_2)\, dx_1 dx_2 = c(q_1, q_2)\, dq_1 dq_2$.

(ii) $w_1(x_1)\, dx_1 = dq_1$ *and* $w_2(x_2)\, dx_2 = dq_2$.

(iii) $\dfrac{w(x_1, x_2)}{w_1(x_1) w_2(x_2)} = c(q_1, q_2)$.

Bivariate value-at-risks and bivariate average value-at-risks are represented by value-at-risks as follows.

Lemma 2.2. *Let a random vector* $(X_1, X_2) \in \mathcal{X}^2$ *and a probability* $p \in (0, 1]$. *Then the following (i) and (ii) hold:*

(i) *The value-at-risk* $\mathrm{VaR}_p(X_1, X_2)$ *is given by*

$$\mathrm{VaR}_p(X_1, X_2) = \{(\mathrm{VaR}_{p_1}(X_1), \mathrm{VaR}_{p_2}(X_2)) \mid (p_1, p_2) \in \Gamma(p)\}. \quad (2.10)$$

(ii) *The average value-at-risk* $\mathrm{AVaR}_p(X_1, X_2)$ *is represented as follows:*

$$\mathrm{AVaR}_p(X_1, X_2)$$
$$= \left\{ \frac{1}{p} \iint_{[0, p_1] \times [0, p_2]} (\mathrm{VaR}_{q_1}(X_1), \mathrm{VaR}_{q_2}(X_2)) \, c(q_1, q_2) \, dq_1 dq_2 \mid (p_1, p_2) \in \Gamma(p) \right\}. \quad (2.11)$$

Hence a case where

$$c(q_1, q_2) = 1 \quad \text{for all } (q_1, q_2) \in (0, 1)^2 \quad (2.12)$$

is equivalent to

$$X_1 \text{ and } X_2 \text{ are independent for all random vectors } (X_1, X_2) \in \mathcal{X}^2$$

(see Table 1). We call this *independent case*. Then (2.11) is reduced to

$$\mathrm{AVaR}_p(X_1, X_2) = \{(\mathrm{AVaR}_{p_1}(X_1), \mathrm{AVaR}_{p_2}(X_2)) \mid (p_1, p_2) \in \Gamma(p)\}. \quad (2.13)$$

Now we investigate properties of bivariate value-at-risks $\mathrm{VaR}_p(\cdot, \cdot)$ and bivariate average value-at-risks $\mathrm{AVaR}_p(\cdot, \cdot)$.

Lemma 2.3. *Let random vectors* $(X_1, X_2), (Y_1, Y_2) \in \mathcal{X}^2$ *and a probability* $p \in (0, 1]$. *For value-at-risk* $\mathrm{VaR}_p(\cdot, \cdot)$, *the following (i)–(iii) hold:*

(i) *If* $(X_1, X_2) \preceq (Y_1, Y_2)$, *then* $\mathrm{VaR}_p(X_1, X_2) \preceq \mathrm{VaR}_p(Y_1, Y_2)$.
(ii) $\mathrm{VaR}_p(\gamma X_1, \gamma X_2) = \gamma \, \mathrm{VaR}_p(X_1, X_2)$ *for* $\gamma > 0$.
(iii) $\mathrm{VaR}_p(X_1 + \gamma_1, X_2 + \gamma_2) = \mathrm{VaR}_p(X_1, X_2) + (\gamma_1, \gamma_2)$ *for* $(\gamma_1, \gamma_2) \in \mathbb{R}^2$.

Lemma 2.4. *Let random vectors* $(X_1, X_2), (Y_1, Y_2) \in \mathcal{X}^2$ *and a probability* $p \in (0, 1]$. *For average value-at-risk* $\mathrm{AVaR}_p(\cdot, \cdot)$, *the following (i)–(iv) hold:*

(i) *If* $(X_1, X_2) \preceq (Y_1, Y_2)$, *then* $\mathrm{AVaR}_p(X_1, X_2) \preceq \mathrm{AVaR}_p(Y_1, Y_2)$.
(ii) $\mathrm{AVaR}_p(\gamma X_1, \gamma X_2) = \gamma \, \mathrm{AVaR}_p(X_1, X_2)$ *for* $\gamma > 0$.
(iii) $\mathrm{AVaR}_p(X_1 + \gamma_1, X_2 + \gamma_2) = \mathrm{AVaR}_p(X_1, X_2) + (\gamma_1, \gamma_2)$ *for* $(\gamma_1, \gamma_2) \in \mathbb{R}^2$.
(iv) *If copula function* C *satisfies*

$$\frac{\partial^2}{\partial q_1^2} C(q_1, q_2) \le 0 \text{ and } \frac{\partial^2}{\partial q_2^2} C(q_1, q_2) \le 0 \quad \text{for all } (q_1, q_2) \in (0, 1)^2, \quad (2.14)$$

then $\mathrm{AVaR}_p(X_1 + Y_1, X_2 + Y_2) \succeq \mathrm{AVaR}_p(X_1, X_2) + \mathrm{AVaR}_p(Y_1, Y_2)$.

It is clear that independent case (2.12) satisfies (2.14). If copula function C satisfies (2.14), $\rho(\cdot, \cdot) = -\mathrm{AVaR}_p(\cdot, \cdot)$ is a coherent risk measure on $\mathcal{X}^2$ for any $p \in (0, 1]$ from Lemma 2.4 and Definition 2.2. Further popular copula functions C satisfying (2.14) are found in Table 1, and a lot of general copulas are also found in Nelson [5].

Table 1. Copula functions $C(p_1, p_2)$

Copula function Θ: Range of parameters θ Θ_1: Range of θ satisfying (2.14)	$C(p_1, p_2)$
Independent case (2.12)	$p_1 p_2$
Farlie-Gumbel-Morgentern family $\Theta = [-1, 1]$, $\Theta_1 = [0, 1]$	$p_1 p_2 (1 + \theta(1 - p_1)(1 - p_2))$
Clayton-Cook-Johnson family $\Theta = [-1, \infty) \setminus \{0\}$, $\Theta_1 = (0, \infty)$	$(p_1^{-\theta} + p_2^{-\theta} - 1)^{-1/\theta}$
Gumbel-Hougaad family $\Theta = [1, \infty)$, $\Theta_1 = [1, \infty)$	$\exp\left(-((-\log p_1)^\theta + (-\log p_2)^\theta)^{1/\theta}\right)$
Frank family $\Theta = \mathbb{R} \setminus \{0\}$, $\Theta_1 = (0, \infty)$	$\frac{1}{\theta} \log\left(1 + \frac{(e^{\theta p_1} - 1)(e^{\theta p_2} - 1)}{e^\theta - 1}\right)$
Gauss family $\Theta = (-1, 1)$, $\Theta_1 = \{0\}$	$\frac{1}{2\pi\sqrt{1 - \theta^2}} \int_{-\infty}^{\Phi^{-1}(p_1)} \int_{-\infty}^{\Phi^{-1}(p_2)} e^{-\frac{z_1^2 + z_2^2 - 2\theta z_1 z_2}{2(1 - \theta^2)}} \, dz_1 dz_2,$

$$\text{where } \Phi(x) = \frac{1}{\sqrt{2\pi}} \int_{-\infty}^{x} e^{-\frac{z^2}{2}} \, dz \quad \text{for } x \in \mathbb{R}.$$

3   Bivariate Value-at-Risks and Stochastic Dominances

In this section, we discuss relations among stochastic dominances, bivariate value-at-risks and weighted quasi-arithmetic means on two-dimensional regions. We say a random variable $X(\in \mathcal{X})$ is dominated by a random variable $Y(\in \mathcal{X})$ in the sense of *the first-order stochastic dominance* [7] if

$$P(X \leq x) \geq P(Y \leq x) \text{ for any } x \in \mathbb{R}. \tag{3.1}$$

Then we write it as $X \preceq_{\text{FSD}} Y$. And we also say a random variable $X(\in \mathcal{X})$ is dominated by a random variable $Y(\in \mathcal{X})$ in the sense of *the second-order stochastic dominance* [7] if

$$\int_{-\infty}^{x} P(X \leq z) \, dz \geq \int_{-\infty}^{x} P(Y \leq z) \, dz \text{ for any } x \in \mathbb{R}. \tag{3.2}$$

Then we write it as $X \preceq_{\text{SSD}} Y$. Hence from (3.1) the first-order stochastic dominance on bivariate space $\mathcal{X}^2$ is given as follows [7].

Definition 3.1. Let $(X_1, X_2), (Y_1, Y_2) \in \mathcal{X}^2$ be random vectors. (X_1, X_2) is dominated by (Y_1, Y_2) in the sense of *the first-order stochastic dominance* if

$$P((X_1, X_2) \preceq (x_1, x_2)) \geq P((Y_1, Y_2) \preceq (x_1, x_2)) \text{ for any } (x_1, x_2) \in \mathbb{R}^2. \text{ (3.3)}$$

Then we write it as $(X_1, X_2) \preceq_{\text{FSD}} (Y_1, Y_2)$.

We denote by $\mathcal{L}^2$ a family of all C^1-class strictly increasing functions f on domain $D = I_1 \times I_2$ with non-empty open intervals I_1 and I_2, i.e. $f_{x_1}(x_1, x_2) > 0$ and $f_{x_2}(x_1, x_2) > 0$ for $(x_1, x_2) \in D$ and $f(x_1, x_2) = 0$ for $(x_1, x_2) \in \mathbb{R}^2 \setminus D$ such that $f(X_1, X_2)$ are integrable for all random vectors $(X_1, X_2) \in \mathcal{X}^2$. We also denote by $\mathcal{L}_c^2$ a family of all C^2-class concave functions $f \in \mathcal{L}^2$. Then the following relations hold for the first-order stochastic dominance.

Theorem 3.1. *Let $(X_1, X_2), (Y_1, Y_2) \in \mathcal{X}^2$ be random vectors. The following (a), (b) and (c) are equivalent:*

(a) $(X_1, X_2) \preceq_{\text{FSD}} (Y_1, Y_2)$.
(b) $X_1 \preceq_{\text{FSD}} X_2$ *and* $Y_1 \preceq_{\text{FSD}} Y_2$.
(c) $E(f(X_1, X_2)) \leq E(f(Y_1, Y_2))$ *for all* $f \in \mathcal{L}^2$.

Next we investigate the relations between the first-order stochastic dominance and weighted quasi-arithmetic means on two-dimensional regions. Denote a family of rectangle regions by $\mathcal{R}(D) = \{R = J_1 \times J_2 \mid J_1 \text{ and } J_2 \text{ are bounded closed intervals and } R \subset D\}$. For a rectangle region $R \in \mathcal{R}(D)$, *weighted quasi-arithmetic means* $M_w^f(R)$ on region R with an increasing function $f \in \mathcal{L}^2$ and a weighting function $w : \mathbb{R}^2 \mapsto [0, \infty)$ are given by

$$\left\{ (\tilde{x}_1, \tilde{x}_2) \in R \mid f(\tilde{x}_1, \tilde{x}_2) \iint_R w(x_1, x_2) \, dx_1 dx_2 = \iint_R f(x_1, x_2) w(x_1, x_2) \, dx_1 dx_2 \right\}. \text{ (3.4)}$$

Then we have $M_w^f(R) \neq \emptyset$ since f is continuous on R. Hence the first-order stochastic dominance is characterized by density functions as follows.

Theorem 3.2. *Let $(X_1, X_2), (Y_1, Y_2) \in \mathcal{X}^2$ be random vectors with density functions w and v respectively.*

(i) *Let $f \in \mathcal{L}^2$. If $M_w^f(R) \preceq M_v^f(R)$ for all rectangle regions $R \in \mathcal{R}(D)$, then $E(f(X_1, X_2)) \preceq E(f(Y_1, Y_2))$.*
(ii) *If $M_w^f(R) \preceq M_v^f(R)$ for all $f \in \mathcal{L}^2$ and all rectangle regions $R \in \mathcal{R}(D)$, then $\frac{w_{x_1}}{w} \leq \frac{v_{x_1}}{v}$ and $\frac{w_{x_2}}{w} \leq \frac{v_{x_2}}{v}$ on D.*
(iii) *In independent case (2.12), the following (a), (b) and (c) are equivalent:*
    (a) $M_w^f(R) \preceq M_v^f(R)$ *for all* $f \in \mathcal{L}^2$ *and all rectangle regions* $R \in \mathcal{R}(D)$.
    (b) $\frac{w_{x_1}}{w} \leq \frac{v_{x_1}}{v}$ *and* $\frac{w_{x_2}}{w} \leq \frac{v_{x_2}}{v}$ *on* D.
    (c) $\frac{w_1'}{w_1} \leq \frac{v_1'}{v_1}$ *on* I_1 *and* $\frac{w_2'}{w_2} \leq \frac{v_2'}{v_2}$ *on* I_2.

Now in this paper we introduce stochastic orders induced from value-at-risks and average value-at-risks.

Definition 3.2

(i) For random variables $X, Y \in \mathcal{X}$, *value-at-risk order* $X \preceq_{\mathrm{VaR}} Y$ implies

$$\mathrm{VaR}_p(X) \preceq \mathrm{VaR}_p(Y) \text{ for all probabilities } p \in (0, 1]. \tag{3.5}$$

(ii) For random variables $X, Y \in \mathcal{X}$, *average value-at-risk order* $X \preceq_{\mathrm{AVaR}} Y$ implies

$$\mathrm{AVaR}_p(X) \preceq \mathrm{AVaR}_p(Y) \text{ for all probabilities } p \in (0, 1]. \tag{3.6}$$

(iii) For random vectors $(X_1, X_2), (Y_1, Y_2) \in \mathcal{X}^2$, we define *value-at-risk order* $(X_1, X_2) \preceq_{\mathrm{VaR}} (Y_1, Y_2)$ as follows:

$$\mathrm{VaR}_p(X_1, X_2) \preceq \mathrm{VaR}_p(Y_1, Y_2) \text{ for all probabilities } p \in (0, 1]. \tag{3.7}$$

(iv) For random vectors $(X_1, X_2), (Y_1, Y_2) \in \mathcal{X}^2$, we define *average value-at-risk order* $(X_1, X_2) \preceq_{\mathrm{AVaR}} (Y_1, Y_2)$ as follows:

$$\mathrm{AVaR}_p(X_1, X_2) \preceq \mathrm{AVaR}_p(Y_1, Y_2) \text{ for all probabilities } p \in (0, 1]. \tag{3.8}$$

Then we obtain the following relations for the first-order stochastic dominance and the value-at-risks.

Theorem 3.3. *Let* $(X_1, X_2), (Y_1, Y_2) \in \mathcal{X}^2$ *be random vectors. The following (a), (b) and (c) are equivalent:*

(a) $(X_1, X_2) \preceq_{\mathrm{FSD}} (Y_1, Y_2)$.
(b) $(X_1, X_2) \preceq_{\mathrm{VaR}} (Y_1, Y_2)$.
(c) $X_1 \preceq_{\mathrm{VaR}} Y_1$ *and* $X_2 \preceq_{\mathrm{VaR}} Y_2$.

4   Second Stochastic Dominances and Bivariate Average Value-at-risk Order

The original study of stochastic orders is the estimation of random risks in economics. The first-order stochastic dominances (3.1) shows comparison of the downside risks. Recently the study of stochastic orders is developed to various fields. It is well-known that the second-order stochastic dominance (3.2) is a weaker stochastic order than the first-order stochastic dominance (3.1), and it is related to the estimation by decision maker's risk averse utility (Levi [4]). Now we introduce a second-order stochastic dominance for random vectors [7].

Definition 4.1. Let $(X_1, X_2), (Y_1, Y_2) \in \mathcal{X}^2$ be random vectors. (X_1, X_2) is dominated by (Y_1, Y_2) in the sense of the (lower-orthant) *second-order stochastic dominance* if

$$\iint_{(-\infty, x_1] \times (-\infty, x_2]} P((X_1, X_2) \preceq (z_1, z_2)) dz_1 dz_2 \geq \iint_{(-\infty, x_1] \times (-\infty, x_2]} P((Y_1, Y_2) \preceq (z_1, z_2)) dz_1 dz_2 \tag{4.1}$$

for any $(x_1, x_2) \in \mathbb{R}^2$. Then we write it as $(X_1, X_2) \preceq_{\mathrm{SSD}} (Y_1, Y_2)$.

When we extend the second-order stochastic dominance (3.2) to a stochastic order for random vectors, there are a lot of approaches in general (See Shaked and Shanthikumar [7, Sect. 7.A.9]). However this paper adopts Definition 4.1 as the second-order stochastic dominance on $\mathcal{X}^2$ because it preserves good properties for estimation of random risks, i.e. monotonicity, positively homogeneity and translation invariance. In the rest of this section, we investigate these properties.

Lemma 4.1. *Let $X, Y \in \mathcal{X}$ be random variables.*

(i) *For stochastic order $\preceq_{\mathrm{FSD}}$, the following (a), (b) and (c) hold:*
   (a) *If $X \preceq Y$, then $X \preceq_{\mathrm{FSD}} Y$. (monotonicity)*
   (b) *If $X \preceq_{\mathrm{FSD}} Y$, then $\gamma X \preceq_{\mathrm{FSD}} \gamma Y$ for all $\gamma > 0$. (positively homogeneity*
   (c) *If $X \preceq_{\mathrm{FSD}} Y$, then $X + \gamma \preceq_{\mathrm{FSD}} Y + \gamma$ for all $\gamma \in \mathbb{R}$. (translation invariance)*
   Further stochastic orders $\preceq_{\mathrm{SSD}}$, $\preceq_{\mathrm{VaR}}$ and $\preceq_{\mathrm{AVaR}}$ also have the same properties, i.e. monotonicity, positively homogeneity and translation invariance.
(ii) *$X \preceq_{\mathrm{FSD}} Y$ if and only if $X \preceq_{\mathrm{VaR}} Y$.*
(iii) *If $X \preceq_{\mathrm{FSD}} Y$, then $X \preceq_{\mathrm{SSD}} Y$.*
(iv) *If $X \preceq_{\mathrm{SSD}} Y$, then $X \preceq_{\mathrm{AVaR}} Y$.*

Hence we obtain the following relations regarding the second-order stochastic dominance for random vectors.

Theorem 4.1. *Let $(X_1, X_2), (Y_1, Y_2) \in \mathcal{X}^2$ be random vectors.*

(i) *If $(X_1, X_2) \preceq_{\mathrm{FSD}} (Y_1, Y_2)$, then $(X_1, X_2) \preceq_{\mathrm{SSD}} (Y_1, Y_2)$.*
(ii) *If $E(f(X_1, X_2)) \le E(f(Y_1, Y_2))$ for all concave $f \in \mathcal{L}_c^2$, then $X_1 \preceq_{\mathrm{SSD}} Y_1$ and $X_2 \preceq_{\mathrm{SSD}} Y_2$.*
(iii) *In independent case (2.12), the following assertion holds: If $X_1 \preceq_{\mathrm{SSD}} Y_1$ and $X_2 \preceq_{\mathrm{SSD}} Y_2$, then $(X_1, X_2) \preceq_{\mathrm{SSD}} (Y_1, Y_2)$.*

Average value-at-risks are known as criteria for random risks. Now the following theorem implies average value-at-risk order $\preceq_{\mathrm{AVaR}}$ is weaker than the second-order stochastic dominance $\preceq_{\mathrm{SSD}}$ for random vectors.

Theorem 4.2. *Let $(X_1, X_2), (Y_1, Y_2) \in \mathcal{X}^2$ be random vectors.*

(i) *Stochastic orders $\preceq_{\mathrm{FSD}}$, $\preceq_{\mathrm{SSD}}$, $\preceq_{\mathrm{VaR}}$ and $\preceq_{\mathrm{AVaR}}$ have the properties of monotonicity, positively homogeneity and translation invariance.*
(ii) *If $(X_1, X_2) \preceq_{\mathrm{VaR}} (Y_1, Y_2)$, then $(X_1, X_2) \preceq_{\mathrm{AVaR}} (Y_1, Y_2)$.*
(iii) *In independent case (2.12), the following (a) and (b) are equivalent:*
   (a) *$(X_1, X_2) \preceq_{\mathrm{AVaR}} (Y_1, Y_2)$*
   (b) *$X_1 \preceq_{\mathrm{AVaR}} Y_1$ and $X_2 \preceq_{\mathrm{AVaR}} Y_2$.*
(iv) *In independent case (2.12), the following assertion holds: If $X_1 \preceq_{\mathrm{SSD}} Y_1$ and $X_2 \preceq_{\mathrm{SSD}} Y_2$, then $(X_1, X_2) \preceq_{\mathrm{AVaR}} (Y_1, Y_2)$.*

5   Conclusion

The first-order stochastic dominance $\preceq_{\mathrm{FSD}}$ and the second-order stochastic dominance $\preceq_{\mathrm{SSD}}$ are well-known as important stochastic orders in economics. In Definition 4.1, we have introduced $\preceq_{\mathrm{SSD}}$ for random vectors $\mathcal{X}^2$ to estimate downside risks. Then value-at-risk order $\preceq_{\mathrm{VaR}}$ is almost equivalent to $\preceq_{\mathrm{FSD}}$, however we have found average value-at-risk order $\preceq_{\mathrm{AVaR}}$ is weaker than $\preceq_{\mathrm{SSD}}$ and it has similar properties. These facts imply that average value-at-risk order $\preceq_{\mathrm{AVaR}}$ is an important stochastic order for economic risks and it is applicable in wider various fields. We expect average value-at-risk order $\preceq_{\mathrm{AVaR}}$ will be utilized in various multi-object case-based reasoning with risk aversity.

Appendix

Proof of Lemma 2.1. (i) For $x_1 = \mathrm{VaR}_{q_1}(X_1)$ and $x_2 = \mathrm{VaR}_{q_2}(X_2)$, from (2.3) and (2.8) we have $c(q_1, q_2) = \frac{\partial^2}{\partial q_1 \partial q_2} P(X_1 \leq x_1, X_2 \leq x_2) = w(x_1, x_2) \frac{dx_1}{dq_1} \frac{dx_2}{dq_2}$. Thus we get (i). (ii) Since $P(X_1 \leq x_1) = q_1$ and $P(X_2 \leq x_2) = q_2$, we get $w_1(x_1)\, dx_1 = dq_1$ and $w_2(x_2)\, dx_2 = dq_2$. (iii) is trivial from (i) and (ii). □

Proof of Lemma 2.2. Let a random $(X_1, X_2) \in \mathcal{X}^2$ and a probability $p \in (0, 1]$. (i) From (2.4), (2.5) and (2.6), we have

$$
\begin{aligned}
&\mathrm{VaR}_p(X_1, X_2) \\
&= \{(x_1, x_2) \in \mathbb{R}^2 \mid P(X_1 \leq x_1) = p_1, P(X_2 \leq x_2) = p_2, C(p_1, p_2) = p\} \\
&= \{(\mathrm{VaR}_{p_1}(X_1), \mathrm{VaR}_{p_2}(X_2)) \mid (p_1, p_2) \in \Gamma(p)\}.
\end{aligned}
$$

Thus we get (i). (ii) From (i), (2.7) and Lemma 2.1(i), we also have

$$
\begin{aligned}
&\mathrm{AVaR}_p(X_1, X_2) \\
&= \{E((X_1, X_2) \mid (X_1, X_2) \preceq \mathrm{VaR}_p(X_1, X_2))\} \\
&= \{E((X_1, X_2) \mid (X_1, X_2) \preceq (\mathrm{VaR}_{p_1}(X_1), \mathrm{VaR}_{p_2}(X_2))) \mid (p_1, p_2) \in \Gamma(p)\} \\
&= \{E((X_1, X_2) \mid X_1 \leq \mathrm{VaR}_{p_1}(X_1), X_2 \leq \mathrm{VaR}_{p_2}(X_2)) \mid (p_1, p_2) \in \Gamma(p)\} \\
&= \left\{ \frac{1}{p} \iiint_{(-\infty, \mathrm{VaR}_{p_1}(X_1)] \times (-\infty, \mathrm{VaR}_{p_2}(X_2)]} (x_1, x_2)\, w(x_1, x_2)\, dx_1 dx_2 \mid (p_1, p_2) \in \Gamma(p) \right\} \\
&= \left\{ \frac{1}{p} \iint_{[0, p_1] \times [0, p_2]} (\mathrm{VaR}_{q_1}(X_1), \mathrm{VaR}_{q_2}(X_2))\, c(q_1, q_2)\, dq_1 dq_2 \mid (p_1, p_2) \in \Gamma(p) \right\}.
\end{aligned}
$$

Therefore this lemma holds. □

Proof of Lemma 2.3. Value-at-risks (2.1) has similar properties (Artzner [2]). Therefore from Lemma 2.2(i) we can easily check (i)–(iii). □

Proof of Lemma 2.4. Value-at-risks (2.2) has similar properties to (i)–(iii) [2]. From Lemma 2.2(ii), we can easily check (i)–(iii). (iv) It is well-known in one-dimensional case (2.2) that average value-at-risks has sub-additivity:

$$\text{AVaR}_p(X+Y) \geq \text{AVaR}_p(X) + \text{AVaR}_p(Y) \tag{A.1}$$

for all $X, Y \in \mathcal{X}$ (Artzner [2] and Yoshida [9]). Let $(X_1, X_2), (Y_1, Y_2) \in \mathcal{X}^2$ and let $p \in (0,1]$ and $(p_1, p_2) \in \Gamma(p)$. From (2.9) and (2.14), maps $q_1 \mapsto \frac{\partial}{\partial q_1} C(q_1, p_2) = \int_0^{p_2} c(q_1, q_2)\, dq_2$ and $q_2 \mapsto \frac{\partial}{\partial q_2} C(p_1, q_2) = \int_0^{p_1} c(q_1, q_2)\, dq_1$ are non-increasing. When we approximate these nonnegative-valued functions by sums of non-increasing step functions on $(0,1]$, $q \mapsto \gamma 1_{[0,\delta]}(q)$ ($\gamma > 0$, $\delta \in (0,1]$), from (A.1) we can check

$$\frac{1}{p} \iint_{[0,p_1] \times [0,p_2]} (\text{VaR}_{q_1}(X_1 + Y_1), \text{VaR}_{q_2}(X_2 + Y_2))\, c(q_1, q_2)\, dq_1 dq_2$$
$$\succeq \frac{1}{p} \iint_{[0,p_1] \times [0,p_2]} (\text{VaR}_{q_1}(X_1), \text{VaR}_{q_2}(X_2))\, c(q_1, q_2)\, dq_1 dq_2$$
$$+ \frac{1}{p} \iint_{[0,p_1] \times [0,p_2]} (\text{VaR}_{q_1}(Y_1), \text{VaR}_{q_2}(Y_2))\, c(q_1, q_2)\, dq_1 dq_2.$$

Therefore we obtain (i) from Lemma 2.2(ii), and this lemma holds.   □

Proof of Theorem 3.1. (a) $\implies$ (b) is trivial when we let $x_1 \to \infty$ or $x_2 \to \infty$ in Definition 3.1. (b) $\implies$ (a): From (2.9), maps $p_1 \mapsto C(p_1, p_2)$ and $p_2 \mapsto C(p_1, p_2)$ are non-decreasing. And we get (a) from (b) with (2.4). (a) $\implies$ (c): Let $f \in \mathcal{L}^2$. We can approximate decreasing function $-f$ by sums of step functions $\gamma 1_{(-\infty, x_1] \times (-\infty, x_2]}$ ($\gamma > 0$) and we get (c) from (3.3). (c) $\implies$ (a): We can easily check (a) when we approximate step functions $1_{(-\infty, x_1] \times (-\infty, x_2]}$ by decreasing functions $-f$ with $f \in \mathcal{L}^2$.   □

Proof of Theorem 3.2. (i) Letting $R \uparrow D$ in $M_w^f(R) \preceq M_v^f(R)$, we get $E(f(X_1, X_2)) \leq E(f(Y_1, Y_2))$. (ii) For simplicity, we use small order notation $o(\cdot)$ with Landau's symbol, i.e. $\phi(t) \in o(\psi(t)) \iff \limsup_{t \downarrow 0} \left| \frac{\varphi(t)}{\psi(t)} \right| = 0$ for functions $\varphi, \psi : (0, \infty) \mapsto \mathbb{R}$. Let $(a, b) \in D$ and take an increasing utility function $f(x_1, x_2) = (x_1 - a) + (x_2 - b)$. Let $R = [a, a+h] \times [b, b+k] \in \mathcal{R}(D)$ with $h > 0$ and $k > 0$, and let $R(t) = [a, a+th] \times [b, b+tk] \subset R$ for $t \in (0,1)$. From $M_w^f(R(t)) \preceq M_v^f(R(t))$, we have

$$\frac{\iint_{R(t)} f(x_1, x_2) w(x_1, x_2)\, dx_1 dx_2}{\iint_{R(t)} w(x_1, x_2)\, dx_1 dx_2} \leq \frac{\iint_{R(t)} f(x_1, x_2) v(x_1, x_2)\, dx_1 dx_2}{\iint_{R(t)} v(x_1, x_2)\, dx_1 dx_2}. \tag{A.2}$$

Hence, by Taylor's theorem, there exists $\theta \in [0, t]$ such that $w(x_1, x_2) = w(a, b) + w_{x_1}(a + \theta h, b + \theta k)(x_1 - a) + w_{x_2}(a + \theta h, b + \theta k)(x_2 - b)$. Then regarding $M_w^f(R(t))$ we have

$$\iint_{R(t)} f(x_1, x_2) w(x_1, x_2)\, dx_1 dx_2 - \frac{(h+k)t}{2} \iint_{R(t)} w(x_1, x_2)\, dx_1 dx_2$$

$$= \frac{h^2 k^2 t^4}{4}(w_{x_1}(a,b) + w_{x_2}(a,b)) + \frac{hkt^4}{3}(h^2 w_{x_1}(a,b) + k^2 w_{x_2}(a,b))$$

$$- \frac{(h+k)hkt^4}{4}(hw_{x_1}(a,b) + kw_{x_2}(a,b)) + o(t^4).$$

We also have similar equation holds for $M_v^f(R(t))$. Applying the equations to (A.1) and letting $t \downarrow 0$, we get $\frac{h^2 w_{x_1}(a,b) + k^2 w_{x_2}(a,b)}{w(a,b)} \leq \frac{h^2 v_{x_1}(a,b) + k^2 v_{x_2}(a,b)}{v(a,b)}$ for all $(a,b) \in D$ and $h >$ and $k > 0$. Thus we obtain $\frac{w_{x_1}}{w} \leq \frac{v_{x_1}}{v}$ and $\frac{w_{x_2}}{w} \leq \frac{v_{x_2}}{v}$ on D. (iii) In case of $c(\cdot, \cdot) = 1$, $w(x_1, x_2) = w_1(x_1)w_2(x_2)$ and $v(x_1, x_2) = v_1(x_1)v_2(x_2)$ for $(x_1, x_2) \in D$. Therefore (b) and (c) are equivalent. In case of $c(\cdot, \cdot) = 1$, (a) and (b) are equivalent from (ii) and the proof in Yoshida [11, Theorem 2.3]. $\square$

Proof of Theorem 3.3. (a) $\Longrightarrow$ (b): From Theorem 3.1 we have $X_1 \preceq_{\mathrm{FSD}} X_2$ and $Y_1 \preceq_{\mathrm{FSD}} Y_2$, and then $P(X_i \leq x_i) \geq P(Y_i \leq x_i)$ for $x_i \in \mathbb{R}$ and $i = 1, 2$. They imply $\mathrm{VaR}_p(X_i) \leq \mathrm{VaR}_p(Y_i)$ for all $p \in (0, 1]$ and $i = 1, 2$. Thus we get (b) by Lemma 2.2(i) since maps $p_1 \mapsto C(p_1, p_2)$ and $p_2 \mapsto C(p_1, p_2)$ are nondecreasing from (2.9). (b) $\Longrightarrow$ (a): From (b) and (2.6), for any $(x_1, x_2) \in \mathbb{R}^2$ there exists $(y_1, y_2) \in \mathbb{R}^2$ such that $(x_1, x_2) \preceq (y_1, y_2)$ and $P((X_1, X_2) \preceq (x_1, x_2)) = p = P((Y_1, Y_2) \preceq (y_1, y_2))$. Then $P((X_1, X_2) \preceq (x_1, x_2)) = P((Y_1, Y_2) \preceq (y_1, y_2)) \geq P((Y_1, Y_2) \preceq (x_1, x_2))$ and we get (a). The equivalence of (b) and (c) is trivial from (2.10). $\square$

Proof of Lemma 4.1. (i) is trivial from the definitions (2.1), (2.2), (3.1) and (3.2). (ii) can be checked easily from the definition (2.1). (iii) is also trivial from the definitions (3.1) and (3.2). (iv) From $X \preceq_{\mathrm{SSD}} Y$ we have

$$\int_{-\infty}^{x} P(X \leq z)\, dz \geq \int_{-\infty}^{x} P(Y \leq z)\, dz \qquad (A.3)$$

for any $x \in \mathbb{R}$. Let $p \in (0, 1]$. Let $Z \in \mathcal{X}$ have a density function u. By the integration of parts, we have $\frac{1}{p}\int_{-\infty}^{\mathrm{VaR}_p(Z)} P(Z \leq z)\, dz = \frac{1}{p} \times \mathrm{VaR}_p(Z)P(Z \leq \mathrm{VaR}_p(Z)) - \frac{1}{p}\int_{-\infty}^{\mathrm{VaR}_p(Z)} zu(z)\, dz = \mathrm{VaR}_p(Z) - \mathrm{AVaR}_p(Z)$. Thus we get

$$\mathrm{AVaR}_p(Z) = \mathrm{VaR}_p(Z) - \frac{1}{p}\int_{-\infty}^{\mathrm{VaR}_p(Z)} P(Z \leq z)\, dz \quad \text{for all } Z \in \mathcal{X}. \quad (A.4)$$

Put a function $g(x) = x - \frac{1}{p}\int_{-\infty}^{x} P(Y \leq z)\, dz$ for $x \in \mathbb{R}$. Then $g'(x) = 1 - \frac{P(Y \leq x)}{p} \gtrless 0 \iff x \lessgtr \mathrm{VaR}_p(Y)$, and $g(x)$ has a maximum at $x = \mathrm{VaR}_p(Y)$. Together with (A.3) and (A.4) we get $\mathrm{AVaR}_p(Y) = g(\mathrm{VaR}_p(Y)) \geq g(\mathrm{VaR}_p(X)) = \mathrm{VaR}_p(X) - \frac{1}{p}\int_{-\infty}^{\mathrm{VaR}_p(X)} P(Y \leq z)\, dz \geq \mathrm{VaR}_p(X) - \frac{1}{p}\int_{-\infty}^{\mathrm{VaR}_p(X)} P(X \leq z)\, dz = \mathrm{AVaR}_p(X)$. Therefore we have $X \preceq_{\mathrm{AVaR}} Y$. $\square$

Proof of Theorem 4.1. (i) is trivial from the definitions (3.1) and (4.1). (ii) Let $f_1 : \mathbb{R} \mapsto \mathbb{R}$ be a C^2-class strictly increasing concave function. Approximating a function $(x_1, x_2)(\in \mathbb{R}^2) \mapsto f_1(x_1)$ by concave functions $f \in \mathcal{L}_c^2$, we get $E(f_1(X_1)) \leq E(f_1(Y_1))$ for all concave functions $f_1 : \mathbb{R} \mapsto \mathbb{R}$. In the same way, we also have $E(f_2(X_2)) \leq E(f_2(Y_2))$ for all concave functions $f_2 : \mathbb{R} \mapsto \mathbb{R}$. It is well-known that these inequalities are equivalent to $X_1 \preceq_{\text{SSD}} Y_1$ and $X_2 \preceq_{\text{SSD}} Y_2$ (Levi [4]). (iii) From $X_1 \preceq_{\text{SSD}} Y_1$ and $X_2 \preceq_{\text{SSD}} Y_2$, we have $\int_{-\infty}^{x_1} P(X_1 \leq z_1) \, dz_1 \geq \int_{-\infty}^{x_1} P((Y_1 \leq z_1) \, dz_1 \geq 0$ and $\int_{-\infty}^{x_2} P(X_2 \leq z_2) \, dz_2 \geq \int_{-\infty}^{x_2} P((Y_2 \leq z_2) \, dz_2 \geq 0$ for $(x_1, x_2) \in \mathbb{R}^2$. Since $c(\cdot, \cdot) = 1$, we get (2.4) and $(X_1, X_2) \preceq_{\text{SSD}} (Y_1, Y_2)$ holds. □

Proof of Theorem 4.2. (i)–(iii) are trivial from Lemma 2.2 and (2.13). (iv) By Lemma 4.1(iv), from $X_1 \preceq_{\text{SSD}} Y_1$ and $X_2 \preceq_{\text{SSD}} Y_2$ we have $X_1 \preceq_{\text{AVaR}} Y_1$ and $X_2 \preceq_{\text{AVaR}} Y_2$. Thus we get (iii) together with (2.13). □

References

1. Arrow, K.J.: Essays in the Theory of Risk-Bearing. Markham, Chicago (1971)
2. Artzner, P., Delbaen, F., Eber, J.-M., Heath, D.: Coherent measures of risk. Math. Finance **9**, 203–228 (1999)
3. Jouini, E., Meddeb, M., Touzi, N.: Vector-valued coherent risk measures. Finance Stoch. **8**, 531–552 (2004)
4. Levy, H.: Stochastic dominance and expected utility: survey and analysis. Manag. Sci. **38**, 555–593 (1992)
5. Nelson, R.B.: An Introduction to Copulas, 2nd edn. Springer, New York (2006). https://doi.org/10.1007/0-387-28678-0
6. Pratt, J.W.: Risk aversion in the small and the large. Econometrica **32**, 122–136 (1964)
7. Shaked, M., Shanthikumar, J.G.: Stochastic Orders. Springer, New York (2007). https://doi.org/10.1007/978-0-387-34675-5
8. Yoshida, Y.: Weighted quasi-arithmetic means and a risk index for stochastic environments. Int. J. Uncertain. Fuzziness Knowl. Based Syst. (IJUFKS) **16**(suppl.), 1–16 (2011)
9. Yoshida, Y.: An ordered weighted average with a truncation weight on intervals. In: Torra, V., Narukawa, Y., López, B., Villaret, M. (eds.) MDAI 2012. LNCS (LNAI), vol. 7647, pp. 45–55. Springer, Heidelberg (2012). https://doi.org/10.1007/978-3-642-34620-0_6
10. Yoshida, Y.: Weighted quasi-arithmetic mean on two-dimensional regions and their applications. In: Torra, V., Narukawa, Y. (eds.) MDAI 2015. LNCS (LNAI), vol. 9321, pp. 42–53. Springer, Cham (2015). https://doi.org/10.1007/978-3-319-23240-9_4
11. Yoshida, Y.: Weighted quasi-arithmetic means on two-dimensional regions: an independent case. In: Torra, V., Narukawa, Y., Navarro-Arribas, G., Yañez, C. (eds.) MDAI 2016. LNCS (LNAI), vol. 9880, pp. 82–93. Springer, Cham (2016). https://doi.org/10.1007/978-3-319-45656-0_7

Stochastic Orders on Two-Dimensional Space: Application to Cross Entropy

Mateu Sbert[1(✉)] and Yuji Yoshida[2]

[1] Graphics and Imaging Laboratory, University of Girona, 17003 Girona, Spain
mateu.sbert@udg.edu
[2] Faculty of Economics, University of Kitakyushu, Kitakyushu 802-8577, Japan
yoshida@kitakyu-u.ac.jp

Abstract. We present in this paper the extension to 2d probability mass functions (PMFs) of the first and likelihood-ratio stochastic orders for 1d PMFs. We show that first stochastic order ensures Kolmogorov mean order invariance. We also review the concept of comonotonic sequences and show its direct relationship with likelihood-ratio order. We give some application examples related to frequency histograms and cross entropy.

1 Introduction

The stochastic orders [1,2] have its origin in economics risk literature [3]. The probably most interesting stochastic orders are the first stochastic order, which is characterized by order invariance of generalized arithmetic mean, or Kolmogorov mean, and its suborder, the likelihood-ratio. This is, if two PMFs are ordered under these orders, all generalized weighted means obtained weighting any increasing sequence with these PMFs are ordered in the same direction. This invariance property has been used for instance in the study of optimal sampling allocation and optimal combination of techniques in Monte Carlo multiple importance sampling technique [4], and in studying cross entropy and likelihood [6]. We present here the extension of the ordering between 1d PMFs to the ordering between 2d PMFs. As 1d PMF gives a 1d random variable probability distribution, the 2d PMFs correspond to the probability distributions of two joint 1d random variables. The 2d PMFs can be represented as 2d histograms, which have applications in the field of image processing [5]. We will study the general case of non independent random variables, and give some particular results for independent ones. We will see also how the order between 2d PMFs submits the same order between the marginal probabilities. The concept of comonotonicity is reviewed and we show its direct relationship with likelihood-ratio order.

The rest of the paper is organized as follows. In Sect. 2 we review the first stochastic and likelihood-ratio between 1d PMFs. We introduce in Section 3 the stochastic order between 2d PMFs. In Sect. 4 we introduce 1d and 2d comonotonic sequences and its relationship to the corresponding 1d or 2d likelihood-ratio order. Finally in Sect. 5 we present our conclusions and future work.

V. Torra et al. (Eds.): MDAI 2020, LNAI 12256, pp. 28–40, 2020.
https://doi.org/10.1007/978-3-030-57524-3_3

2   Stochastic Order in 1d

We review in this section stochastic orders in 1 dimension. Be $\mathcal{N} = \{1, 2, \ldots, N\}$. Consider a random variable X' with states in $\mathcal{N}$ with PMF (probability mass function) $\{q_i\}_{i=1}^{i=N}$ and random variable X with PMF $\{p_i\}_{i=1}^{i=N}$, this is, $\forall i : p_i, q_i \geq 0$ and $\sum_{i=1}^{N} p_i = \sum_{i=1}^{N} q_i = 1$.

Definition 1 (First stochastic order in 1d).   *We say that the PMF $\{q_i\}$ precedes in first stochastic order or first stochastic dominance (FSD) the PMF $\{p_i\}$, and we write $\{q_i\} \prec_{FSD} \{p_i\}$, when for all $n \in \mathcal{N}$*

$$\sum_{i=1}^{n} q_i \geq \sum_{i=1}^{n} p_i. \tag{1}$$

Observe that, geometrically, Eq. 1 means that the cumulative distribution function for X' will always be below the one for X.

Using the fact that $\sum_{i=1}^{N} p_i = \sum_{i=1}^{N} q_i = 1$, condition in Eq. 1 can be written in a different way:

Lemma 1. *If $\{q_i\}$ and $\{p_i\}$ are PMFs, the following conditions are equivalent:*

1. $\{q_i\} \prec_{FSD} \{p_i\}$
2. For all $n \in \mathcal{N}$,

$$\sum_{i=n}^{N} q_i \leq \sum_{i=n}^{N} p_i. \tag{2}$$

Corollary 1. *If $\{q_i\} \prec_{FSD} \{p_i\}$ then $q_1 \geq p_1$ and $q_N \leq p_N$.*

Proof. It comes directly from Eq. 1 and Eq. 2.   □

Definition 2 (Likelihood-ratio order in 1d).   *We say that the PMF $\{q_i\}$ precedes in likelihood-ratio order (LR) the PMF $\{p_i\}$, and we write $\{q_i\} \prec_{LR} \{p_i\}$, when for all i, j, $\frac{q_i}{p_i} \geq \frac{q_j}{p_j}$, this is, the sequence $\{\frac{q_i}{p_i}\}$ is decreasing.*

Note 1. Both $\prec_{FSD}$ and $\prec_{LR}$ can be shown to be partial orders.

Several indirect proofs can be found in the literature [1,2,7] for the following result:

Lemma 2. *Given PMFs $\{q_i\}$, $\{p_i\}$, $\{q_i\} \prec_{LR} \{p_i\} \implies \{q_i\} \prec_{FSD} \{p_i\}$.*

Note 2. Observe that the reverse of Lemma 2 is not true. Consider the PMFs: $\{1/6, 2/6, 1/6, 2/6\}$, $\{2/6, 1/6, 2/6, 1/6\}$. We have that $\{2/6, 1/6, 2/6, 1/6\}$ $\prec_{FSD} \{1/6, 2/6, 1/6, 2/6\}$ but $\{2/6, 1/6, 2/6, 1/6\} \not\prec_{LR} \{1/6, 2/6, 1/6, 2/6\}$.

Let us remember the concept of quasi-arithmetic mean:

Definition 3 (Kolmogorov or quasi-arithmetic mean). *Be* $g(x)$ *any strictly monotonic function (this is, strictly increasing or decreasing),* $\{x_i\}$ *a sequence of numbers and* $\{p_i\}$ *a PMF (that we also call weights), then we define the Kolmogorov or quasi-arithmetic mean* $M_g(\{p_i\}, \{x_i\})$ *as*

$$M_g(\{p_i\}, \{x_i\}) = g^{-1}\Big(\sum_{i=1}^{N} p_i g(x_i)\Big). \tag{3}$$

For instance arithmetic mean corresponds to $g(x) = x$, harmonic mean to $g(x) = 1/x$, power mean with index r to $g(x) = x^r$.

The following invariance theorem has been proven in [7] (more precisely, that Eq. 1 is equivalent to condition 2 of Lemma 3).

Lemma 3. *Be* $\{p_i\}, \{q_i\}$ *PMFs. Then, the following conditions are equivalent:*

1. $\{q_i\} \prec_{FSD} \{p_i\}$
2. *For any strictly monotonic function* $g(x)$ *and for any increasing sequence* $\{x_i\}$,

$$M_g(\{q_i\}, \{x_i\}) \leq M_g(\{p_i\}, \{x_i\}) \tag{4}$$

Example. The fact that the theorem applies only to increasing sequences of numbers seems to be a restriction to its usefulness. However, there is a ubiquitous scenario where all sequences are increasing, which is the case of frequency histograms, as a histogram can be considered an ordered range of bins with the corresponding frequencies and increasing representative values $\{x_i\}$. If we have two histograms with the same bins and same representative values $\{x_i\}$, if the frequencies of the two histograms are related by $\prec_{FSD}$ order, all means will be ordered in the same direction. Histograms with same $\{x_i\}$ values could represent for instance a time series, or different images where the intensity values have been discretized in the same number of bins.

Corollary 2. *Be* $\{x_i\}$ *any increasing sequence, and be* $\{p_i\}, \{q_i\}$ *PMFs. Be* $g(x)$ *a strictly monotonic function. Then, condition 1 implies condition 2 but not viceversa:*

1. $\{q_i\} \prec_{LR} \{p_i\}$
2. $M_g(\{q_i\}, \{x_i\}) \leq M_g(\{p_i\}, \{x_i\})$

Proof. That 1 implies 2 comes from that $\{q_i\} \prec_{LR} \{p_i\} \implies \{q_i\} \prec_{FSD} \{p_i\}$ and Lemma 3. The reverse is not true as $\prec_{LR}$ is a strict suborder of $\prec_{FSD}$. □

3   Stochastic Order in 2d

Stochastic orders in 2d have been introduced in [8]. Be $\mathcal{M} = \{1, 2, \ldots, M\}$. Let us consider now joint random variables (X, Y), (X', Y') with states in $\mathcal{N} \times \mathcal{M}$, and be $\{q_{ij}\}$ and $\{p_{ij}\}$ their joint probabilities. We will generalize first stochastic order in Definition 1 to 2 dimensions:

Definition 4 (First stochastic order in 2d). *We say that the 2d PMF* $\{q_{ij}\}$ *precedes in first stochastic order the 2d PMF* $\{p_{ij}\}$, *and we write that* $\{q_{ij}\} \prec_{\textbf{FSD}} \{p_{ij}\}$ *when for all* $(n, m) \in \mathcal{N} \times \mathcal{M}$ *we have*

$$\sum_{i=1}^{n} \sum_{j=1}^{m} q_{ij} \geq \sum_{i=1}^{n} \sum_{j=1}^{m} p_{ij}. \tag{5}$$

Observe that, geometrically, Eq. 5 means that the cumulative distribution function for (X, Y) will always be below the one for (X', Y') for all $\{1, 2 \ldots, n\} \times \{1, 2, \cdots, m\} \subset \mathcal{N} \times \mathcal{M}$.

The following theorem tells us that if two 2d PMFs are ordered their marginal distributions are ordered too.

Theorem 1. $\{q_{ij}\} \prec_{\textbf{FSD}} \{p_{ij}\} \implies (\{q_{i\cdot}\} \prec_{FSD} \{p_{i\cdot}\} \& \{q_{\cdot j}\} \prec_{FSD} \{p_{\cdot j}\})$, *where* $q_{i\cdot} = \sum_{j=1}^{M} q_{ij}$ *and* $q_{\cdot j} = \sum_{i=1}^{N} q_{ij}$, *and* $p_{i\cdot} = \sum_{j=1}^{M} p_{ij}$ *and* $p_{\cdot j} = \sum_{i=1}^{N} p_{ij}$.

Proof. From Definition 4 we have that for all $1 \leq n \leq N$ and $1 \leq m \leq M$

$$\sum_{i=1}^{n} \sum_{j=1}^{M} q_{ij} \geq \sum_{i=1}^{n} \sum_{j=1}^{M} p_{ij}, \tag{6}$$

and

$$\sum_{i=1}^{m} \sum_{j=1}^{N} q_{ij} \geq \sum_{i=1}^{m} \sum_{j=1}^{N} p_{ij}, \tag{7}$$

and the result comes from the definition of $\prec_{FSD}$.   □

Corollary 3. *If* $\{q_{ij}\} \prec_{\textbf{FSD}} \{p_{ij}\}$ *then* $q_{11} \geq p_{11}$ *and for all* i, j, $q_{N\cdot} \leq p_{N\cdot}, q_{\cdot M} \leq p_{\cdot M}$.

Proof. $q_{11} \geq p_{11}$ comes from Eq. 5 taking $n, m = 1$. $q_{N\cdot} \leq p_{N\cdot}, q_{\cdot M} \leq p_{\cdot M}$ comes from Theorem 1 and Corollary 1.   □

Theorem 2. *Be* $\{p_i\}, \{q_i\}, \{r_i\}, \{s_i\}$ *PMFs. Then,* $\{q_i s_j\} \prec_{\textbf{FSD}} \{p_i r_j\} \iff (\{q_i\} \prec_{FSD} \{p_i\} \& \{s_j\} \prec_{FSD} \{r_j\})$

Proof. The $\implies$ implication is a direct application of Theorem 1 with independent random variables. The $\impliedby$ implication comes from the fact that condition in Definition 4 can be written because of the hypothesis as $\sum_{i=1}^{n} q_i \sum_{j=1}^{m} s_j \geq \sum_{i=1}^{n} p_i \sum_{j=1}^{m} r_j$, which results immediately from multiplying both members of the inequalities $\sum_{i=1}^{n} q_i \geq \sum_{i=1}^{n} p_i$ and $\sum_{j=1}^{m} q_j \geq \sum_{j=1}^{m} r_j$, which are true by the definition of $\prec_{FSD}$.   □

Theorem 3. *The following conditions are equivalent:*

1. $\{q_i s_j\} \prec_{\textbf{FSD}} \{p_i r_j\}$
2. For all $(n, m) \in \mathcal{N} \times \mathcal{M}$

$$\sum_{i=n}^{N}\sum_{j=m}^{M}q_i s_j \le \sum_{i=n}^{N}\sum_{j=m}^{M}p_i r_j. \tag{8}$$

Proof. From Theorem 2 and Lemma 1, condition 1 is equivalent to: for all n, m,

$$\Big(\sum_{i=n}^{N}q_i \le \sum_{i=n}^{N}p_i\Big)\&\Big(\sum_{j=m}^{M}s_j \le \sum_{j=m}^{M}r_j\Big), \tag{9}$$

thus $1 \implies 2$ by multiplying both terms of the inequalities in Eq. 9, and $2 \implies 1$ by setting $n = 1, m = 1$ in Eq. 8. $\qquad\square$

Let us extend the quasi-arithmetic mean to 2d.

Definition 5 (quasi-arithmetic mean extended to 2d arrays). *The quasi-arithmetic or Kolmogorov mean $M_g(\{p_{ij}\}, \{x_{ij}\})$ of 2d array $\{x_{ij}\}$ with PMF (or weights) $\{p_{ij}\}$, is defined as*

$$M_g(\{p_{ij}\}, \{x_{ij}\}) = g^{-1}\Big(\sum_{i=1}^{N}\sum_{j=1}^{M}p_{ij}g(x_{ij})\Big).$$

Theorem 4. *Be $\{p_{ij}\}, \{q_{ij}\}$ 2d PMFs, and $\{q_{ij}\} \prec_{\mathbf{FSD}} \{p_{ij}\}$. Then, for any 2d array $\{x_{ij}\}$ increasing in i and j we have:*

$$\sum_{i=1}^{N}\sum_{j=1}^{M}q_{ij}x_{ij} \le \sum_{i=1}^{N}\sum_{j=1}^{M}p_{ij}x_{ij}. \tag{10}$$

Proof. Define $A_{kl} = \sum_{i=1}^{N}p_{il}, A'_{kl} = \sum_{i=1}^{N}q_{il}$, and $A_{0l} = A'_{0l} = 0$. Then $p_{kl} = (A_{kl} - A_{k-1,l}), q_{kl} = (A'_{kl} - A'_{k-1,l})$, and

$$\sum_{l=1}^{M}\sum_{k=1}^{N}p_{kl}x_{kl} - \sum_{l=1}^{M}\sum_{k=1}^{N}q_{kl}x_{kl} \tag{11}$$

$$= \sum_{l=1}^{M}\sum_{k=1}^{N}(A_{kl} - A_{k-1,l})x_{kl} - \sum_{l=1}^{M}\sum_{k=1}^{N}(A'_{kl} - A'_{k-1,l})x_{kl}$$

$$= \sum_{l=1}^{M}\sum_{k=1}^{N}(A_{kl} - A'_{kl})x_{kl} - \sum_{l=1}^{M}\sum_{k=1}^{N}(A_{k-1,l} - A'_{k-1,l})x_{kl}$$

$$= \sum_{l=1}^{M}(A_{Nl} - A'_{Nl})x_{Nl} + \sum_{l=1}^{M}\sum_{k=1}^{N-1}(A_{kl} - A'_{kl})x_{kl} - \sum_{l=1}^{M}\sum_{k=1}^{N-1}(A_{kl} - A'_{kl})x_{k+1,l}$$

$$= \sum_{l=1}^{M}(A_{Nl} - A'_{Nl})x_{Nl} + \sum_{l=1}^{M}\sum_{k=1}^{N-1}(A_{kl} - A'_{kl})(x_{kl} - x_{k+1,l}).$$

But the second sum is positive because $(A_{kl} - A'_{kl}) \leq 0$ by Definition 4 and $(x_{kl} - x_{k+1,l}) \leq 0$ by being x_{ij} increasing in both i, j. About the first term:

$$\sum_{l=1}^{M}(A_{Nl} - A'_{Nl})x_{Nl} \geq x_{NM} \sum_{l=1}^{M}(A_{Nl} - A'_{Nl}) = x_{NM}(1 - 1) = 0. \quad (12)$$

Therefore this theorem holds.   □

Theorem 5. *If $\{q_{ij}\} \prec_{\textbf{FSD}} \{p_{ij}\}$ then for all $(n, m) \in \mathcal{N} \times \mathcal{M}$*

$$\sum_{i=n,j=m}^{N,M} q_{ij} \leq \sum_{i=n,j=m}^{N,M} p_{ij}. \quad (13)$$

Proof. The results follow from Eq. 10 when we take $x_{ij} = 1$ if $n \leq i \leq N$ and $m \leq j \leq M$ and we also take $x_{ij} = 0$ otherwise.   □

From Theorem 5 and Definition 4 the following corollary is immediate:

Corollary 4. *Be $\{p_{ij}\}, \{q_{ij}\}$ 2d PMFs. Then*

$$\{q_{ij}\} \prec_{\textbf{FSD}} \{p_{ij}\} \implies \{p_{N-i+1,M-j+1}\} \prec_{\textbf{FSD}} \{q_{N-i+1,M-j+1}\}.$$

Now we are in condition to prove the following theorem:

Theorem 6. *Be $\{p_{ij}\}, \{q_{ij}\}$ 2d PMFs. Then, the following conditions are equivalent:*

1. *$\{q_{ij}\} \prec_{\textbf{FSD}} \{p_{ij}\}$*
2. *For any any strictly monotonic function $g(x)$ and 2d array $\{x_{ij}\}$ increasing in i and j, $M_g(\{q_{ij}\}, \{x_{ij}\}) \leq M_g(\{p_{ij}\}, \{x_{ij}\})$.*

Proof. 1 $\implies$ 2: Consider first $g(x)$ increasing, then $g_{ij} = g(x_{ij})$ is increasing in both i, j, thus Eq. 10 holds, and it is enough to apply $g^{-1}(x)$, which is increasing, to both terms of Eq. 10. Suppose now $g(x)$ decreasing, then $g(x_{N-i+1,M-j+1})$ is increasing, and from Corollary 4 and Theorem 4 we have that

$$\sum_{i=1}^{N}\sum_{j=1}^{M} p_{ij}g(x_{ij}) = \sum_{i=1}^{N}\sum_{j=1}^{M} p_{N-i+1,M-j+1}g(x_{N-i+1,M-j+1})$$

$$\leq \sum_{i=1}^{N}\sum_{j=1}^{M} q_{N-i+1,M-j+1}g(x_{N-i+1,M-j+1}) = \sum_{i=1}^{N}\sum_{j=1}^{M} q_{ij}g(x_{ij}),$$

and applying $g(x)$, a decreasing function, to both terms of the above equation it reverses direction and we obtain the desired result.

2 $\implies$ 1: taking $g(x) = x$, and we take $x_{ij} = 1$ if $n \leq i \leq N$ and $m \leq j \leq M$ and we also take $x_{ij} = 0$ otherwise, we obtain Eq. 13, and thus by Definition 4, $\{p_{N-i+1,M-j+1}\} \prec_{\textbf{FSD}} \{q_{N-i+1,M-j+1}\}$, and then we apply Corollary 4.   □

Example. A digital image is an array of pixel values. These values can be discretized into N bins, with bins assigned increasing values $\{x_i\}$. A second image is discretized into M bins, with bins assigned increasing values $\{x_j\}_{j=1}^m$. The joint histogram is an $N \times M$ array where in position (i, j) it contains the number of pixels that a) occupy the same position in both images, and b) the value of the pixel intensity in the first image corresponds to x_i and in the second image to x_j. Their joint normalized histogram will be a 2d PMF $\{p_{ij}\}$. If one or both images change then we have another normalized histogram $\{q_{ij}\}$. For instance, in image registration [5] one of the images is moved to obtain the maximum alignment with the other image. Theorem 4 tells us that if we assign to the bins a doubly increasing sequence of values $\{x_{ij}\}$ (for instance $x_{ij} = x_i x_j$, or simply $x_{ij} = x_i$ or $x_{ij} = x_j$), then all means will change in the same direction when $\{q_{ij}\} \prec_{\mathbf{FSD}} \{p_{ij}\}$ or $\{p_{ij}\} \prec_{\mathbf{FSD}} \{q_{ij}\}$. 2d histograms are used too to represent a two-channel image, see https://svi.nl/TwoChannelHistogram. We can also generalize the likelihood-ratio order, and we have

Definition 6 (Likelihood-ratio order in 2d). *We say that the 2d PMF $\{q_{ij}\}$ precedes in likelihood-ratio order the 2d PMF $\{p_{ij}\}$ when the 2d array $\{\frac{q_{ij}}{p_{ij}}\}$ is both decreasing in i and j, and then we write that $\{q_{ij}\} \prec_{LR} \{p_{ij}\}$.*

Theorem 7. $\{q_i s_j\} \prec_{LR} \{p_i r_j\} \iff (\{q_i\} \prec_{LR} \{p_i\} \& \{s_j\} \prec_{LR} \{r_j\})$.

Proof. Let see first the $\impliedby$ implication. By hypothesis, for all $i_1 \leq j_1, i_2 \leq j_2$, $\frac{q_{i_1}}{q_{j_1}} \geq \frac{p_{i_1}}{p_{j_1}}$ and $\frac{s_{i_1}}{s_{j_1}} \geq \frac{r_{i_1}}{r_{j_1}}$, and multiplying left and right terms of those two inequalities we obtain the result. Now, for the $\implies$ implication, without loss of generality suppose that $\{q_i\} \not\prec_{LR} \{p_i\}$. It means there are indexes $i_1 \leq j_1$ such that $\frac{q_{i_1}}{q_{j_1}} < \frac{p_{i_1}}{p_{j_1}}$. By hypothesis, if $i_1 \leq j_1, i_2 \leq j_2$, $\frac{q_{i_1} s_{i_1}}{q_{j_1} s_{j_1}} \geq \frac{p_{i_1} r_{i_1}}{p_{j_1} r_{j_1}}$, but then we obtain $\frac{q_{i_1} s_{i_1}}{q_{j_1} s_{j_1}} \geq \frac{p_{i_1} r_{i_1}}{p_{j_1} r_{j_1}} > \frac{q_{i_1} s_{i_1}}{q_{j_1} s_{j_1}}$. $\qquad\square$

Theorem 8. $\{q_i s_j\} \prec_{LR} \{p_i r_j\} \implies \{q_i s_j\} \prec_{FSD} \{p_i r_j\}$.

Proof. It comes from successive application of Theorem 7, Lemma 2 and Theorem 2. $\qquad\square$

We will show later in Theorem 15 that this result generalizes to non independent distributions.

4   Comonotonic and Countermonotonic sequences

Comonotonicity was introduced in [9,10].

Definition 7 (Comonotonic sequences). *We say that sequences $\{f_i\}$ and $\{h_i\}$ are comonotonic when for all $i, j \in [1, \ldots, N], f_i > f_j \implies h_i \geq h_j$.*

Definition 8 (Countermonotonic sequences). *We say that sequences $\{f_i\}$ and $\{h_i\}$ are countermonotonic when for all $i, j \in [1, \ldots, N], f_i > f_j \implies h_i \leq h_j$.*

Note 3. Observe that if for all i, $h_i \neq 0$, if $\{f_i\}$ and $\{h_i\}$ are comonotonic then $\{f_i\}$ and $\{1/h_i\}$ are countermonotonic and viceversa.

Theorem 9. *If $\{f_i\}$ is comonotonic with $\{h_i\}$ and for all i, $h_i > 0$, then for all strictly monotonous function $g(x)$,*

$$M_g(\{1/N\}, \{f_i\}) \leq M_g(\{h_i / \sum_{k=1}^{N} h_k\}, \{f_i\}). \tag{14}$$

Proof. We reorder the sequences $\{f_i\}$ and $\{h_i\}$ in increasing order of $\{f_i\}$, obtaining sequence $\{f_i^\star\}$ and $\{h_i^\star\}$. By the comonotonicity, $\{h_i^\star\}$ is also in increasing order. As the sequence $\frac{1/N}{\frac{\{h_i^\star\}}{\sum_{k=1}^{N} h_k}}$ is decreasing, $\frac{1}{N} \prec_{LR} \frac{\{h_i^\star\}}{\sum_{k=1}^{N} h_k}$, and as $\prec_{LR}$ order implies $\prec_{FSD}$ order, applying Lemma 3 we obtain Eq. 14: $M_g(\{1/N\}, \{f_i\}) = M_g(\{1/N\}, \{f_i^\star\}) \leq M_g(\{h_i^\star / \sum_{k=1}^{N} h_k\}, \{f_i^\star\}) = M_g(\{h_i / \sum_{k=1}^{N} h_k\}, \{f_i\})$. □

Corollary 5. *If $\{f_i\}$ is comonotonic with $\{h_i\}$ and for all i, $h_i > 0$, then*

$$\frac{1}{N}(\sum_{i=1}^{N} h_i)(\sum_{i=1}^{N} f_i) \leq \sum_{i=1}^{N} h_i f_i \tag{15}$$

Proof. It is enough to apply Theorem 9 with the function $g(x) = x$, this is, the arithmetic mean. □

Theorem 10. *If $\{f_i\}$ is comonotonic with $\{p_i/q_i\}$ where $\{p_i\}$, $\{q_i\}$ are PMFs, and for all i, $p_i, q_i > 0$, then for all strictly monotonous function $g(x)$,*

$$M_g(\{q_i\}, \{f_i\}) \leq M_g(\{p_i\}, \{f_i\}). \tag{16}$$

In particular,

$$\sum_{i=1}^{N} q_i f_i \leq \sum_{i=1}^{N} p_i f_i. \tag{17}$$

Proof. Reordering $\{f_i\}$ in increasing order, the sequence $\{q_i/p_i\}$ is decreasing and thus $\{q_i\} \prec_{LR} \{p_i\}$, and the result is obtained applying Corollary 2. Taking $g(x) = x$ we obtain Eq. 17. □

We give now a direct proof of Eq. 17, for the purpose of its 2d generalization:

Theorem 11. *Let $\{p_i\}$ and $\{q_i\}$ be sequences of positive PMF. If two sequences $\{f_i\}$ and $\{\frac{p_i}{q_i}\}$ are comonotonic, then it holds that*

$$\sum_{i=1}^{N} f_i q_i \leq \sum_{i=1}^{N} f_i p_i.$$

Proof. This proof is derived from the proof of Yoshida [11, Theorem 3.1]. Let $H = \sum_{i=1}^{N} f_i p_i - \sum_{j=1}^{N} f_j q_j = \sum_{i=1}^{N} f_i p_i \sum_{j=1}^{N} q_j - \sum_{j=1}^{N} f_j q_j \sum_{i=1}^{N} p_i$. Then

$$H = \sum_{i=1}^{N} \sum_{j=1}^{N} p_i q_j (f_i - f_j)$$

$$= \sum_{i=1}^{N} \sum_{j=1}^{i-1} p_i q_j (f_i - f_j) + \sum_{i=1}^{N} \sum_{j=i+1}^{N} p_i q_j (f_i - f_j)$$

$$= \sum_{i=1}^{N} \sum_{j=1}^{i-1} p_i q_j (f_i - f_j) + \sum_{i=1}^{N} \sum_{j=1}^{i-1} p_j q_i (f_j - f_i)$$

$$= \sum_{i=1}^{N} \sum_{j=1}^{i-1} (p_i q_j - p_j q_i)(f_i - f_j) = \sum_{i=1}^{N} \sum_{j=1}^{i-1} q_i q_j \left(\frac{p_i}{q_i} - \frac{p_j}{q_j} \right) (f_i - f_j).$$

Since functions $\{f_i\}$ and $\{\frac{p_i}{q_i}\}$ are comonotonic, we have $\left(\frac{p_i}{q_i} - \frac{p_j}{q_j} \right) (f_i - f_j) \geq 0$ for i, j. Thus we get $H \geq 0$ and this theorem holds. $\qquad\square$

Corollary 6. *If PMF $\{\alpha_i\}$ is comonotonic with $\{p_i/q_i\}$, where $\{p_i\}$, $\{q_i\}$ are PMFs, and for all i, $p_i, q_i > 0$, then*

$$-\sum_{i=1}^{N} p_i \log \alpha_i = CE(\{p_i\}, \{\alpha_i\}) \leq CE(\{q_i\}, \{\alpha_i\}) = -\sum_{i=1}^{N} q_i \log \alpha_i,$$

where CE stands for cross entropy.

Observe that this result complements the results in [6], where $\{\alpha_i\}$ had to be given in increasing order.

Proof. Applying Theorem 10 with function $g(x) = \log x$, and applying $\log x$ to both sides of the inequality in Eq. 16, as $\log x$ is increasing, we obtain $\sum_{i=1}^{N} q_i \log \alpha_i \leq \sum_{i=1}^{N} p_i \log \alpha_i$, and multiplying by -1 both sides we obtain the result. $\qquad\square$

Similar theorems can be proved for countermonotonicity.

Theorem 12. *Be $\{p_i\}$ and $\{q_i\}$ PMFs satisfying $p_i, q_i > 0$ for all i. If $\{f_i\}$ is countermonotonic with $\{p_i/q_i\}$, then for all strictly monotonous function $g(x)$*

$$M_g(\{p_i\}, \{f_i\}) \leq M_g(\{q_i\}, \{f_i\}).$$

Corollary 7. *Be $\{p_i\}$ and $\{q_i\}$ PMFs satisfying $p_i, q_i > 0$ for all i. If PMF $\{\alpha_i\}$ is countermonotonic with $\{p_i/q_i\}$, then*

$$-\sum_{i=1}^{N} q_i \log \alpha_i = CE(\{q_i\}, \{\alpha_i\}) \leq CE(\{p_i\}, \{\alpha_i\}) = -\sum_{i=1}^{N} p_i \log \alpha_i.$$

Definition 9 (Comonotonic 2d arrays). *We say that 2d arrays $\{f_{ij}\}$ and $\{h_{ij}\}$ are comonotonic when for all $i, j, k, l \in [1, \ldots, N], f_{ij} > f_{kl} \implies h_{ij} \geq h_{kl}$.*

Theorem 13. *Let $\{p_{ij}\}$ and $\{q_{ij}\}$ be sequences of positive PMF. If two sequences $\{f_{ij}\}$ and $\{\frac{p_{ij}}{q_{ij}}\}$ are comonotonic, then it holds that*

$$\sum_{i=1}^{M}\sum_{j=1}^{N} f_{ij}q_{ij} \leq \sum_{i=1}^{M}\sum_{j=1}^{N} f_{ij}p_{ij}.$$

Proof. Let $H = \sum_{j=1}^{N}\sum_{l=1}^{N}\sum_{i=1}^{M}\sum_{k=1}^{M} q_{ij}p_{kl}(f_{ij} - f_{kl})$. In the same way as the proof in Theorem 12, we have

$$H = \sum_{j=1}^{N}\sum_{l=1}^{N}\sum_{i=1}^{M}\sum_{k=1}^{i-1} (q_{ij}p_{kl} - q_{kl}p_{ij})(f_{ij} - f_{kl}) + \sum_{j=1}^{N}\sum_{l=1}^{N}\sum_{i=1}^{M} q_{ij}p_{il}(f_{ij} - f_{il}).$$

Hence we exchange sums $\sum_j\sum_l$ and $\sum_i\sum_k$ in the term, and in the same way as the proof in Theorem 12, we get

$$H = -\sum_{i=1}^{M}\sum_{k=1}^{i-1}\sum_{j=1}^{N}\sum_{l=1}^{j-1} q_{ij}q_{kl}\left(\frac{p_{ij}}{q_{ij}} - \frac{p_{kl}}{q_{kl}}\right)(f_{ij} - f_{kl})$$

$$-\sum_{i=1}^{M}\sum_{k=1}^{i-1}\sum_{j=1}^{N}\sum_{l=1}^{j-1} q_{il}q_{kj}\left(\frac{p_{il}}{q_{il}} - \frac{p_{kj}}{q_{kj}}\right)(f_{il} - f_{kj})$$

$$-\sum_{i=1}^{M}\sum_{j=1}^{N}\sum_{l=1}^{j-1} q_{il}q_{ij}\left(\frac{p_{il}}{q_{il}} - \frac{p_{ij}}{q_{ij}}\right)(f_{il} - f_{ij})$$

$$-\sum_{i=1}^{M}\sum_{k=1}^{i-1}\sum_{j=1}^{N} q_{ij}q_{kj}\left(\frac{p_{ij}}{q_{ij}} - \frac{p_{kj}}{q_{kj}}\right)(f_{ij} - f_{kj}).$$

We obtain $H \leq 0$ since $\{f_{ij}\}$ and $\{\frac{p_{ij}}{q_{ij}}\}$ are comonotonic. Therefore this theorem holds. $\qquad\square$

Theorem 14. *If $\{q_{ij}\} \prec_{\mathbf{LR}} \{p_{ij}\}$ then for all 2d array $\{x_{ij}\}$ increasing both in i and j and for all strictly monotonous function $g(x)$ we have:*

$$M_g(\{q_{ij}\}, \{x_{ij}\}) \leq M_g(\{p_{ij}\}, \{x_{ij}\}). \tag{18}$$

Proof. Suppose $g(x)$ is increasing. Observe that $\{g_{ij} = g(x_{ij})\}$ is increasing in i and j, thus $\{g_{ij}\}$ is comonotonic with $\{\frac{p_{ij}}{q_{ij}}\}$, and applying Theorem 13 we obtain

$$\sum_{i=1}^{N}\sum_{j=1}^{M} q_{ij}g_{ij} \leq \sum_{i=1}^{N}\sum_{j=1}^{M} p_{ij}g_{ij}. \tag{19}$$

Then observe that $g^{-1}(x)$ exists and is increasing because $g(x)$ is strictly increasing, and now applying $g^{-1}(x)$ to both members of Eq. 19 the result comes from Definition 5. Suppose now $g(x)$ is strictly decreasing. Then $\{g_{N-i+1,M-j+1}\}$, where $\{g_{ij} = g(x_{ij})\}$, is increasing in both i,j and comonotonic with $\{\frac{q_{N-i+1,M-j+1}}{p_{N-i+1,M-j+1}}\}$. Thus applying Theorem 13 we obtain

$$\sum_{i=1}^{N}\sum_{j=1}^{M} p_{ij}g_{ij} = \sum_{i=1}^{N}\sum_{j=1}^{M} p_{N-i+1,M-j+1}g_{N-i+1M-j+1}$$

$$\leq \sum_{i=1}^{N}\sum_{j=1}^{M} q_{N-i+1,M-j+1}g_{N-i+1,M-j+1} = \sum_{i=1}^{N}\sum_{j=1}^{M} q_{ij}g_{ij}, \qquad (20)$$

and applying $g^{-1}(x)$ which exists and is strictly decreasing because $g(x)$ is strictly decreasing, to both members of the inequality its direction changes and we obtain the expected result. □

Theorem 15. $\{q_{ij}\} \prec_{\mathbf{LR}} \{p_{ij}\} \implies \{q_{ij}\} \prec_{\mathbf{FSD}} \{p_{ij}\}$.

Proof. It is a direct consequence of Theorem 14 and Theorem 6. □

Examples

(i) Observe that for $\{p_{ij}\}$ decreasing in both i,j, $\{p_{ij}\} \prec_{\mathbf{LR}} \{1/NM\}$, and if increasing $\{1/NM\} \prec_{\mathbf{LR}} \{p_{ij}\}$. Thus for liability (or cost) function $L(i,j)$ increasing in both i,j, any $\{p_{ij}\}$ distribution decreasing in both i,j is better than uniform distribution, while for a utility function $U(i,j)$ increasing in both i,j any $\{p_{ij}\}$ distribution of shares increasing in both i,j is better than uniform distribution.

(ii) Consider $\{p_{ij} = \frac{1}{M^2N^2}2M(i-1)+(2j-1)\}$, and $\{q_{ij} = \frac{1}{M^2N^2+MN}2M(i-1)+2j\}$, odd and even numbers organized as an $N \times M$ array. It is easy to check that q_{ij}/p_{ij} is decreasing in both i and j, and thus $\{q_{ij}\} \prec_{\mathbf{LR}} \{p_{ij}\}$. Thus any Kolmogorov mean of 2d array $\{f_{ij}\}$ increasing on both i and j will be less for weights q_{ij} than for weights p_{ij}. For instance, taking $f_{ij} = q_{ij}$, increasing in both i and j, and $g(x) = 1/x$, we obtain, for all positive integers M,N, the following inequality: $\dfrac{1}{2} + \dfrac{3}{4} + \dfrac{5}{6} + \ldots + \dfrac{2M(N-1)+2M-1}{2M(N-1)+2M} \leq \dfrac{M^2N^2}{MN+1}$. Taking $f_{ij} = p_{ij}q_{ij}$, increasing in both i and j, and $g(x) = 1/x$, we obtain $\dfrac{1}{2} + \dfrac{1}{4} + \dfrac{1}{6} + \ldots + \dfrac{1}{2M(N-1)+2M} \leq \left(\dfrac{NM}{NM+1}\right)1 + \dfrac{1}{3} + \dfrac{1}{5} + \ldots + \dfrac{1}{2M(N-1)+2M-1}$.

Theorem 16. *For all $1 \leq i,j \leq N$, be for fixed i, $\{q_{j|i}\} = \{q_{ij}/q_{i\cdot}\}$, $\{p_{j|i}\} = \{p_{ij}/p_{i\cdot}\}$, and for fixed j, $\{q_{i|j}\} = \{q_{ij}/q_{\cdot j}\}$, $\{p_{i|j}\} = \{p_{ij}/p_{\cdot j}\}$, i.e., the row and column conditional probabilities, then $\{q_{ij}\} \prec_{\mathbf{LR}} \{p_{ij}\} \iff (\{q_{j|i}\} \prec_{LR} \{p_{j|i}\})\&(\{q_{i|j}\} \prec_{LR} \{p_{i|j}\})$.*

Proof. For fixed i, for $j \leq l$, $\frac{q_{j|i}}{p_{j|i}} \geq \frac{q_{l|i}}{p_{l|i}} \iff \frac{q_{ij}/q_{i.}}{p_{ij}/p_{i.}} \geq \frac{q_{il}/q_{i.}}{p_{il}/q_{i.}} \iff \frac{q_{ij}}{p_{ij}} \geq \frac{q_{il}}{p_{il}}$,

and for fixed j, for $i \leq k$, $\frac{q_{i|j}}{p_{i|j}} \geq \frac{q_{k|j}}{p_{k|j}} \iff \frac{q_{ij}/q_{.j}}{p_{ij}/p_{.j}} \geq \frac{q_{kj}/q_{.j}}{p_{kj}/q_{.j}} \iff \frac{q_{ij}}{p_{ij}} \geq \frac{q_{kj}}{p_{kj}}$ □

Corollary 8. *If $\{q_{ij}\} \prec_{\mathbf{LR}} \{p_{ij}\}$ and $\{\alpha_k\}$ is an increasing 1d PMF, then for all $1 \leq i,j \leq N$, for fixed i, $CE(\{p_{j|i}\}, \{\alpha_j\} \leq \{q_{j|i}\}, \{\alpha_j\})$, and for fixed j $CE(\{p_{i|j}\}, \{\alpha_i\} \leq \{q_{i|j}\}, \{\alpha_i\})$.*

Proof. From Theorem 16 we obtain the 1d orderings between the conditional probabilities rows and columns, apply to these orderings Corollary 2 with function $g(x) = \log x$, then apply $g^{-1}(x)$ to both members of resulting inequality and finally multiply by -1 to switch the inequality. □

5   Conclusions and Future Work

We have extended in this paper first stochastic order and likelihood-ratio order between 1d PMFs to 2d PMFs. The fundamental result is that the extended orders guarantee Kolmogorov mean order invariance, as in 1d. We have hinted to applications of these orders, such as comparing 2d histograms. We have also shown the relationship of likelihood ratio order with comonotonic sequences, and given some examples of application to cross-entropy. In the future we will study whether the results presented here generalize to any dimension, and we will consider other orders such as second stochastic and increase convex orders.

Acknowledgments. M.S. has been funded by grant TIN2016-75866-C3-3-R from Spanish Government.

References

1. Shaked, M., Shanthikumar, J.G.: Stochastic Orders. Springer, New York (2007). https://doi.org/10.1007/978-0-387-34675-5
2. Belzunce, F., Martinez-Riquelme, C., Mulero, J.: An Introduction to Stochastic Orders. Academic Press, Elsevier (2016)
3. Hadar, J., Russell, W.: Rules for ordering uncertain prospects. Am. Econ. Rev. **59**(1), 25–34 (1969)
4. Sbert, M., Havran, V., Szirmay-Kalos, L., Elvira, V.: Multiple importance sampling characterization by weighted mean invariance. Vis. Comput. **34**, 843–852 (2018). https://doi.org/10.1007/s00371-018-1522-x
5. Maes, F., Collignon, A., Vandermeulen, D., Marchal, G., Suetens, P.: Multi-modality image registration by maximization of mutual information. IEEE Trans. Med. Imaging **16**(2), 187–198 (1997)
6. Sbert, M., Poch, J., Chen, M., Bardera, A.: Some order preserving inequalities for cross entropy and Kullback-Leibler divergence. Entropy **20**(959), 1–10 (2018)
7. Sbert, M., Poch, J.: A necessary and sufficient condition for the inequality of generalized weighted means. J. Inequalities Appl. **2016**(2), 292 (2016)
8. Yoshida, Y.: Weighted quasi-arithmetic means on two-dimensional regions: an independent case. In: Torra, V., Narukawa, Y., Navarro-Arribas, G., Yañez, C. (eds.) MDAI 2016. LNCS (LNAI), vol. 9880, pp. 82–93. Springer, Cham (2016). https://doi.org/10.1007/978-3-319-45656-0_7

9. Renneberg, D.: Non Additive Measure and Integral. Kluwer Academic Publ., Dordrecht (1994)
10. Dellacherie, C.: Quelques commentarires sur les prolongements de capacités, Séminare de Probabilites 1969/1970, Strasbourg. In: LNAI, vol. 191, pp. 77–81. Springer (1971)
11. Yoshida, Y.: Weighted quasi-arithmetic means and a risk index for stochastic environments. Int. J. Uncertain. Fuzziness Knowl. Based Syst. (IJUFKS) **16**(suppl.), 1–16 (2011)

Modeling Decisions in AI: Re-thinking Linda in Terms of Coherent Lower and Upper Conditional Previsions

Serena Doria[1]([⊠]) [iD] and Alessandra Cenci[2] [iD]

[1] Department of Engineering and Geology, University G.d'Annunzio,
67100 Chieti, Italy
serena.doria@unich.it
[2] Institute for the Study of Culture, Department of Philosophy,
University of Southern Denmark, Campusvej 55, 5230 Odense M, Denmark
alessandra@sdu.dk

Abstract. In this paper, the model of coherent upper and lower conditional previsions is proposed to represent preference orderings and equivalences between random variables that human beings consider consciously or unconsciously when making decisions. To solve the contradiction, Linda's Problem (i.e., conjunction fallacy) is re-interpreted in terms of the probabilistic model based on coherent upper and lower conditional probabilities. Main insights of this mathematical solution for modeling decisions in AI are evidenced accordingly.

Keywords: Coherent upper and lower conditional previsions · Linda's Problem · Human decision making · Conscious and unconscious thought

1 Introduction

Whether and how rational decisions are/should be taken under risk or uncertainty are central questions but explained rather differently in several different disciplines (philosophy, economics, sociology, cognitive psychology, probability theory and so on). Recently, these same issues became crucial in AI as well. Traditionally, mathematical models aim to illustrate how optimal decisions under certain constraints could be achieved. However, several studies in different disciplines have demonstrated that decision-making models based on standard rational behavior (Rational Choice Theory - RCT; Expected Utility Theory - EUT) diverge substantially from how human beings, cognitively imperfect rational agents form their preference orderings and take decisions in real contexts. Reasons behind foremost deviations have been explained in different ways. Undoubtedly, one of the most popular interpretations is represented by the overall work of Daniel Kahneman and Amos Tversky (and others) in cognitive psychology and behavioral economics. Overall, they present a critique of EUT in

V. Torra et al. (Eds.): MDAI 2020, LNAI 12256, pp. 41–52, 2020.
https://doi.org/10.1007/978-3-030-57524-3_4

terms of a descriptive model of rational behavior; namely, the Prospect Theory [12] which became one of the main founding theories when using experimental methods. By means of different experiments, a number (48) of representativeness heuristics (i.e., biases of human intuition) have been identified and used to explore and explain the functioning of human reasoning, human mind and brain. A compendium of these ideas is used to build the two systems account (System 1 and 2) which, by further relying on the dual process theory, depicts the brain activity as regulated by two rather different ways of thinking; namely, fast and slow thinking [13]. System 1 regulates intuitive, involuntary, unconscious and effortless activities (driving, reading angry facial expressions etc.) and is mainly based on immediate heuristics. System 2 is the conscious part of the brain in charge of logical reasoning, articulates judgments, making choices, endorses or rationalizes ideas and feelings, thus, it requires slowing down and not jumping to quick conclusions, deliberating, solving problems, reasoning, computing. This approach is applied in computer science: a heuristic is a technique designed for solving a problem more quickly, when classic optimizing methods are too slow, or for finding an approximate result when classic methods fail to find any exact solution. One of the most discussed heuristics (the n. 15), is the so-called "conjunction fallacy", also known as Linda's problem [21] that exemplifies how real human decision-making under risk or uncertainty constantly violates the logic and most fundamental rules of probability calculus - which is vital when making predictions (also in AI). Briefly, after hearing priming details about a made-up person (Linda), people choose a plausible story over a probable story. Logically, it is more likely that a person will have one characteristic than two characteristics. However, after reading a priming description of Linda most of respondents are more likely to give her two characteristics. Indeed, the more details one adds to a description, forecast, or judgment the less likely they are to be probable. Kahneman [13] newly explains the fallacy by assuming the two systems thinking: the fallacy occurs since System 1/the unconscious overlooks logic in favour of a plausible story. Here, intuition favours what is plausible but improbable over what is implausible and probable since logical reasoning in System 2/the conscious does not intervene at all. Fundamentally, heuristics-based, descriptive interpretations of rational behaviour, although no more rational in the traditional sense of normative, are known to increase realism thus, can be more exploratory and explanatory of both RCT and EUT axioms since tackles on real choices made by real agents. Even so, the increased complexity of real human behaviour and shortages of information unavoidably suffered by rational decision-makers are significant challenges to reliable predictive analysis. They evaded in classical models, precisely, by the espousal of restrictive notion of rationality (RCT) that also underlie the unconditional probability (EUT) those, in view of obtaining optimal results, had the function of limiting the information considered valid, both its nature and amount. In this paper, a mathematical re-interpretation of Linda's dilemma - that avoids the fallacy but involving contexts complexity, partial knowledge - is proposed as an alternative to both standard decision models and heuristics-based explanations. Its insights for modeling decisions in

AI are stressed in the conclusions. Another related matter which arises in the study of human decision-making is the role of emotions and unconscious thinking. According to Matte Blanco [14] conscious and unconscious activities are two different modes of being. Specifically, he draws a distinction between logical conscious/asymmetrical thought - structured on the categories of time and space and ruled by Aristotle's principle of non-contradiction and unconscious/symmetrical thought - based upon the principles of symmetry and generalization. Both types of thoughts are supposed to combine in different human thinking experiences since they yield a bi-logic asset. Emotions are ways to reach and decode the unconscious since they function at the same way. Both scenarios stress the importance of emotion-driven, unconscious thinking in rational decision-making and inspired the main question that this paper attempts to answer: *can more realistic and unconscious choices - embodying complexity and partial knowledge - be mathematically described and are there partial orderings and relations that manage to better represent human preference orderings and equivalences between random variables?* In the attempt of providing a positive answer to the above, this paper, differently from EUT concepts criticized by Kahneman and Tversky, proposes adopting two different measures of uncertainty such as coherent lower and upper conditional probabilities to represent partial knowledge and coherent lower and upper previsions to represent partial orderings and equivalence between random variables [10]. The model of coherent upper and lower conditional probabilities based on Hausdorff outer and inner measures allows to assess conditional probability according to the complexity of the conditioning event that represent the given information. Here, the two non-linear functionals represent different binary relations between random variables since preference orderings described in terms of coherent lower conditional previsions satisfy the antisymmetric property which is not satisfied by the binary relation represented by their conjugate coherent upper conditional previsions. Unlike predictive models based on EUT unconditional probabilities, causing Linda's problem, the mathematical model of coherent upper and lower conditional probabilities most profitably deal with the complexity of real choice environments, partial knowledge (usually considered detrimental or even precluding for predicting events) can solve the fallacy. Accordingly, the mathematical solution based on conditional probability offers a re-interpretation of Kahneman's System 1–2 concerning the role of emotions and unconscious thinking in rational decisions: they can be considered in ways analogous to the modalities described by Matte Blanco.

2   Representation of Partial Knowledge by Means of Coherent Upper and Lower Conditional Probabilities

Let Ω be a non-empty set let $\mathbf{B}$ be a partition of Ω. A bounded random variable is a function $X : \Omega \to \Re$ and $L(\Omega)$ is the class of all bounded random variables defined on Ω; for every $B \in \mathbf{B}$ denote by $X|B$ the restriction of X to B and by $\sup(X|B)$ the supremum value that X assumes on B. Let $L(B)$ be the class of all bounded random variables $X|B$. Denote by I_A the indicator function of any event $A \in \wp(B)$, i.e. $I_A(\omega) = 1$ if $\omega \in A$ and $I_A(\omega) = 0$ if $\omega \in A^c$. For every $B \in \mathbf{B}$ coherent upper conditional previsions $\overline{P}(\cdot|B)$ are functionals defined on $L(B)$ [22].

Definition 1. *Coherent upper conditional previsions are functionals* $\overline{P}(\cdot|B)$ *defined on* $L(B)$, *such that the following conditions hold for every* X *and* Y *in* $L(B)$ *and every strictly positive constant* λ:

1) $\overline{P}(X|B) \leq \sup(X|B)$;
2) $\overline{P}(\lambda X|B) = \lambda\overline{P}(X|B)$ *(positive homogeneity)*;
3) $\overline{P}(X + Y)|B) \leq \overline{P}(X|B) + \overline{P}(Y|B)$ *(subadditivity)*;
4) $\overline{P}(I_B|B) = 1$.

Suppose that $\overline{P}(X|B)$ is a coherent upper conditional prevision on $L(B)$ then its conjugate coherent lower conditional prevision is defined by

$$\underline{P}(X|B) = -\overline{P}(-X|B).$$

Let K be a linear space contained in $L(B)$; if for every X belonging to K we have $P(X|B) = \underline{P}(X|B) = \overline{P}(X|B)$ then $P(X|B)$ is called a coherent *linear* conditional prevision [2, 9, 16] and it is a linear, positive and positively homogenous functional on $L(B)$. The unconditional coherent upper prevision $\overline{P} = \overline{P}(\cdot|\Omega)$ is obtained as a particular case when the conditioning event is Ω. Coherent upper conditional probabilities are obtained when only 0–1 valued random variables are considered. An upper prevision is a real-valued function defined on some class of bounded random variables K. A necessary and sufficient condition for an upper prevision $\overline{P}$ to be coherent is to be the *upper envelope* of linear previsions, i.e. there is a class M of linear previsions such that $\overline{P} = \sup\{P : P \in M\}$.

2.1 Partial Preference Orderings and Equivalence Between Random Variables Represented, Respectively, by Lower and Upper Conditional Previsions

Let $\overline{P}(X|B)$ be a coherent upper conditional prevision and let $\underline{P}(X|B)$ its conjugate lower conditional prevision. A partial strict order, which is an antisymmetric and transitive binary relation between random variables, can be represented by the lower conditional prevision $\underline{P}(X|B)$.

Definition 2. *We say that* X *is preferable to* Y *given* B *with respect to* $\underline{P}$, *i.e.* $X \succ_* Y$ *in* B *if and only if*

$$\underline{P}((X - Y)|B) > 0$$

Here, the binary relation $\succ_*$ is proven to satisfy the antisymmetric property.

Proposition 1. *Let $X, Y \in L(B)$ two random variables such that X is preferable to Y given B with respect to $\underline{P}$, i.e. $X \succ_* Y$ in B, then Y is not preferable to X given B. Proof. We have that*

$$X \succ_* Y \Longleftrightarrow \underline{P}((X - Y)|B) > 0 \Longrightarrow$$
$$\underline{P}((Y - X|B) \leq 0 \Longleftrightarrow Y \, not \succ_* X.$$

In fact

$$0 < \underline{P}((X - Y)|B) < \overline{P}((X - Y)|B) \Longrightarrow$$
$$\overline{P}((X - Y)|B) = -\underline{P}((Y - X)|B) > 0$$

so that $\overline{P}((Y - X)|B) < 0$ that is $Y \, not \succ_* X$.

Remark 1. *Two random variables which have previsions equal to zero cannot be compared by the ordering $\succ_*$.*

A binary relation $\propto$ can be defined on $L(B)$ with respect to $\overline{P}$ but it cannot represent a strict preference ordering because it does not satisfy the antisymmetric property.

Definition 3. *We say that $X \propto Y$ given B if and only if $\overline{P}((X - Y)|B) > 0$.*

The following example show that the relation $\propto$ does not satisfy the antisymmetric property.

Example 1. *Let $X, Y \in L(B)$ such that $\overline{P}((X-Y)|B) > 0$ and $\underline{P}((X-Y)|B) < 0$; then*

$$\overline{P}((X - Y)|B) > 0 \ does \ not \ imply \ \overline{P}((Y - X)|B) < 0$$

since

$$\overline{P}((Y - X)|B) < 0 \Longleftrightarrow -\underline{P}((X - Y)|B) < 0 \Longleftrightarrow \underline{P}((X - Y)|B) > 0.$$

Two complete equivalence relations, which are reflexive, symmetric and transitive binary relations on $L(B)$ can be represented by the coherent upper conditional prevision $\overline{P}(X|B)$.

Definition 4. *Two random variables X and $Y \in L(B)$ are equivalent given B with respect to $\overline{P}$ if and only if $\overline{P}(X|B) = \overline{P}(Y|B)$.*

If two random variables are equivalent given B with respect to the coherent upper conditional prevision, by the conjugacy property the random variable $-X$ and $-Y$ are equivalent given B with respect to the lower conditional prevision. In fact, by the conjugacy property we have

$$\overline{P}(X|B) = \overline{P}(Y|B) \Longleftrightarrow -\underline{P}(-X|B) = -\underline{P}(-Y|B) \Longleftrightarrow \underline{P}(-X)|B) = \underline{P}(-Y|B).$$

Definition 5. *We say that X and Y are indifferent given B with respect to $\overline{P}$, i.e. $X \approx Y$ in B if and only if*

$$\overline{P}((X - Y)|B) = \overline{P}((Y - X)|B) = 0.$$

By coherence, we have that $|\overline{P}(X) - \overline{P}(Y)| \leq \overline{P}(|X - Y|)$ (Walley 2.6.1.) so that if two random variables are indifferent then they are equivalent. In the next section the previous results are considered to analyze Linda's Problem.

3   Linda's Problem Revisited

In this section, the experiment by Tversky and Kahneman [21], that violates the logic of probability calculus is examined and revisited. Instead, we propose coherent upper and lower conditional probabilities to represent both unconscious and conscious thought (i.e., Kahneman's System 1 and System 2). System 1 heuristics (i.e., mental techniques leading to fast conclusions with minimal effort) are described by equivalences between events represented by a coherent upper conditional probability such that the conjugate coherent lower conditional probability assesses zero probability to the same events. Following this paper approach, these events cannot be compared by the conscious thought (System 2). Let's start with the original experimental tests in which participants were presented with the following dilemma: Linda is 31 years old, single, outspoken, and very bright. She majored in philosophy. As a student, she was deeply concerned with issues of discrimination and social justice, and also participated in anti-nuclear demonstrations. Which event about Linda is currently more likely?

(1) Linda is a bank teller.
(2) Linda is a bank teller and is active in the feminist movement.

85% of those asked chose option (2), but the correct answer is option (1). If we represent partial knowledge by a unique probability as in EUT is required that the probability of two events occurring together (in "conjunction") is always less than or equal to the probability of either one occurring alone. Most people get this problem wrong because they rely on representativeness heuristics (the n. 15) to make this kind of judgment: option (2) seems more "representative" of Linda based on the description of her, even though it is clearly mathematically less likely. It might occur since people perceive the option (2) as closer to the complexity of her description than the less complex but most probable option (1). A simpler, more general proposition has more chances of being true, but seen through our biased mental lens, more detailed and specific propositions seem more probable. The conjunction fallacy arose when espousing EUT notion of unconditional probability i.e., standard probability calculus states that the probability that Linda is a bank teller has to be at least "as great as" the probability that she is a bank teller and a feminist. Conversely, in the case represented in Linda's experiment, respondents consider the probability that Linda is bank teller and a feminist greater than the probability she is a bank teller. Thus, heuristic-based explanation is used to describe how decisions are taken in similar cases.

If we analyze the Linda's Problem in terms of coherent upper and lower conditional probabilities unconscious answers given by people can be represented by the coherent upper conditional probability $\overline{P}$ while conscious answers can be represented by coherent lower probability $\underline{P}$. Consider the events

E: "'Linda is a bank teller."'
F: "'Linda is active in the feminist movement."'
$E \cap F$: "'Linda is a bank teller and is active in the feminist movement."'
B: "'Linda is 31 years old, single, outspoken, and very bright. She majored in philosophy. As a student, she was deeply concerned with issues of discrimination and social justice, and also participated in anti-nuclear demonstrations."'

Let $S = \{\varnothing, \Omega, E \cap F|B, E|B, F|B\}$; let μ be the monotone set function on S such that $\mu(E \cap F|B) = \frac{1}{3}, \mu(E|B) = \frac{1}{3}, \mu(F|B) = \frac{3}{4}, \mu(\varnothing) = 0, \mu(\Omega) = 1$. Then, we extend the monotone set function μ to any subset A of Ω by its outer and inner measures

$$\overline{P}(A) = inf\left\{\mu(H) : H \supseteq A; H \in S\right\} \forall A \in \wp(\Omega)$$

and

$$\underline{P}(A) = sup\left\{\mu(M) : M \subset A; M \in S\right\} \forall A \in \wp(\Omega)$$

The outer measure is a coherent upper prevision and the inner measure is a coherent lower probability. Hence, the following assessment can be considered:

	$E \cap F|B$	$E|B$	$F|B$	$(E \cap F)^c|B$	$E^c|B$	$F^c|B$
$\overline{P}$	$\frac{1}{3}$	$\frac{1}{3}$	$\frac{3}{4}$	1	1	1
$\underline{P}$	0	0	0	$\frac{2}{3}$	$\frac{2}{3}$	$\frac{1}{4}$

The conjunction fallacy does not occur because

$$\underline{P}(E \cap F|B) \leq \underline{P}(E|B) \text{ and } \overline{P}(E \cap F|B) \leq \overline{P}(E|B).$$

Moreover, we obtain

$$\overline{P}(E \cap F|B) = \frac{1}{3} = \overline{P}(E|B) \leq \overline{P}(F|B)$$

and

$$\overline{P}(F^c|B) < \overline{P}(F|B)$$

which represents the preference ordering assigned by System 1 in Linda experiment, that is the event $E \cap F$ and E are equivalent but the majority of the participants assess a greater probability to the event F (Linda is feminist) than to the event F^c given the information B they receive about Linda. Since

$$\underline{P}(E|B) = \underline{P}(F|B) = \underline{P}(E \cap F|B) = 0$$

the events $E \cap F$, E and F cannot be compared with respect to the lower conditional probability $\underline{P}(\cdot|B)$ because they have probability zero given B. This last result describes a situation in which System 2 is not yet involved in the assessment of a preference ordering on the events $E \cap F$, E and F. This result points up the problem in analyzing how events with zero upper and/or lower conditional probability should be compared. In the next section, we describe how the model of coherent upper and lower conditional previsions based on Hausdoff outer and inner measures can describe the knowledge updating and the process of awareness of human being when new information is available and also, how the assessments on the class S can be given to take the information about Linda into account.

4   Knowledge Updating

A new model of coherent upper conditional probability based on Hausdorff outer measures is introduced in [3–8]. For the definition of Hausdorff outer measure and its basic properties see [17] and [10]. For every $B \in \mathbf{B}$ denote by s the Hausdorff dimension of B and let h^s be the Hausdorff s-dimensional Hausdorff outer measure associated to the coherent upper conditional prevision. For every bounded random variable X a coherent upper conditional prevision $\overline{P}(X|B)$ is defined by the Choquet integral with respect to its associated Hausdorff outer measure if the conditioning event has positive and finite Hausdorff outer measure in its Hausdorff dimension. Otherwise if the conditioning event has Hausdorff outer measure in its Hausdorff dimension equal to zero or infinity it is defined by a 0–1 valued finitely, but not countably, additive probability.

Theorem 1. *Let (Ω, d) be a metric space and let $\mathbf{B}$ be a partition of Ω. For every $B \in \mathbf{B}$ denote by s the Hausdorff dimension of the conditioning event B and by h^s the Hausdorff s-dimensional outer measure. Let m_B be a 0-1 valued finitely additive, but not countably additive, probability on $\wp(B)$. Then for each $B \in \mathbf{B}$ the functional $\overline{P}(X|B)$ defined on $L(B)$ by*

$$\overline{P}(X|B) = \begin{cases} \frac{1}{h^s(B)} \int_B X dh^s & \text{if } 0 < h^s(B) < +\infty \\ m_B & \text{if } h^s(B) \in \{0, +\infty\} \end{cases}$$

is a coherent upper conditional prevision.

When the conditioning event B has Hausdorff outer measure in its Hausdorff dimension equal to zero or infinity, an additive conditional probability is coherent if and only if it takes only 0–1 values. Because linear previsions on $L(B)$ are uniquely determined by their restrictions to events, the class of linear previsions on $L(B)$ whose restrictions to events take only the values 0 and 1 can be identified with the class of 0–1 valued additive probability defined on all subsets of B.

Theorem 2. *Let (Ω, d) be a metric space and let $\boldsymbol{B}$ be a partition of Ω. For every $B \in \boldsymbol{B}$ denote by s the Hausdorff dimension of the conditioning event B and by h^s the Hausdorff s-dimensional outer measure. Let m_B be a 0-1 valued finitely additive, but not countably additive, probability on $\wp(B)$. Thus, for each $B \in \boldsymbol{B}$, the function defined on $\wp(B)$ by*

$$\overline{P}(A|B) = \begin{cases} \frac{h^s(A \cap B)}{h^s(B)} & if\ 0 < h^s(B) < +\infty \\ m_B & if\ h^s(B) \in \{0, +\infty\} \end{cases}$$

is a coherent upper conditional probability.

The innovative aspect of the proposed model consists on the fact that the measure used to define conditional upper probability depends on the complexity of the conditioning event, given in terms of Hausdorff dimension of the set B. Therefore, the events with a zero-value a priori probability determine the change of the measure of uncertainty that represents the level of knowledge of the subject. In this section, we analyze how the process of upgrading the level of knowledge can be described by the model of coherent upper conditional probabilities defined by Hausdorff outer measures. This model describes variations of the uncertainty measure adopted by a subject when new information illustrated by the B_1 set is assumed. The update procedure is equal for both coherent lower conditional probability and coherent upper conditional probability. The set B depicted the subject's initial information. The degree of confidence of a subject on an event A is represented with an interval, which represents the belief interval that an individual gives to the event A based on one's own experience

$$[\underline{P}(A|B); \overline{P}(A|B)].$$

If the set B has positive and finite Hausdorff measure in its Hausdorff dimension t, then the a priori probability of an event A is defined by choosing the set B as a conditioning set. The a priori probability is defined by

$$\overline{P}(A|B) = \frac{h^t(A \cap B)}{h^t(B)} \text{ if } 0 < h^t(B) < +\infty.$$

The initial measure of uncertainty, representing the level of knowledge of the subject, is defined by means of the Hausdorff measure h^t, where t is the Hausdorff dimension of B. Let the set B_1 a new piece of information. If we denote by s the Hausdorff dimension of the set B_1 and by h^s the s-dimensional Hausdorff measure, thus we can define the conditional or a posteriori probability of an event A given B_1 in the following way

$$\overline{P}(A|B_1) = \frac{h^s(A \cap B_1)}{h^s(B_1)} \text{ if } 0 < h^s(B_1) < +\infty.$$

The innovative measure used to define the conditional probability depends on the complexity of the conditioning event measures in terms of Hausdorff dimension. It is noteworthy to note that if the Hausdorff dimension of the set $A \cap B$ is lower than s, that is the Hausdorff dimension of B_1, we have: $h^s(A \cap B) = 0$ and therefore $\overline{P}(A|B_1) = 0$. The a priori probability of the event B_1 is defined similarly, although by using B as the conditioning set. Denoted with t the Hausdorff dimension of B, the a priori probability is defined by

$$\overline{P}(B_1|B) = \frac{h^t(B_1 \cap B)}{h^t(B)} \text{ if } 0 < h^t(B) < +\infty.$$

If the upper a priori probability assigned to B_1 is positive, that is $\overline{P}(B_1|B) > 0$, it implies that the Hausdorff dimension of B^1 is equal to t, that is it is equal to the Hausdorff dimension of B. In this case, the measurement of a posteriori uncertainty of information B_1 is always defined by the outer t-dimensional Hausdorff outer measure. Then, no change occurs in the assessment of the events by the subject. Otherwise, if the new piece of information B_1 is represented by a set which has Hausdorff dimension equal to $s < t$, that is the a priori probability of the event B_1 is zero, arises the problem to update the information conditionally on the event B_1. If B_1 has positive and finite Hausdorff outer measure in its Hausdorff dimension s then the conditional probability will be defined by the s-dimensional Hausdorff measure, no longer by the t-dimensional Hausdorff outer measure. Then, another measure needs to be used to represent the subject's level of knowledge conditioned to the event B_1 thus, the subject updates his knowledge.

5   Conclusions

The main contribution of this paper is that, by adopting the concept of conditional probability instead of the traditional unconditional probability (EUT), the fallacy is resolved. Two states of belief are simultaneously held thus, differently from the original experiment, Linda can be both a bank teller and a feminist. What is important to remark is that the model presented in his paper offers a logical escape to the fallacy - instead of the merely descriptive explanation grounded on heuristics. Why is this result so important to advance AI modeling? Which problems, frequently related to the growing AI impact on future societies, is able to better tackles on? A recent report written by a multidisciplinary expertise of over 150 scholars (computer scientists, sociologist, philosophers, policy makers etc.) indicates that most pressing AI challenges are the lack of agency, freedom, democracy associated with the opacity of algorithmic-based decision making and correlated processes [15]. In our (and other's) view, it has very much to do with how choices are conceptualized in AI related disciplines such as computational social choice [1], a discipline in between computer science and economics social choice theory, when modeling algorithms-based predictive systems decision with the aim of handling risk or uncertainty in ways that optimal decisions - under certain constraints - can always be achieved. Yet, resulting predictive models are known to be highly idealized, reductive and biased concerning the working ideals of "good system performance" assumed (only optimal outcomes are acceptable). Most negative effects are for the reliability of predictions but also for AI practical applications since applied works regularly deal with non-optimal situations and solutions. Thus, increasing the realism of conventional predictive models - by augmenting the complexity of models' assumptions thus, the amount of heterogeneous information used in predictive exercises – seems pivotal. The most important contribution of the outlined model for advancing current debates in machines decision making, machines learning etc. is about depicting a logic, precise, accurate rational decision-making procedure under risk or uncertainty that can be more similar to human real decisions avoiding biased results usually connected with augmented information, contexts complexity. In other words, the mathematical solution adopted offers an analytical decision-making for AI modeling that can be more similar to multifaceted human decisions (perspectival, value-based, emotion-driven etc.) while evading the impossibility of getting cogent results. It could positively increase AI modeling concern for human agency, free-will, positive freedom without undermining the consistency or the possibility of getting exact predictive outcomes. As a matter of fact, our model - based on conditional probability rather than on EUT unconditional probability - can most profitably handle structural lack of information that real rational agents encounter in real choices (we say: complexity and partial knowledge). These aspects are neglected in standard decision models by supposing restrictive notions of rationality (RCT axioms of completeness, transitivity) which are known to negatively constrain predictive models' informative basis (see [18]). Important limitations collapse in our solution to the conjunction fallacy since the underlying model is able to illustrate - mathematically - how rational choices under risk or uncertainty (we say: involving complexity and partial knowledge) can take place without violating the logic of probability calculus (differently understood). Our model major contribution (i.e., the mathematical solution to the conjunction fallacy) is that it meaningfully embodies and handles intuitions bias, influence of the contexts, asymmetrical or missing information (we say: complexity and partial knowledge) formerly addressed only descriptively by models of bounded rationality or heuristics-based explanations of rational behavior. It is important to remark that the logic behind the mathematical solution proposed copes with key aspects of [11] critique of Kahneman's and Tversky's interpretation of Linda. Similarly, it naturally connects with latest normative - not descriptive - advancements in theory of rationality, rational social choice theory (see [19,20]) in which incompleteness, incomplete preference orderings are no longer considered detrimental for a rational social choice - even though often imply renouncing to optimality in favor of maximal choices. Major advantages are modeling collective decision-making optimizing procedures in which human diversity, value pluralism (i.e., complexity) and/or unavoidable disagreements, lack of information (i.e., partial knowledge) do not impede a social choice (as in axiomatic accounts like Kenneth Arrow's impossibility theorem and other later impossibility results). To conclude, the mathematical model proposed, by assuming two non-additive conditional probabilities, can help in modeling strategies better handling what is normally considered reasoning bias (heuristics) or rationality failures (incompleteness). It demonstrates that both aspects can be converted in logical reasoning thus, always delivering exact and reliable predictive results. Thus, once compared to former accounts, our model offers several new insights that open the way to apply most recent Bayesian advances to modeling decisions in AI.

References

1. Brandt, F., Conitzer, V., Endriss, U., Lang, J., Procaccia, A.D.: Handbook of Computational Social Choice. Cambridge University Press, Cambridge (2016)
2. de Finetti, B.: Theory of Probability. Wiley, London (1974)
3. Doria, S.: Probabilistic independence with respect to upper and lower conditional probabilities assigned by Hausdorff outer and inner measures. Int. J. Approx. Reason. **46**, 617–635 (2007)
4. Doria, S.: Characterization of a coherent upper conditional prevision as the Choquet integral with respect to its associated Hausdorff outer measure. Ann. Oper. Res. **195**, 33–48 (2012). https://doi.org/10.1007/s10479-011-0899-y
5. Doria, S.: Symmetric coherent upper conditional prevision by the Choquet integral with respect to Hausdorff outer measure. Ann. Oper. Res. **229**, 377–396 (2014)
6. Doria, S.: On the disintegration property of a coherent upper conditional prevision by the Choquet integral with respect to its associated Hausdorff outer measure. Ann. Oper. Res. **216**(2), 253–269 (2017)
7. Doria, S., Dutta, B., Mesiar, R.: Integral representation of coherent upper conditional prevision with respect to its associated Hausdorff outer measure: a comparison among the Choquet integral, the pan-integral and the concave integral. Int. J. Gen. Syst. **216**(2), 569–592 (2018)
8. Doria, S.: Preference orderings represented by coherent upper and lower previsions. Theory Decis. **87**, 233–259 (2019)
9. Dubins, L.E.: Finitely additive conditional probabilities, conglomerability and disintegrations. Ann. Probab. **3**, 89–99 (1975)
10. Falconer, K.J.: The Geometry of Fractal Sets. Cambridge University Press, Cambridge (1986)
11. Hertwig, R., Gigerenzer, G.: The 'conjunction fallacy' revisited: how intelligent inferences look like reasoning errors. J. Behav. Decis. Mak. **12**, 275–305 (1999)
12. Kahneman, D., Tversky, A.: Prospect theory: an analysis of decision under risk. Econometrica **47**(2), 263–291 (1979)
13. Kahneman, D.: Thinking, Fast and Slow, Farrar, Straus and Giroux (2011)
14. Matte Blanco, I.: The Unconscious as Infinite Sets: An Essay on Bi-logic. Gerald Duckworth, London (1975)
15. Peer Research Centre: Artificial intelligence and the future of humans (2018). www.pewresearch.org
16. Regazzini, E.: De Finetti's coherence and statistical inference. Ann. Stat. **15**(2), 845–864 (1987)
17. Rogers, C.A.: Hausdorff Measures. Cambridge University Press, Cambridge (1970)
18. Sen, A.: Prediction and economic theory. Proc. R. Soc. Lond. A **407**, 3–23 (1986)
19. Sen, A.: Reasoning and justice: the maximal and the optimal. Philosophy **92**, 5–19 (2017)
20. Sen, A.: The importance of incompleteness. Int. J. Econ. Theory **14**, 9–20 (2018)
21. Tversky, A., Kahnemann, D.: Extensional versus intuitive reasoning: the conjunction fallacy in probability judgment. Psycological Rev. **90**(4), 293 (1983)
22. Walley, P.: Statistical Reasoning with Imprecise Probabilities. Chapman and Hall, London (1991)

Ensemble Learning, Social Choice and Collective Intelligence
An Experimental Comparison of Aggregation Techniques

Andrea Campagner[(✉)], Davide Ciucci, and Federico Cabitza

Dipartimento di Informatica, Sistemistica e Comunicazione,
University of Milano–Bicocca, viale Sarca 336, 20126 Milan, Italy
a.campagner@campus.unimib.it

Abstract. Ensemble learning provides a theoretically well-founded approach to address the bias-variance trade-off by combining many learners
to obtain an aggregated model with reduced bias or variance. This same
idea of extracting knowledge from the predictions or choices of individuals has been also studied under different perspectives in the domains
of *social choice theory* and *collective intelligence*. Despite this similarity,
there has been little research comparing and relating the aggregation
strategies proposed in these different domains. In this article, we aim to
bridge the gap between these disciplines by means of an experimental
evaluation, done on a set of standard datasets, of different aggregation
criteria in the context of the training of ensembles of decision trees. We
show that a social-science method known as *surprisingly popular decision* and the three-way reduction, achieve the best performance, while
both bagging and boosting outperform social choice-based Borda and
Copeland methods.

Keywords: Ensemble learning · Social choice theory · Collective
intelligence · Aggregation criteria

1 Introduction

Ensemble learning, currently one of the most successful Machine Learning (ML)
paradigms, emerged from Kearns and Valiant question [26,27] of whether it
would be possible to aggregate a set of high-bias models into a single low-bias
one or, less technically put, if it would be possible to obtain a very accurate predictor from a set of predictors slightly better than chance. Since then, ensemble
learning, mainly through its two main variants *bagging* [5] and *boosting* [19], has
grown into a theoretically well-founded [37] and empirically successful approach
to tackle the bias-variance trade-off.

Interestingly, this same idea of extracting *knowledge* from a collection of
not necessarily well-performing individual agents have been fruitfully explored
also in other scientific disciplines by scholars who focus on *social choice theory*,
a theoretical framework to study how to aggregate individual preferences or

© Springer Nature Switzerland AG 2020
V. Torra et al. (Eds.): MDAI 2020, LNAI 12256, pp. 53–65, 2020.
https://doi.org/10.1007/978-3-030-57524-3_5

opinions to reach collective welfare, and on *collective intelligence*, an umbrella term to account for the decision-making capabilities of groups and collectives.

Despite these similarities, the cross-fertilization between these disciplines have, until recently, been limited, with each community focusing on different aggregation methods and focusing on different concerns and priorities, with little or no interaction. However, in the recent years, several studies have tried to bridge this disciplinary gap, by acknowledging the manifest similarities and proposing some sort of integration of their methodologies. In this recent mould, most works have focused on the application of voting mechanisms for model aggregation in ensemble learning [12,29,30], or the application of ensemble learning, and more in general Machine Learning (ML), approaches in collective intelligence and crowdsourcing scenarios [2,8,23,36,42]. Other works also considered the development of ensemble learning strategies based on collective intelligence methods [31], or the study of ensemble learning methodologies from the perspective of social choice theory [10,33].

This article represents a contribution in this strand of research, by empirically evaluating different aggregation techniques that have been proposed in these different fields as model combination techniques in ensemble learning. Compared to recent works in this strand of research [12,29,30], we provide a more extensive comparison of different ensembling techniques in that we both compare a larger number of aggregation methods and also evaluate them on a larger collection of benchmark datasets, of varying complexity. Our work also provides a first comparison, at least to the authors' knowledge, of collective intelligence techniques in the field of ensemble methods. Particularly, we present novel ensembling algorithms based on the recently proposed "surprisingly popular decision" algorithm [35] and on crowdsourcing strategies [8] based on possibility theory [15,41] and three-way decision [40].

2 Methods

2.1 Background

In this section, we recall some basic notions about ensemble learning methods, collective intelligence techniques and voting rules in social choice theory.

Ensemble Methods. Ensemble learning refers to a set of techniques to aggregate different ML models into a single model with reduced bias and/or variance. Formally, given a set of predictors (regression models or classifiers) $h_1, ..., h_n$, the goal is to obtain a combined predictor h_{ens} such that its bias and/or variance is lower than each of the h_i. Among many different variants, the most popular methodologies are *boosting* and *bagging*.

Bagging, which is at the heart of the popular *random forest* algorithm [6], revolves around the idea of training a set of classifiers on a collection of different datasets obtained by boostrapping (i.e. sampling with replacement) of the training set. Specifically, given an initial dataset D, a number n of synthetic datasets

$D_1, ..., D_n$ are obtained by sampling uniformly with replacement from D (i.e., $\forall i, |D_i| = |D|$). A classifier h_i is trained on each dataset D_i and the *bagging* model is obtained by combination of $h_1, ..., h_n$ through simple *plurality voting* (usually weighted by the prediction score of each model, in order to take into account the confidence in the predictions).

Boosting techniques, on the other hand, are based on an iterative procedure: a initial model h_0 is trained and evaluated on the given training set D_0. The dataset D_0 is then modified by a weighting mechanism that puts greater weights on the instances $x \in D$ on which h_0 made an error. This weighting mechanism can be implemented by either attaching a weight to the instances (requiring a modification of the training algorithm to take into account differently weighted instances) or by an oversampling procedure. In so doing, from the dataset D_i a new dataset D_{i+1} is obtained, on which a new classifier h_{i+1} is trained, so that the general classifier $h_{boost} = \sum_{j=1}^{i+1} \alpha_j * h_j$ has minimum error. A particularly popular implementation of the boosting procedure is the Gradient Boosting [32] algorithm, which is based on the explicit minimization of a loss function through gradient descent.

Social Choice and Voting Rules. Social choice theory, which dates from the work of Condorcet [13] in the 18th century and finds its modern form in the work by Arrow [1], studies how to best aggregate preferences and opinions by many individuals into social welfare decisions. The central notion in social choice theory is that of a *voting rule* $f : \mathcal{P}^n \mapsto \mathcal{A}$, where $\mathcal{A}$ is the set of *alternatives*, $\{1, ..., n\}$ is the set of *voters* and $\mathcal{P}$ is the set of total orders over $\mathcal{A}$, which represents the preferences of the voters. The most basic voting rule is the plurality rule, which simply returns the alternatives that has been ranked first by the greatest number of voters. While this rule is optimal, from several perspectives, in the binary case, it presents many disadvantages when it is extended to the multi-alternative case [4], such as its weakness to strategic manipulation, or the fact that it could declare as loser an alternative which however would be preferred by a majority of voters (i.e. it fails to elect the Condorcet winner, see below). Thus, several alternative methods or generalizations have been developed to overcome these shortcomings: in this article we will consider the *Borda count*, as an example of a scoring method, and the *Copeland method*, as an instance of a *Condorcet method*.

The Borda voting rule [16] is defined via a weighting vector $v \in \mathbb{R}_+^{|\mathcal{A}|}$ s.t. $\forall i, v_i > v_{i+1}$. Given the preferences $\langle a_1, ..., a_{|\mathcal{A}|} \rangle$ of a voter, a weight of v_i is assigned to the alternative a_i ranked in the i-th position. The *Borda winner* is the alternative with the greatest associated count. In the rest of this article, we will consider the Dowdall weighting vector [18] defined by $v_i = \frac{1}{i}$.

The Copeland method [11] is based on the notion of *pairwise majority*. Given two alternatives a_i, a_j we have that $a_i >_{pm} a_j$ iff a_i is preferred over a_j by a majority of voters. The Copeland score of an alternative a is defined as $Copeland(a) = |\{a_i : a >_{pm} a_i\}| - |\{a_i : a_i >_{pm} a\}|$. The Copeland winner is the alternative with the greatest Copeland score, if $Copeland(a) = |\mathcal{A}|$ then a is said to be a *Condorcet winner*.

Another approach in social choice theory, which does not require each voter to provide a full ranking of the alternatives, is approval voting [3]. In this case each voter provides a set of approved alternatives $appr(i) \subseteq \mathcal{A}$ and the score of alternative a is $score(a) = |\{i : a \in appr(i)\}|$ and the winning alternative is the one with maximum score.

Collective Intelligence. Collective intelligence, also studied under the popular name of *Wisdom of Crowds* [39], regards the study of phenomena that can be traced back to the notion of "intelligence" in group decision making, and the development of mechanisms through which this intelligence can emerge. While most of the research in the collective intelligence literature has focused on the simple plurality rule for aggregation, an increasing interest has been placed onto mechanisms to identify high-performing individuals [28] or, more in general, to correct possible errors due to the outcome of the majority [38], by drawing techniques also from the learning theory [2,23] and the multi-armed bandit [7] literature. A simple example of these mechanisms (which however also has its own shortcomings) is based on a confidence weighting [24], in which each member of the group is asked to both provide an answer (or choose an option) and also rate their own confidence in their answer: answers whose confidence are higher are associated with a greater weight in the aggregation (thus implementing a mechanism similar to the probability score weighting in bagging ensemble algorithms).

An alternative to this scheme, which is based on a Bayesian perspective [34], is the *surprisingly popular algorithm* [35], which was shown to empirically outperform both simple plurality aggregation and confidence-weighted approaches [21]. This method is based on asking to each individual i not only their answer $p(i)$ but also the answer that they predict will be the most voted one $v(i)$. The score of each alternative is then computed as $SP(a) = |\{i : p(i) = a\}| - |\{i : v(i) = a\}|$, and the winning alternative is then the one with greatest $SP(a)$.

A different perspective, based on uncertainty management theories such as possibility theory and three-way decision, was recently proposed in [8]. In this approach each individual i is asked to provide, for each alternative a, their subjective probability $p_i(a)$ that the answer is the correct one: we denote as $\pi(i) = \langle p_i(a_1), ..., p_i(a_n) \rangle$ the ranking of alternatives in decreasing probability order for individual i. Then the *possibilistic reduction* is defined by

$$poss(i) = \langle 1, \frac{p_i(a_2)}{p_i(a_1)}, ..., \frac{p_i(a_n)}{p_i(a_1)} \rangle \tag{1}$$

while the *three-way reduction* is defined by:

$$three\text{-}way(i) = \{a_1, ..., a_m\} \text{ s.t. } \sum_{j=1}^{m} p_i(a_j) \geq \tau > \sum_{j=1}^{m-1} p_i(a_j) \tag{2}$$

where $\tau > \frac{1}{n}$ is a threshold. For both techniques, the score of alternative a is obtained by summing the weights for each individual (in the three-way reduction the weight of alternative a for individual i is 1 if $a \in three\text{-}way(i)$, 0 otherwise). It is not hard to show that the possibilistic reduction represents a variation on the probability score weighted ensemble scheme [9], while the aggregation mechanism for the three-way reduction corresponds to approval voting [3] in social choice theory: indeed, for each individual i, the result of $three\text{-}way(i)$ is a set of approved alternatives and then the combination is simply performed via approval voting.

2.2   Experimental Design

The goal of this work, as already stated in Sect. 1, is to contribute to bridging the gap between those interested in ML ensemble methods, and those interested in collective intelligence and social choice theory. To this aim, we provide an experimental comparison of the aggregation mechanisms proposed in these different communities with respect to their capability to accurately extract knowledge from individual predictions, expanding on the initial results reported in [12, 29, 30]. In particular, we evaluated a set of different aggregation mechanisms as ensemble combination techniques on a variety of standard datasets from the UCI database [14]. We compared 7 different aggregation criteria: namely, bagging with probability score weighting (Bagging); Gradient Boosting (Boost); the Surprisingly Popular Algorithm (Sur. Pop.); the Borda voting rule; the Copeland voting rule (Copeland); the Possibilistic reduction and the three-way one (Three-way). In order to implement a fair comparison, we employed the same underlying ensemble algorithm with fixed hyperparameter settings for all of the considered aggregation mechanisms: all algorithms were based on an ensemble of 100 Decision Trees. For Boosting and Bagging we employed the default implementations provided by the Python scikit-learn library. All the other algorithms were implemented relying on bagging: we trained multiple Decision Trees on different and independent samples, drawn with replacement, of the original dataset. The predictions of the single Decision Trees were combined according to each specific aggregation criteria: for the Copeland and Borda methods only the ordinal ranking of each class label (w.r.t. the prediction scores given by each tree) was employed in the final aggregation, while the probability scores were employed to define the Three-way reduction for each of the trees, and similarly for the Possibilistic reduction. As the Surprisingly Popular algorithm requires a prediction (for each model) of what the most popular prediction will be within the ensemble (besides the top-rated answer for each model in the ensemble), we implemented Algorithm 1, for ensemble construction and prediction.

In regards to the three-way ensemble algorithm, we set $\tau = 0.75$ for binary classification problems, and $\tau = 0.5$ for multi-class ones. These two values were chosen in order to constraint the models in the ensemble to provide precise predictions in favor of one class (w.r.t. the other) only when: the evidence in favor of that class was statistically significant, in the binary case; the probability score assigned to the top-ranked class label was greater than the sum of all other probability score, in the multi-class case. The algorithms were compared on the following list of datasets:

Algorithm 1. The ensembling algorithm for the surprisingly popular aggregation method.

---

procedure SURPRISINGLY POPULAR ENSEMBLING($D = \langle X, y \rangle$: dataset, n: models to ensemble)

    Generate $D_1, ..., D_n$ sampling with replacement from D

    for all $i = 1 ... n$ **do**

        Train decision tree T_i on D_i

    end for

    $E = Bagging\text{-}Ensemble(T_1, ..., T_n)$

    $\hat{y} = E.predict(X)$

    for all $i = 1 ... n$ **do**

        Train decision tree P_i on $\langle X, \hat{y} \rangle$

    end for

    return $\langle T_1, ..., T_n \rangle, \langle P_1, ..., P_n \rangle$

end procedure

procedure SURPRISINGLY POPULAR PREDICT($Ens = \langle \langle T_1, ..., T_n \rangle, \langle P_1, ..., P_n \rangle \rangle$: ensemble, x: instance)

    $c = \langle 0, ..., 0 \rangle$ s.t. $|c| = |Y|$

    $p = \langle 0, ..., 0 \rangle$ s.t. $|c| = |Y|$

    for all $i = 1 ... n$ **do**

        $c = c + T_i.predict(x)$

        $p = p + P_i.predict(x)$

    end for

    $\hat{y} = argmax_i\{c[i] - p[i]\}$

    return $\hat{y}$

end procedure

- **Iris**: 4 features, 150 instances, 3 classes;
- **Wine**: 13 features, 178 instance, 2 classes;
- **Digits**: 64 features, 1797 instances, 10 classes;
- **Breast Cancer** (Cancer): 30 features, 569 instances, 2 classes;
- **Covtype**: 50 features, 581012 instances, 7 classes;
- **Diabetes**: 8 features, 768 instances, 2 classes;
- **Heart**: 75 features, 303 instances, 2 classes;
- **Arrhythmia**: 279 features, 459 instances, 16 classes;
- **Olivetti Faces** (Faces): 4096 features, 400 instances, 40 classes;
- **20 Newsgroups** (News): 130107 features, 18846 instances, 20 classes.

For dataset **20 Newsgroups**, feature reduction was performed in order to select the most important 10000 principal components using PCA. In order to compare the algorithms, we computed their average accuracy and 95% empirical confidence intervals on the different datasets using a 10-fold stratified nested cross-validation procedure. In order to assess if the measured differences in performance among the algorithms were statistically significant we performed the Friedman test [20] with Nemenyi post-hoc test [25], which already accounts for family-wise error and thus does not require p-value corrections.

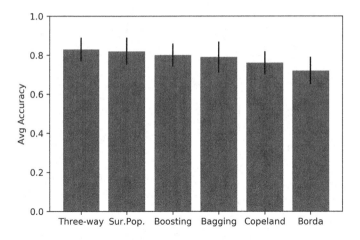

Fig. 1. Average accuracy and 95% boostrap confidence intervals for all algorithms.

Table 1. Accuracy of the tested ensembling methods in terms of the average ± width of the 95% confidence interval.

	Bagging	Sur. Pop.	Borda	Copeland	Boost	Three-way
Iris	0.95 ± 0.03	**0.96 ± 0.04**	0.81 ± 0.03	0.82 ± 0.10	**0.96 ± 0.03**	0.95 ± 0.02
Wine	**0.96 ± 0.02**	**0.96 ± 0.02**	0.88 ± 0.04	0.93 ± 0.01	0.86 ± 0.03	**0.96 ± 0.02**
Digits	0.93 ± 0.03	**0.94 ± 0.03**	0.87 ± 0.03	0.81 ± 0.03	0.84 ± 0.04	0.91 ± 0.02
Cancer	0.93 ± 0.02	**0.96 ± 0.03**	0.86 ± 0.05	0.90 ± 0.04	0.94 ± 0.02	0.95 ± 0.02
Covtype	0.82 ± 0.02	0.83 ± 0.02	0.75 ± 0.01	0.78 ± 0.05	0.81 ± 0.02	**0.85 ± 0.02**
Diabetes	0.71 ± 0.04	0.75 ± 0.03	0.63 ± 0.06	0.67 ± 0.03	0.74 ± 0.02	**0.77 ± 0.03**
Heart	0.72 ± 0.04	0.71 ± 0.02	0.67 ± 0.6	0.70 ± 0.01	0.80 ± 0.03	**0.81 ± 0.02**
Arrhythmia	0.68 ±0.03	0.73 ± 0.04	0.71 ± 0.04	**0.77 ± 0.02**	0.72 ± 0.02	0.75 ± 0.03
Faces	0.75 ± 0.01	**0.81 ± 0.01**	0.71 ± 0.01	0.77 ± 0.02	0.73 ± 0.02	0.80 ± 0.01
News	0.73 ± 0.02	0.78 ± 0.02	0.71 ± 0.02	0.75 ± 0.03	**0.81 ± 0.02**	**0.81 ± 0.01**

3   Results

The obtained average accuracy scores and their 95% confidence intervals are reported in Table 1. The average accuracy across the datasets, and relative 95% bootstrap confidence intervals, are shown Fig. 1.

The results for the possibilistic reduction are not reported in the Table as they were identical to the ones for the Bagging method. The ranks of the algorithms, needed to perform the Friedman test, are reported in Table 2 and Fig. 2. The Friedman statistics was $Q = 23.22$. We obtained a *p-value* $\ll 0.001$ which was found to be significant even after a Bonferroni correction (corrected $\hat{\alpha} = 0.003$), thus we also performed the post-hoc pairwise comparisons.

The p-values computed according to Nemenyi test for each pairwise comparison are reported in Table 3. The pairwise difference between the Surprisingly Popular (resp., Three-way) and all other methods was statistically significant

Table 2. Ranks of the ensembling methods for Friedman test.

	Bagging	Sur. Pop.	Borda	Copeland	Boost	Three-way
Iris	3.5	1	6	5	1	3.5
Wine	2	2	5	4	6	2
Digits	2	1	4	6	5	3
Breast cancer	4	1	6	5	3	2
Covtype	3	2	5	6	4	1
Diabetes	4	2	6	5	3	1
Heart	3	4	6	5	2	1
Arrhythmia	6	3	5	1	4	2
Olivetti faces	4	1	6	3	5	2
20 Newsgroups	5	1	6	4	3	2
Average rank	3.65	1.8	5.5	4.1	3.3	1.95

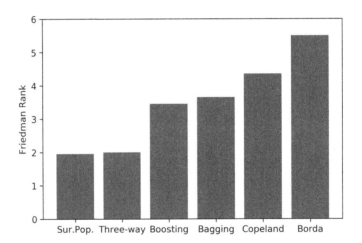

Fig. 2. Friedman ranks for all algorithms (lower is better).

Table 3. P-values for the pairwise comparisons according to Nemenyi test.

	Sur. Pop.	Borda	Copeland	Boost	Three-way
Bagging	0.014	0.014	0.295	0.663	0.02
Sur. Pop.	–	$\ll 0.001$	0.003	0.047	0.57
Borda	–	–	0.057	0.004	$\ll 0.001$
Copeland	–	–	–	0.17	0.005
Boost	–	–	–	–	0.044

at $\alpha = 0.05$, while the difference between the Surprisingly Popular method and the three-way reduction was found not to be significant (p-value = 0.57). The difference between the Bagging (resp. Boosting, possibilistic reduction) methods and Borda voting was also found to be statistically significant. All the other differences were found not to be statistically significant at $\alpha = 0.05$.

4 Discussion

From the results reported in the previous section, we can see that two of the criteria derived from the collective intelligence and crowdsourcing (i.e. the Surprisingly popular algorithm and the Three-way one) literature were found to significantly outperform all the other considered criteria. We conjecture that this is the case for two main reasons (but others can apply):

– The advantage of the surprisingly popular algorithm is given by the implemented ensembling methods, which requires each model T_i to also predict which answer will be the most popular within the ensemble (see Algorithm 1). As the training of the model P_i for providing this second answer is based on the same bootstrapped dataset D_i, the accuracy of P_i is directly connected to the representativeness of D_i with respect to the whole D, that is it implies that both T_i and P_i are trained on a distribution that well approximates the real population distribution. Hence, as models that are good at correctly predicting the most popular answer will be weighted more in the ensembling process, the algorithm implements an effective weighting scheme to discover the more reliable classifiers in the ensemble;
– The advantage of the three-way reduction method is due to both its noise-reducing effect, as only alternatives that are assigned large probabilities are considered in the aggregation process and weighted equally in order to remove fluctuations in the probability scores that may be due to chance.

Furthermore, the difference between these two methods was found to be statistically not significant. This suggests that ensemble methods based on the three-way reduction can be seen as more cost-effective than those based on the surprisingly popular algorithm, as they would provide similar performance but at a reduced computational cost: in fact, the surprisingly popular ensemble method requires to train $2n$ models instead of n. While the computational complexity is usually not considered relevant for the training of decision tree ensembles (as induction methods for decision trees is very efficient), this complexity may have an impact in the case of large datasets or if more time-consuming algorithms were employed in the ensemble.

The ensemble method based on the Borda voting rule was found to be the worst performing one (albeit the difference between Borda and Copeland was not statistically significant) and, more in general, both social choice-based criteria performed worse than all other methods. Interestingly, this result differs from the one reported in [29] where both Borda and Copeland were found to outperform the simple majority method: we notice, however, that the differences reported in that work were evident only for the k-NN algorithm (and very less so for the Naive Bayes algorithm) and statistical significance was never assessed. Furthermore, simple plurality was not directly assessed in this article because probability score weighting is known to be more efficient [22]. We conjecture that the lower performance for the social choice-based methods in our study is due to the facts that those methods simply consider the ordinal component of each classifier's ranking, and completely ignore the probability scores which are nevertheless important in that they allow to discriminate cases in which the top-ranked alternatives have been associated with low confidence.

The standard ensemble techniques, Bagging and Boosting, were found to have similar performance and to be intermediate in accuracy between collective intelligence-based and social choice-based criteria. Interestingly, the possibilistic reduction ensembling had the same performance as standard Bagging for all considered datasets. While these two ensemble strategies are relatively similar (since they essentially differ only for a normalization factor, and a strong relationship between the two methods has been established [8]), that the two approaches give the same results is not an obvious outcome of our experiment, and this suggests that further investigation should be done to this respect.

5   Conclusion

In this article we reported about a comparative study of different aggregation techniques, which were originally proposed in ensemble learning, collective intelligence and social choice theory, and considered them as ensembling criteria. In doing so, we aim to bridge the gaps between the above scholarly communities, and contribute to building an understanding of the properties of the considered aggregation methods.

To this aim, we performed an experimental evaluation of different methods (namely, Bagging, Boosting, Borda, Copeland, Three-way reduction and the Surprisingly popular algorithm) and proposed two novel ensemble algorithms. The findings show that collective intelligence-based criteria seem to be very effective as ensembling strategies, while social choice-based methods were associated with a worse performance than the other evaluated methods, probably due to their reliance only on ordinal information. In particular we think that both the Surprisingly Popular and the three-way ensembling criteria could be recommended for use in Machine Learning applications (e.g. in Random Forest, which is similar to the setting we evaluated in this work), especially so for the three-way algorithm which was also more computationally efficient.

While in this article we mainly considered the effectiveness of criteria drawn from collective intelligence and social choice theory in ensemble learning, also other directions should be explored: for example, studying the normative character or formal properties of ensemble learning or collective intelligence methods using techniques from social choice [33]. Another possible direction of research would involve studying the formal properties of these methods, from the perspective of learning theory [37], in order to verify our conjectures about the observed performance differences of the methods considered.

Furthermore, while in this article we have only focused on classic social choice voting rules, it would also be interesting to consider randomized voting rules [17], which have been studied in probabilistic social choice, and their connections with ensemble methods and multi-armed bandit theory.

References

1. Arrow, K.J.: Social Choice and Individual Values. Wiley, New York (1951)
2. Awasthi, P., Blum, A., Haghtalab, N., Mansour, Y.: Efficient PAC learning from the crowd. arXiv preprint arXiv:1703.07432 (2017)
3. Brams, S., Fishburn, P.C.: Approval Voting. Springer, New York (2007). https://doi.org/10.1007/978-0-387-49896-6
4. Brandt, F., Conitzer, V., Endriss, U., Lang, J., Procaccia, A.D.: Handbook of Computational Social Choice. Cambridge University Press, Cambridge (2016)
5. Breiman, L.: Bagging predictors. Mach. Learn. **24**(2), 123–140 (1996)
6. Breiman, L.: Random forests. Mach. Learn. **45**(1), 5–32 (2001)
7. Bubeck, S., Cesa-Bianchi, N., et al.: Regret analysis of stochastic and nonstochastic multi-armed bandit problems. Found. Trends® Mach. Learn. **5**(1), 1–122 (2012)
8. Cabitza, F., Campagner, A., Ciucci, D.: New frontiers in explainable AI: understanding the GI to interpret the GO. In: Holzinger, A., Kieseberg, P., Tjoa, A.M., Weippl, E. (eds.) CD-MAKE 2019. LNCS, vol. 11713, pp. 27–47. Springer, Cham (2019). https://doi.org/10.1007/978-3-030-29726-8_3
9. Campagner, A., Ciucci, D., Svensson, C.M., Figge, T., Cabitza, F.: Ground truthing from multi-rater labelling with three-way decisions and possibility theory. Inf. Sci. (2020, submitted)
10. Chourasia, R., Singla, A.: Unifying ensemble methods for q-learning via social choice theory. CoRR abs/1902.10646 (2019)
11. Colomer, J.M.: Ramon Llull: from 'Ars electionis' to social choice theory. Soc. Choice Welf. **40**(2), 317–328 (2013). https://doi.org/10.1007/s00355-011-0598-2
12. Cornelio, C., Donini, M., Loreggia, A., Pini, M.S., Rossi, F.: Voting with random classifiers (VORACE). arXiv preprint arXiv:1909.08996 (2019)
13. de Condorcet, J.: Essai sur l'application de l'analyse à la probabilité des decisions rendues à la pluralité des voix. Imprimerie Royale, Paris (1785)
14. Dua, D., Graff, C.: UCI machine learning repository (2017). http://archive.ics.uci.edu/ml
15. Dubois, D., Prade, H.: Possibility Theory: An Approach to Computerized Processing of Uncertainty. Springer, Boston (2012). https://doi.org/10.1007/978-1-4684-5287-7
16. Emerson, P.: The original Borda count and partial voting. Soc. Choice Welf. **40**(2), 353–358 (2013). https://doi.org/10.1007/s00355-011-0603-9
17. Endriss, U.: Trends in Computational Social Choice. Lulu.com, Morrisville (2017)
18. Fraenkel, J., Grofman, B.: The Borda count and its real-world alternatives: comparing scoring rules in Nauru and Slovenia. Aust. J. Polit. Sci. **49**(2), 186–205 (2014)
19. Freund, Y., Schapire, R.E.: A desicion-theoretic generalization of on-line learning and an application to boosting. In: Vitányi, P. (ed.) EuroCOLT 1995. LNCS, vol. 904, pp. 23–37. Springer, Heidelberg (1995). https://doi.org/10.1007/3-540-59119-2_166

20. Friedman, M.: A comparison of alternative tests of significance for the problem of m rankings. Ann. Math. Stat. **11**(1), 86–92 (1940)
21. Görzen, T., Laux, F.: Extracting the wisdom from the crowd: a comparison of approaches to aggregating collective intelligence. Paderborn University, Faculty of Business Administration and Economics, Technical report (2019)
22. Hastie, T., Tibshirani, R., Friedman, J.: The Elements of Statistical Learning: Data Mining, Inference, and Prediction. Springer, New York (2009). https://doi.org/10.1007/978-0-387-84858-7
23. Heinecke, S., Reyzin, L.: Crowdsourced PAC learning under classification noise. In: Proceedings of the AAAI Conference on Human Computation and Crowdsourcing, vol. 7, pp. 41–49 (2019)
24. Hertwig, R.: Tapping into the wisdom of the crowd-with confidence. Science **336**(6079), 303–304 (2012)
25. Hollander, M., Wolfe, D.A., Chicken, E.: Nonparametric Statistical Methods, vol. 751. Wiley, Hoboken (2013)
26. Kearns, M.: Thoughts on hypothesis boosting. Unpublished manuscript 45, 105 (1988)
27. Kearns, M., Valiant, L.: Cryptographic limitations on learning Boolean formulae and finite automata. J. ACM (JACM) **41**(1), 67–95 (1994)
28. Lee, M.D., Steyvers, M., De Young, M., Miller, B.: Inferring expertise in knowledge and prediction ranking tasks. Top. Cogn. Sci. **4**(1), 151–163 (2012)
29. Leon, F., Floria, S.A., Bădică, C.: Evaluating the effect of voting methods on ensemble-based classification. In: IEEE INISTA, vol. 2017, pp. 1–6 (2017)
30. Leung, K.T., Parker, D.S.: Empirical comparisons of various voting methods in bagging. In: Proceedings of the Ninth ACM SIGKDD International Conference on Knowledge Discovery and Data Mining, pp. 595–600 (2003)
31. Luo, T., Liu, Y.: Machine truth serum. arXiv preprint arXiv:1909.13004 (2019)
32. Mason, L., Baxter, J., Bartlett, P.L., Frean, M.R.: Boosting algorithms as gradient descent. In: Advances in Neural Information Processing Systems, pp. 512–518 (2000)
33. Pennock, D.M., Maynard-Reid II, P., Giles, C.L., Horvitz, E.: A normative examination of ensemble learning algorithms. In: ICML, pp. 735–742 (2000)
34. Prelec, D.: A Bayesian truth serum for subjective data. Science **306**(5695), 462–466 (2004)
35. Prelec, D., Seung, H.S., McCoy, J.: A solution to the single-question crowd wisdom problem. Nature **541**(7638), 532–535 (2017)
36. Rangi, A., Franceschetti, M.: Multi-armed bandit algorithms for crowdsourcing systems with online estimation of workers' ability. In: Proceedings of ICAAMAS, vol. 2018, pp. 1345–1352 (2018)
37. Shalev-Shwartz, S., Ben-David, S.: Understanding Machine Learning: From Theory to Algorithms. Cambridge University Press, Cambridge (2014)
38. Simmons, J.P., Nelson, L.D., Galak, J., Frederick, S.: Intuitive biases in choice versus estimation: implications for the wisdom of crowds. J. Consum. Res. **38**(1), 1–15 (2011)
39. Surowiecki, J.: The Wisdom of Crowds. Anchor (2005)
40. Yao, Y.: An outline of a theory of three-way decisions. In: Yao, J.T., et al. (eds.) RSCTC 2012. LNCS (LNAI), vol. 7413, pp. 1–17. Springer, Heidelberg (2012). https://doi.org/10.1007/978-3-642-32115-3_1

41. Zadeh, L.: Fuzzy sets as the basis for a theory of possibility. Fuzzy Sets Syst. **1**, 3–28 (1978)
42. Zhang, H., Conitzer, V.: A PAC framework for aggregating agents' judgments. In: Proceedings of the AAAI Conference on Artificial Intelligence, vol. 33, pp. 2237–2244 (2019)

.

An Unsupervised Capacity Identification Approach Based on Sobol' Indices

Guilherme Dean Pelegrina[1,3](✉) [ID], Leonardo Tomazeli Duarte[2] [ID],
Michel Grabisch[3], and João Marcos Travassos Romano[1]

[1] School of Electrical and Computer Engineering, University of Campinas,
400 Albert Einstein Avenue, Campinas 13083-852, Brazil
pelegrina@decom.fee.unicamp.br, romano@dmo.fee.unicamp.br
[2] School of Applied Sciences, University of Campinas, 1300 Pedro Zaccaria Street,
Limeira 13484-350, Brazil
leonardo.duarte@fca.unicamp.br
[3] Centre d'Économie de la Sorbonne, Université Paris I Panthéon-Sorbonne,
106-112 Boulevard de l'Hôpital, 75647 Paris Cedex 13, France
michel.grabisch@univ-paris1.fr

Abstract. In many ranking problems, some particular aspects of the
addressed situation should be taken into account in the aggregation process. An example is the presence of correlations between criteria, which
may introduce bias in the derived ranking. In these cases, aggregation
functions based on a capacity may be used to overcome this inconve-
nience, such as the Choquet integral or the multilinear model. The adop-
tion of such strategies requires a stage to estimate the parameters of
these aggregation operators. This task may be difficult in situations in
which we do not have either further information about these parameters
or preferences given by the decision maker. Therefore, the aim of this
paper is to deal with such situations through an unsupervised approach
for capacity identification based on the multilinear model. Our goal is to
estimate a capacity that can mitigate the bias introduced by correlations
in the decision data and, therefore, to provide a fairer result. The viability of our proposal is attested by numerical experiments with synthetic
data.

Keywords: Multicriteria decision making · Multilinear model ·
Unsupervised capacity identification · Sobol' index

1 Introduction

In multicriteria decision making (MCDM) [3], a typical problem consists in
obtaining a ranking of a set of alternatives (candidates, projects, cars, ...) based

This work was supported by the São Paulo Research Foundation (FAPESP, grant
numbers 2016/21571-4 and 2017/23879-9) and the National Council for Scientific and
Technological Development (CNPq, grant number 311357/2017-2).

V. Torra et al. (Eds.): MDAI 2020, LNAI 12256, pp. 66–77, 2020.
https://doi.org/10.1007/978-3-030-57524-3_6

on their evaluations in a set of decision criteria. These evaluations are generally aggregated in order to achieve overall values for the alternatives and, therefore, to define the ranking. In the literature [8], one may find several aggregation functions that can be used to deal with such problems. A simple example is the weighted arithmetic mean (WAM), which comprises a linear aggregation and is based on parameters representing weight factors associated to each criterion. Although largely used, there are some characteristics about the addressed decision problem that the WAM cannot deal with. An example is the interaction among criteria, which should be modelled in order to overcome biased results originated from the correlation structure of the decision data [9].

Different aggregation functions have been developed to model interactions among criteria. For instance, one may cite the well-known Choquet integral [1, 4], which derives the overall evaluations through a piecewise linear function. Moreover, although less used in comparison with the Choquet integral, one also may consider the multilinear model [10]. In this case, we aggregate the set of evaluations through a polynomial function.

A drawback of both Choquet integral and multilinear model, in comparison with the WAM, is that we need many more parameters (the capacity coefficients) to model the interactions among criteria. Therefore, the task of capacity identification is an important issue to be addressed when considering these functions. In the literature, one may find some supervised approaches (i.e., based on the information about both criteria evaluations and overall values to be used as learning data) for both Choquet integral [6] and multilinear model [11]. Moreover, one also may find unsupervised approaches (i.e., based only on the information about the decision data) to estimate the parameters of the Choquet integral [2].

Since in the unsupervised approaches one does not have access to the overall evaluations as learning data, one needs to assume a characteristic about the decision problem that we would like to deal with. This characteristic will be considered when implementing the capacity identification model. For instance, [2] associates some Choquet integral parameters to similarity measures of pairs of criteria in order to deal with the bias provided by correlations in the decision data. Therefore, the goal is the estimation of a capacity that leads to fairer overall evaluations in the sense that this bias is mitigated.

Motivated by the interesting results obtained by [2], in this paper, we tackle the unsupervised capacity identification problem in the context of the multilinear model, which remains largely unknown in the literature. However, instead of using the similarity measures, we deal with correlations by considering the Sobol' indices of coalitions of criteria, which can be directly associated with the multilinear model [7]. In order to attest the efficacy of the proposal, we apply our approach in scenarios with different numbers of alternatives and different degrees of correlation.

The rest of this paper is organized as follows. Section 2 describes the underlying theoretical concepts of this paper. In Sect. 3, we present the proposed unsupervised approach to deal with the problem of capacity identification. Numerical experiments are conducted in Sect. 4. Finally, in Sect. 5, we present our conclusions and future perspectives.

2 Theoretical Background

This section presents the theoretical aspects associated with our proposal, mainly the multilinear model and the Sobol' indices.

2.1 Multicriteria Decision Making and the Multilinear Model

The MCDM problem addressed in this paper comprises the ranking of n alternatives $a_1, a_2, \ldots, a_n$ based on their evaluations with respect to a set C of m criteria. Generally, we represent the decision data in a matrix $\mathbf{V}$, defined by

$$
\mathbf{V} = \begin{bmatrix} v_{1,1} & v_{1,2} & \cdots & v_{1,m} \\ v_{2,1} & v_{2,2} & \cdots & v_{2,m} \\ \vdots & \vdots & \ddots & \vdots \\ v_{n,1} & v_{n,2} & \cdots & v_{n,m} \end{bmatrix},
\tag{1}
$$

where $v_{i,j}$ is the evaluation of alternative a_i with respect to the criterion j. Therefore, in order to obtain the ranking, for each alternative a_i we aggregate the criteria evaluations through an aggregation function $F(\cdot)$ and order the alternatives based on the overall evaluations $r_i = F(v_{i,1}, \ldots, v_{i,m})$.

As mentioned in Sect. 1, candidates for $F(\cdot)$ are the WAM and the multilinear model. The WAM is defined as

$$
F_{WAM}(v_{i,1}, \ldots, v_{i,m}) = \sum_{j=1}^{m} w_j v_{i,j},
\tag{2}
$$

where w_j ($w_j \geq 0$, for all $j = 1, \ldots, m$, and $\sum_j^m w_j = 1$) represents the weight factor associated to criterion j. On the other hand, the multilinear model [10] is defined as

$$
F_{ML}(v_{i,1}, \ldots, v_{i,m}) = \sum_{A \subseteq C} \mu(A) \prod_{j \in A} v_{i,j} \prod_{j \in \overline{A}} (1 - v_{i,j}),
\tag{3}
$$

where $v_{i,j} \in [0,1]$, $\overline{A}$ is the complement set of A and the parameters $\mu = [\mu(\emptyset), \mu(\{1\}), \ldots, \mu(\{m\}), \mu(\{1,2\}), \ldots, \mu(\{m-1, m\}), \ldots, \mu(C)]$, called capacity [1], is a set function $\mu : 2^C \to \mathbb{R}$ satisfying the following axioms[1]:

- $\mu(\emptyset) = 0$ and $\mu(C) = 1$ (boundedness),
- for all $A \subseteq B \subseteq C$, $\mu(A) \leq \mu(B) \leq \mu(C)$ (monotonicity).

---

[1] It is worth mentioning that the multilinear model generalizes the WAM, i.e., if we consider an additive capacity, $F_{ML}(\cdot)$ is equivalent to $F_{WAM}(\cdot)$.

Let us illustrate the application of the considered aggregation functions in the problem of ranking a set of students based on their grades in a set of subjects (we adapted this example from [4]). The decision data as well as the overall evaluations and the ranking positions for both WAM and multilinear model are described in Table 1. For instance, we consider that $w_1 = w_2 = w_3 = 1/3$ and $\mu = [0, 1/3, 1/3, 1/3, 2/5, 2/3, 2/3, 1]$.

Table 1. Illustrative example.

Students	Grades			WAM		Multilinear model	
	Mathematics	Physics	Literature	r_i	Position	r_i	Position
Student 1	1.00	0.94	0.67	0.8700	1	0.7874	2
Student 2	0.67	0.72	0.94	0.7767	3	0.7703	3
Student 3	0.83	0.89	0.83	0.8500	2	0.8172	1

One may note that student 1 has an excellent performance on both mathematics and physics, but a lower grade in literature in comparison with the other students. Student 2 has the lowest grades in mathematics and physics and a very good one in literature. Finally, student 3 has an equilibrated performance, with good grades in all disciplines. With respect to the ranking, by applying the WAM, student 1 achieves the first position. However, one may consider that both mathematics and physics are disciplines that are associated with the same latent factor, i.e., they are correlated. Therefore, an aggregation that takes into account this interaction may lead to fairer results by avoiding this bias. That is the case of the ranking provided by the multilinear model, in which the student 3, the one with equilibrated grades, achieves the first position.

2.2   Interaction Indices and 2-Additive Capacity

In the previous section, we defined the multilinear model in terms of a capacity μ. However, the capacity coefficients do not have a clear interpretation. Therefore, we normally use an alternative representation of μ, called interaction index [5]. In the context of the multilinear model, the Banzhaf interaction index [12] is suitable (see, e.g., [11]), and is defined as follows:

$$I^{\mathcal{B}}(A) = \frac{1}{2^{|C|-|A|}} \sum_{D \subseteq C \backslash A} \sum_{D' \subseteq A} (-1)^{|A|-|D'|} \mu(D' \cup D), \forall A \subseteq C, \qquad (4)$$

where $|A|$ represents the cardinality of the subset A. If we consider, for example, a singleton j, we obtain the Banzhaf power index $\phi_j^{\mathcal{B}} \in [0, 1]$, given by

$$\phi_j^{\mathcal{B}} = \frac{1}{2^{|C|-1}} \sum_{D \subseteq C \backslash \{j\}} [\mu(D \cup \{j\}) - \mu(D)], \qquad (5)$$

which can be interpreted as the marginal contribution of criterion i alone taking into account all coalitions. Moreover, if we consider a pair of criteria j, j', we obtain the interaction index $I_{j,j'}^{\mathcal{B}}$ expressed by

$$I_{j,j'}^{\mathcal{B}} = \frac{1}{2^{|C|-2}} \sum_{D \subseteq C \setminus \{j,j'\}} [\mu(D \cup \{j,j'\}) - \mu(D \cup \{j\}) - \mu(D \cup \{j'\}) + \mu(D)].$$

$$(6)$$

In this case, the interpretation is the following:

- if $I_{j,j'}^{\mathcal{B}} < 0$, there is a negative interaction between criteria j, j', which models a redundant effect between them,
- if $I_{j,j'}^{\mathcal{B}} > 0$, there is a positive interaction between criteria j, j', which models a complementary effect between them,
- if $I_{j,j'}^{\mathcal{B}} = 0$, there is no interaction between criteria j, j', which means that they act independently.

In the multilinear model used to deal with the example presented in the last section, the interaction indices associated with the considered μ are $I^{\mathcal{B}} = [0.4675, 0.2667, 0.2667, 0.4017, -0.1367, 0.1333, 0.1333, 0.2600]$. Therefore, we have a negative interaction between criteria 1 and 2, positive interactions between the others pairs of criteria and a higher power index for criterion 3 in comparison to the other ones. It is worth mentioning that, given $I^{\mathcal{B}}(A), \forall A \subseteq C$, one may retrieve $\mu(A)$ through the following transform

$$\mu(A) = \sum_{D \subseteq C} \left(\frac{1}{2}\right)^{|D|} (-1)^{|D \setminus A|} I^{\mathcal{B}}(D), \forall A \subseteq C. \qquad (7)$$

By using either the capacity μ or the interaction index $I^{\mathcal{B}}$, we have $2^m - 2$ parameters to be determined in order to use the multilinear model. This may pose a problem in some situations, since the number of parameters increases exponentially with the number of criteria. In that respect, one may adopt a specific capacity, called 2-additive [5], which reduces the number of unknown parameters to $m(m + 1)/2 - 1$. We say that a capacity μ is 2-additive if the interaction index $I^{\mathcal{B}}(A) = 0$ for all A such that $|A| \geq 3$. In this paper, we consider such a capacity in the multilinear model.

2.3 Sobol' Indices

In several applications it is useful to analyse the sensitivity of a model output given a subset of all input factors. For this purpose, it is usual to carry out a decomposition of the model into terms associated with different input variables. For instance, consider the high-dimensional model representation (HDMR) of a model $Y = f(Z_1, Z_2, \ldots, Z_m)$ given by

$$Y = f_\emptyset + \sum_{j=1}^{m} f_j(Z_j) + \sum_{j<j'}^{m} f_{j,j'}(Z_j, Z_{j'}) + \ldots + f_C(Z_C), \qquad (8)$$

where $f_\emptyset, f_j, f_{j,j'}, \ldots, f_C, \forall j, j' \subseteq C$, are terms of increasing dimensions. A possible function f that can be used is the following:

$$f_A(Z_A) = \sum_{D \subseteq A} (-1)^{|A \setminus D|} \mathbb{E}[Y|Z_D], \tag{9}$$

where $\mathbb{E}[Y|Z_D]$ denotes the conditional expectation of Y given the variables Z_j, for all $j \in D$. In particular, $f_\emptyset = \mathbb{E}[Y]$, $f_j(Z_j) = \mathbb{E}[Y|Z_j] - f_\emptyset$ and $f_{j,j'}(Z_j, Z_{j'}) = \mathbb{E}[Y|Z_j, Z_{j'}] - f_j(Z_j) - f_j(Z_j) - f_\emptyset$.

As mentioned in [13], one may estimate $\mathbb{E}[Y|Z_j]$ by cutting the Z_j domain into slices and calculating the average value for each slice. Therefore, if these values have a pattern, $\mathbb{E}[Y|Z_j]$ has a large variation across Z_j, which means that this variable is "important" (or has a high "impact") in the output model. On the other hand, if no pattern is found, the variation across Z_j is small and this input is "less important" (or has a "less impact") in the model output.

By using the function f, one may calculate the variation across a variable through the variance of $f_j(Z_j)$, i.e., $\mathrm{Var}[f_j(Z_j)] = \mathrm{Var}[\mathbb{E}[Y|Z_j]]$. For instance, if we consider a single variable Z_j and normalize it, one obtains the first-order Sobol' sensitivity index, given by

$$S_j = \frac{\mathrm{Var}[\mathbb{E}[Y|Z_j]]}{\mathrm{Var}[Y]}, \tag{10}$$

which is a measure that indicates the degree of the aforementioned "importance" (or the "impact") that the input variable Z_j has in the model output. The same conclusions can be obtained for higher-order terms, i.e., for any coalition $A \subseteq C$,

$$S_A = \frac{\mathrm{Var}[f_A(Z_A)]}{\mathrm{Var}[Y]} \tag{11}$$

represents a measure of the "importance" (or the "impact") that the coalition of variables Z_A have in the model output. Under the assumption that the input variables (criteria, in the addressed decision problem) are independent and that they follow a uniform distribution on $[0, 1]$, [7] presented an interesting result:

Theorem 1. *(Grabisch and Labreuche [7])   Consider the multilinear model F_{ML} of a capacity μ. The (nonnormalized) Sobol' index of a subset $\emptyset \neq A \subseteq C$ is given by*

$$\mathrm{Var}[(F_{ML})_A] = \frac{1}{3^{|A|}} (\hat{\mu}(A))^2, \tag{12}$$

where $\hat{\mu}$, defined by

$$\hat{\mu}(A) = \left(\frac{-1}{2}\right)^{|A|} I^{\mathcal{B}}(A), \tag{13}$$

is the Fourier transform of μ.

Therefore, one may remark that the Sobol' index is associated with the Banzhaf interaction index through the equation

$$\mathrm{Var}[(F_{ML})_A] = \frac{1}{12^{|A|}} \left(I^{\mathcal{B}}(A)\right)^2. \tag{14}$$

This relation will be used in our proposal to deal with the unsupervised capacity identification problem.

3   The Proposed Unsupervised Approach for Capacity Identification

In order to present the motivations for our proposal, let us consider a decision problem composed by 5000 alternatives and 3 criteria (generated according to a uniform distribution on $[0, 1]$). For instance, this situation may comprise the problem of ranking the set of students described in Sect. 2.1. Figure 1 presents the scatter plot among pairs of criteria. One may note that criteria 1 and 2 are correlated (with Pearson correlation coefficient $\rho_{1,2} = 0.6768$).

Suppose that we want to analyse the impact that the criteria 1 and 3 alone have in the output model. Moreover, assume an additive capacity, i.e., $\mu = [0, 1/3, 1/3, 1/3, 2/3, 2/3, 2/3, 1]$ (in this case, $\phi_1^{\mathcal{B}} = \phi_2^{\mathcal{B}} = \phi_3^{\mathcal{B}} = 1/3$). If we calculate the nonnormalized Sobol' index according to Eq. (14), one obtains $\mathrm{Var}[(F_{ML})_1)] = \mathrm{Var}[(F_{ML})_3)] = (1/12)\left(\phi_3^{\mathcal{B}}\right)^2 \approx 0.0093$. On the other hand, if we consider Eq. (10) (with Y being the multilinear model and without normalization), one obtains $\mathrm{Var}[\mathbb{E}[F_{ML}|\mathbf{v}_1]] \approx 0.0263$ and $\mathrm{Var}[\mathbb{E}[F_{ML}|\mathbf{v}_3]] \approx 0.0091$, where $\mathbf{v}_j = [v_{1,j}, v_{2,j}, \ldots, v_{n,j}]$. Therefore, one may remark that we achieved similar values for both $\mathrm{Var}[(F_{ML})_3)]$ and $\mathrm{Var}[\mathbb{E}[F_{ML}|\mathbf{v}_3]]$, but very different ones comparing $\mathrm{Var}[(F_{ML})_1)]$ and $\mathrm{Var}[\mathbb{E}[F_{ML}|\mathbf{v}_1]]$. This difference is due to the existing correlation between criteria 1 and 2, which violates the hypothesis about the independence between the input variables assumed by Theorem 1. As a consequence, since these criteria are positively correlated, $\mathrm{Var}[\mathbb{E}[F_{ML}|\mathbf{v}_1]]$ is also influenced by criterion 2, which increases the impact of criterion 1 in the output model. With respect to $\mathrm{Var}[\mathbb{E}[F_{ML}|\mathbf{v}_3]]$, since criterion 3 has no correlation with any other criterion, its value remains independent on the other inputs.

Based on this discussion, our hypothesis is that the difference that we achieved with respect to the Sobol' indices may be reduced by applying a capacity that takes into account interactions among criteria. Suppose a scenario in which we have neither further information about the capacity coefficients nor overall evaluations provided by the decision maker. In this situation, it may be interesting to adopt a capacity that is able to compensate the bias provided by the correlations in the decision data and lead to fairer overall evaluations for the alternatives. In this context, this can be achieved by a capacity that leads to a multilinear model in which all coalitions of criteria with the same cardinality have similar impacts on the obtained overall evaluations. Therefore, the aim in this paper is to adjust a capacity μ in order to minimize the difference between the Sobol' indices for subsets of criteria with the same cardinality. Mathematically, the optimization problem is given by[2]

$$\min_{\mu} \sum_{\substack{A \subset C, \\ A \neq \emptyset}} \sum_{\substack{D \subset C, \\ |D| = |A|}} (S_A - S_D)^2. \tag{15}$$

---

[2] It is worth mentioning that we must satisfy the axioms of a capacity.

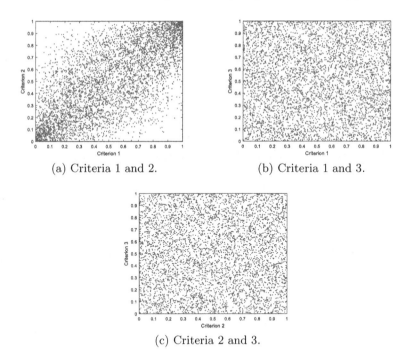

(a) Criteria 1 and 2.   (b) Criteria 1 and 3.

(c) Criteria 2 and 3.

Fig. 1. Scatter plot of pairs of criteria.

4   Numerical Experiments

This section presents the numerical experiments and the obtained results.

4.1   Application of Our Proposal in the Illustrative Example

As a first experiment, let us apply the proposed approach in the decision problem addressed in Sect. 3. In order to reduce the number of parameters to be estimated, but keeping a flexibility to model interactions, we considered the 2-additive multilinear model and a capacity μ such that $\mu(\{1\}) = \mu(\{2\}) = \mu(\{3\}) = 1/3$ (i.e., in the absence of further information about the capacity coefficients, we predefine the same value for all $\mu(\{j\})$). Therefore, one also need to estimate the capacity coefficients associated with pairs of criteria.

With respect to the optimization problem, other than the axioms of capacity presented in Sect. 2.1, we must also guarantee that $I^{\mathcal{B}}(C) = 0$, which leads to the following condition:

$$\mu(\{1,2\}) + \mu(\{1,3\}) + \mu(\{2,3\}) = 2. \tag{16}$$

Moreover, aiming at achieving an aggregation function whose individual criteria have similar impacts on the overall evaluations, we only considered the minimization of the difference between first-order Sobol' indices, i.e., the subsets A

in (15) are restricted to singletons j. In order to solve the optimization problem, we adopted a simple iterative heuristic method based on the golden section search [14]. For instance, we started with the additive capacity and selected at random a $\mu(\{j,j'\})$ to be fixed ($\mu(\{1,3\}) = 2/3$, for example). Thereafter, we selected another $\mu(\{j'',j'''\})$ ($\mu(\{1,2\})$, for example) to be optimized and applied the golden section search in the associated dimension to deal with the addressed optimization problem. In other words, by fixing $\mu(\{j,j'\})$, we perform a one-dimensional search on $\mu(\{j'',j'''\})$ that solves (15). Since we must satisfy (16), one may note that, when applying the golden section search on $\mu(\{j'',j'''\})$, the other capacity coefficient that was not selected so far is automatically adjusted ($\mu(\{2,3\}) = 2 - 2/3 - \mu(\{1,2\})$, in this case). This procedure is repeated until the convergence to the minimum (possibly a local minimum).

Based on the aforementioned assumptions, the application of the proposed approach in the illustrative example led to the capacity and the associated interaction indices presented in Table 2.

Table 2. Achieve capacity and interaction indices.

	A							
	$\emptyset$	$\{1\}$	$\{2\}$	$\{3\}$	$\{1,2\}$	$\{1,3\}$	$\{2,3\}$	C
$\mu(A)$	0	0.3333	0.3333	0.3333	0.4190	0.7778	0.8033	1
$I^B(A)$	0.5000	0.2650	0.2778	0.4572	-0.2477	0.1111	0.1366	0

One may note that we achieved $I^B_{1,2} < 0$, which was expected since both criteria 1 and 2 are correlated. Moreover, the obtained Banzhaf power index ϕ^B_3 was higher in comparison to the other ones, which also contributes to increase the impact of criterion 3 (the independent one) in the output model. With respect to the first-order Sobol' indices, we achieved $S_1 \approx S_2 \approx S_3 \approx 0.0173$.

If we consider the problem of ranking the students, which can be configured as the considered decision problem, the estimated capacity leads to the overall evaluations $r_1 = 0.7976$, $r_2 = 0.8196$ and $r_3 = 0.8445$. Therefore, the student 3 will also be the first one in the ranking.

4.2   Experiments Varying the Number of Alternatives and the Degree of the Correlation

In order to further investigate our proposal, we considered several different scenarios, varying the number of alternatives and the degree of the correlation between criteria 1 and 2. In all cases we considered 3 decision criteria and generated the evaluations according to a uniform distribution on $[0,1]$. Based on the same assumptions considered in the last experiment, the obtained interaction indices (averaged over 100 simulations) for decision problems with $\rho_{1,2} \approx 0.75$, $\rho_{1,2} \approx 0$ and $\rho_{1,2} \approx -0.75$ are presented in Figs. 2, 3 and 4, respectively.

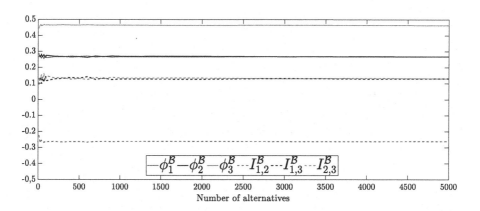

Fig. 2. Results for $\rho_{1,2} \approx 0.75$.

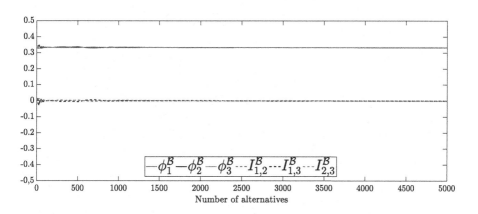

Fig. 3. Results for $\rho_{1,2} \approx 0$.

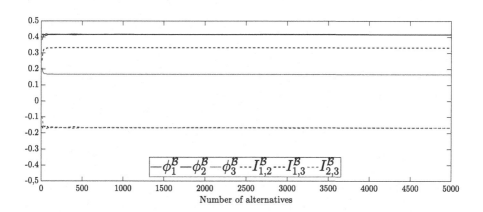

Fig. 4. Results for $\rho_{1,2} \approx -0.75$.

One may note that, in all cases, the stability in the obtained capacity increases as the number of alternatives also increases. This is due to the statistics involved in the Sobol' indices calculation, which require more data to be well estimated.

In Fig. 2, we clearly see that, in order to compensate the positive correlation between criteria 1 and 2, we achieved $I_{1,2}^{\mathcal{B}} < 0$ (redundant effect for the correlated criteria), both $I_{1,3}^{\mathcal{B}}$ and $I_{2,3}^{\mathcal{B}} > 0$ (positive interaction for the independent criteria) and a marginal contribution of criterion 3 ($\phi_3^{\mathcal{B}}$) greater than the other ones. Conversely, in Fig. 4, the negative correlation between criteria 1 and 2 led to $I_{1,2}^{\mathcal{B}} > 0$ (complementary effect for the correlated criteria), both $I_{1,3}^{\mathcal{B}}$ and $I_{2,3}^{\mathcal{B}} < 0$ (negative interaction for the independent criteria) and a marginal contribution of criterion 3 ($\phi_3^{\mathcal{B}}$) lower than the other ones.

With respect to Fig. 3, which contains the results when all the criteria are independent, one may see that the obtained capacity is an additive one. Therefore, we do not need to model interactions or increase marginal contributions to equilibrate the Sobol' indices, since they are already similar.

5   Conclusions

It is usual to observe the presence of correlations between criteria in multicriteria decision making problems. In these cases, the obtained ranking may be biased towards alternatives that have good evaluations in correlated criteria, i.e., that measure the same latent factor. Even with a worst performance in the other criteria, these alternatives can achieve better positions compared to the ones whose evaluations are more equilibrated.

In order to deal with these situations by modelling interactions among criteria, one may use aggregation functions such as the Choquet integral or the multilinear model. These functions are based on a capacity, i.e., a set of parameters associated with all possible coalitions of criteria. Therefore, the number of parameters increases with the number of criteria, making it difficult to define or estimate these values.

In this paper, we addressed the problem of capacity identification in an unsupervised fashion. Differently from our previous work [11], we do not consider any further information about the parameters or overall values provided by the decision maker. Our approach aims at extracting information contained in the decision data and estimating a capacity that can compensate the correlations among criteria. For instance, we assumed that all singletons should have the same impact on the output model and used the Sobol' indices as a means of comparison.

The obtained results attested the application of the proposed approach. In situations with positive (resp. negative) correlation between a pair of criteria, the achieved associated interaction index was negative (resp. positive), which models a redundancy (resp. complementarity) effect. Moreover, the power indices were also adjusted in order to balance the Sobol' indices.

For future perspectives, we would like to analyse the heuristic used to deal with the optimization problem. Other assumptions about the capacities as well as different search algorithms may be investigated. Moreover, we here addressed the situation in which the criteria are not independent. Therefore, future works can be conducted to verify the impact that distributions different from the uniform may have in the Sobol' indices. Finally, we also intend to apply the proposal in real datasets.

References

1. Choquet, G.: Theory of capacities. Annales de l'Institut Fourier **5**, 131–295 (1954)
2. Duarte, L.T.: A novel multicriteria decision aiding method based on unsupervised aggregation via the Choquet integral. IEEE Trans. Eng. Manage. **65**(2), 293–302 (2018)
3. Figueira, J., Greco, S., Ehrgott, M. (eds.): Multiple Criteria Decision Analysis: State of the Art Survey. International Series in Operations Research & Management Science, 2nd edn. Springer, New York (2016). https://doi.org/10.1007/978-1-4939-3094-4
4. Grabisch, M.: The application of fuzzy integrals in multicriteria decision making. Eur. J. Oper. Res. **89**, 445–456 (1996)
5. Grabisch, M.: Alternative representations of discrete fuzzy measures for decision making. Int. J. Uncertainty Fuzziness Knowl. Based Syst. **5**(5), 587–607 (1997)
6. Grabisch, M., Kojadinovic, I., Meyer, P.: A review of methods for capacity identification in Choquet integral based multi-attribute utility theory: applications of the Kappalab R package. Eur. J. Oper. Res. **186**, 766–785 (2008)
7. Grabisch, M., Labreuche, C.: A note on the Sobol' indices and interactive criteria. Fuzzy Sets Syst. **315**, 99–108 (2017)
8. Grabisch, M., Marichal, J.L., Mesiar, R., Pap, E.: Aggregation Functions. Cambridge University Press, New York (2009)
9. Marichal, J.L.: An axiomatic approach of the discrete Choquet integral as a tool to aggregate interacting criteria. IEEE Trans. Fuzzy Syst. **8**, 800–807 (2000)
10. Owen, G.: Multilinear extensions of games. Manage. Sci. Part 2 **18**(5), 64–79 (1972)
11. Pelegrina, G.D., Duarte, L.T., Grabisch, M., Romano, J.M.T.: The multilinear model in multicriteria decision making: the case of 2-additive capacities and contributions to parameter identification. Eur. J. Oper. Res. **282**, 945–956 (2020)
12. Roubens, M.: Interaction between criteria through the use of fuzzy measures. In: 44th Meeting of the European Working Group "Multicriteria Aid for Decisions", Brussels, Belgium (1996)
13. Saltelli, A., et al.: Global Sensitivity Analysis: The Primer. Wiley, Chichester (2008)
14. Vajda, S.: Fibonacci and Lucas Numbers, and the Golden Section: Theory and Applications. Ellis Horword Limited, Chichester (1989)

Probabilistic Measures and Integrals: How to Aggregate Imprecise Data

Michał Boczek[1], Lenka Halčinová[2(✉)], Ondrej Hutník[2], and Marek Kaluszka[1]

[1] Institute of Mathematics, Lodz University of Technology,
ul. Wólczańska 215, 90-924 Lodz, Poland
{michal.boczek.1,marek.kaluszka}@p.lodz.pl
[2] Institute of Mathematics, Faculty of Science, Pavol Jozef Šafárik University
in Košice, Jesenná 5, 040 01 Košice, Slovakia
{lenka.halcinova,ondrej.hutnik}@upjs.sk

Abstract. This paper develops the theory of probabilistic-valued measures and integrals as a suitable aggregation tool for dealing with certain types of imprecise information. The motivation comes from Moore's interval mathematics, where the use of intervals in data processing is due to measurement inaccuracy errors. In case of rounding, the intervals can be considered in distribution function form linked to random variables uniformly distributed over the relevant intervals. We demonstrate how the convolution of distribution functions is taken into account, and integration with respect to probabilistic-valued measures is converted into convolving certain distribution functions. We also improve some existing features of the integral and investigate its convergence properties.

Keywords: Interval-valued aggregation · Distribution function · Moore's interval mathematics · Random variable · Probabilistic integral

1 Introduction

In practice, many physical measurements can be best modelled by the concept of real-valued measures. Indeed, every physical quantity is assigned a specific value by measurement. But each experiment, and each physical quantity is prone to measurement errors, which may be minimalized, but not eliminated. What is more, in accordance with the famous Heisenberg's uncertainty principle, certain pairs of physical quantities cannot be measured simultaneously with arbitrary accuracy. For this reason, it is deemed appropriate to describe the values of physical quantities (that are identified by means of experimental measurements) with the *probabilistic-valued measures*: set functions assigning to each measurable subset a distance distribution function. The value of such function $F(x)$ naturally

This work was supported by the Slovak Research and Development Agency under the contract No. APVV-16-0337 and also cofinanced by bilateral call Slovak-Poland grant scheme No. SK-PL-18-0032 with the Polish National Agency for Academic Exchange PPN/BIL/2018/1/00049/U/00001.

V. Torra et al. (Eds.): MDAI 2020, LNAI 12256, pp. 78–91, 2020.
https://doi.org/10.1007/978-3-030-57524-3_7

describes the probability that the value of measurable physical quantity is less than x. Thus, we achieve more reliable and more accurate results.

Furthermore, there are several other practical examples which attest to the fact that it is not possible to describe phenomena around us by only one exact number value, i.e., we do not know the measure of a set exactly, but we may have certain probabilistic information about it. Examples of the former include, but are not limited to, placing a wager on horse racing, predicting sport results, or results in lottery. Here, only a probabilistic information about the (counting) measure of possibilities to win the prize is known. The importance of such measures can also be traced in other domains, e.g. stochastic differential equations. In fact, the above mentioned examples resemble the original Menger's ideas of probabilistic metric space, see [12], whose introduction was motivated by thinking of situations where the exact distance between two objects is not provided. Thus, the probabilistic point of view comes again into the game.

The above mentioned ideas led us to introduce the concept of probabilistic-valued measures, see [6,7,9]. In general, a set function γ defined on a σ-algebra Σ of subsets of a nonempty set Ω and taking values in the set Δ^+ of distribution functions of nonnegative random variables, is said to be a τ-*decomposable measure* if the "initia" condition $\gamma_\emptyset = \varepsilon_0$ is satisfied and $\gamma_{E \cup F} = \tau(\gamma_E, \gamma_F)$ for each choice of disjoint sets $E, F \in \Sigma$ with ε_0 being the distribution function of Dirac random variable concentrated at point 0, and τ being a triangle function on Δ^+. In fact, this definition resembles the well-known definitions of S-decomposable and $\oplus$-decomposable numerical measures studied in the classical and pseudo-analysis theory by many authors, see e.g. [4,15,19].

An essential part of the global measure theory is the corresponding integration theory. Or, in other words, having certain imprecise information about measure of a set and a function acting on it, we have a natural task how to "aggregate" the information contained in the probabilistic measure and the function. This yields to the construction of a *probabilistic integral* $\int_E f \, d\gamma$ of a nonnegative function $f \colon \Omega \to [0, \infty[$ with respect to a τ-decomposable measure $\gamma \colon \Sigma \to \Delta^+$ with τ being an arbitrary distributive triangle function on Δ^+ and $E \in \Sigma$, see [1,5]. The definition of the above mentioned integral follows the Lebesgue approach to integration.

The aim of this text is to provide a short description of the main features glued in the notion of probabilistic decomposable measure as well as the corresponding integrals, and to demonstrate their applicability in examples from Moore's interval analysis.

2   Basic Ingredients

Distribution and Triangle Functions. A *distance distribution function* is a distribution function whose support is a subset of $\mathbb{R}_+ = [0, \infty[$, see [16, Section 4.3]. The class of all distance distribution functions will be denoted by Δ^+. A special subclass of Δ^+ is the class of unit steps. More precisely, for $a \in \mathbb{R}_+$ we put

$$\varepsilon_a(x) := \begin{cases} 1 & \text{for } x > a, \\ 0 & \text{otherwise.} \end{cases}$$

A *triangle function* is a binary operation $\tau \colon \Delta^+ \times \Delta^+ \to \Delta^+$ which is commutative, associative, nondecreasing in each variable and has ε_0 as the neutral element. One of the most important triangle functions are those obtained from certain aggregation functions, especially triangular norms. A *triangular norm*, shortly a t-norm, is a binary operation $T \colon [0,1] \times [0,1] \to [0,1]$ which is commutative, associative, nondecreasing in each variable and has 1 as the identity element, see [10].

Operations on Δ^+. For a distance distribution function G and a nonnegative constant $c \in \mathbb{R}_+$, we define the $\odot$-*multiplication* of G by a constant c as follows

$$(c \odot G)(x) := \begin{cases} \varepsilon_0(x), & c = 0, \\ G\left(\frac{x}{c}\right), & \text{otherwise.} \end{cases}$$

Clearly, $c \odot G \in \Delta^+$. An *addition* $\oplus_\tau$ in Δ^+ is defined in the sense of addition via triangle function τ, i.e., we put $(G \oplus_\tau H)(x) := \tau(G, H)(x)$. Clearly, $G \oplus_\tau H \in \Delta^+$ for each triangle function τ. We usually omit a triangle function from the subscript $\oplus_\tau$ and write just $\oplus$ when no possible confusion may arise. By associativity of τ, we introduce the $\oplus$-addition of n functions $G_1, \dots, G_n \in \Delta^+$ as

$$\bigoplus_{k=1}^n G_k := \tau\left(G_1, \bigoplus_{k=2}^n G_k\right).$$

Let us denote by $\mathfrak{D}(\Delta^+)$ a set of all *distributive triangle functions*, i.e., for each $c \in \mathbb{R}_+$ and each $G, H \in \Delta^+$ it holds

$$c \odot (G \oplus_\tau H) = (c \odot G) \oplus_\tau (c \odot H). \tag{1}$$

For instance, the binary operation $(G \oplus_{\tau_T} H)(x) = \sup_{u+v=x} T(G(u), H(v))$, with T being a left-continuous t-norm is a distributive triangle function. For M being the minimum t-norm we get the distributive triangle function τ_M used in most cases. Also, $(G \oplus_{\tau_\top} H)(t) = \top(G(t), H(t))$ with $\top$ being a t-norm, is a distributive triangle function.

Metric Structure of Δ^+. The set Δ^+ will be endowed with the modified Lévy metric d_L, see [16, Section 4.2], which metrizes the topology of weak convergence of distribution functions, i.e., a sequence of distribution functions $(G_n)_1^\infty$ converges *weakly* to a distribution function G (shortly, $G_n \xrightarrow{w} G$) if and only if $G_n(t) \to G(t)$ at each point t, where G is continuous. Note that convergence at each point of continuity of G is not equivalent to convergence at every point of $\mathbb{R}$. For the latter convergence (at every point) we use the standard limit notation in the whole text. A triangle function is *continuous* if it is continuous in the complete metric space (Δ^+, d_L). The set of all continuous distributive triangle functions on Δ^+ will be denoted by $\mathfrak{CD}(\Delta^+)$. For instance, $\oplus_{\tau_T} \in \mathfrak{CD}(\Delta^+)$

whenever T is a continuous t-norm, see [16, Theorem 7.2.8]. Furthermore, consider the functional Sugeno integral of $[0, 1]$-valued functions and capacities, i.e.,

$$\mathrm{Su}(x) = \sup_{t \in [0,x]} \{t \wedge \mu(\{1 - t \geq 1 - f\})\}, \ x \in [0, 1].$$

Putting $t = u$ and $1 - t = v$, we get

$$\mathrm{Su}(x) = \sup_{u+v=x} \{u \wedge H(v)\} = \sup_{u+v=x} \{G(u) \wedge H(v)\} = (G \oplus_{\tau_M} H)(x),$$

where $H(v) = \mu(\{1 - f \leq v\})$ and $G(u) = u$. This equality describes a connection between the Sugeno integral and continuous distributive triangle functions on Δ^+. Immediately, this can be extended to the Sugeno-Weber integral with a continuous t-norm T.

Probabilistic Metric Spaces. The original definition of probabilistic metric space introduced by Menger was redefined by many authors over the course of the 20th century. Šertnev in [17] defined the *probabilistic metric space* (PM-space, for short) as a nonempty set Ω together with a triangle function τ and a family $\mathcal{F}$ of distribution functions $F_{p,q} \in \Delta^+$ satisfying the "initial" condition $F_{p,q} = \varepsilon_0$ if and only if $p = q$, the "symmetry" $F_{p,q} = F_{q,p}$, and the "probabilistic analogue" of the triangle inequality expressed by $F_{p,r} \geq \tau(F_{p,q}, F_{q,r})$, which holds for all $p, q, r \in \Omega$. Clearly, this inequality depends on a triangle function. Different triangle functions lead to PM-spaces with different geometric and topological properties. Nowadays, PM-spaces found its application in data analysis. Indeed, in paper [18] authors used PM-spaces for modeling the relationships between machine learning models and statistics. The reason for their usage in this context is that these spaces define metrics by distance distribution functions that allow to represent randomness.

Probabilistic-Valued Decomposable Measures. The concept of probabilistic-valued measure is an extension of a nonnegative measure (not necessarily finite, in general) to its probabilistic case. A mapping $\gamma \colon \Sigma \to \Delta^+$ with $\gamma_\emptyset = \varepsilon_0$ is said to be a τ-*decomposable measure*, if $\gamma_{E \cup F} = \tau(\gamma_E, \gamma_F)$ for each choice of disjoint sets $E, F \in \Sigma$. For a probabilistic-valued measure $\gamma \colon \Sigma \to \Delta^+$ we prefer to write γ_E and $\gamma_E(x)$ instead of $\gamma(E)$ and $\gamma(E)(x)$. Since Δ^+ is the set of all distribution functions with the support $\mathbb{R}_+$, the expression for such mapping γ is necessary to be described just for positive values of x. In case $x \leq 0$ we always suppose $\gamma.(x) = 0$.

3   Probabilistic Integral in Action

There is a variety of well-known various numerical-valued integrals: the Riemann integral, the Lebesgue integral, the so-called fuzzy integrals such as the Choquet integral, the Sugeno integral, etc. They act on numerical (not necessarily additive) measures. Here again the question arises how to handle with cases when we do not know precise values of a measure, but some probabilistic assignment still could be done. Clearly, numerical-valued integrals are not applicable and there is a possibility to introduce a probabilistic-valued integral. Also, in this section we will deal with the issue of describing values of a certain physical quantity which is linked together with other quantities by means of a relation, provided that a probabilistic information about their values is known.

Now, a Lebesgue-type approach to the integration of nonnegative real-valued functions with respect to a τ-decomposable measure will be explained. More precisely, we study the γ-*integral*

$$\int_E f \, d\gamma,$$

where $E \in \Sigma$ with Σ being a σ-algebra of subsets of a nonempty set Ω, $f \colon \Omega \to \mathbb{R}_+$ is a measurable function (in the sense described below), and $\gamma \colon \Sigma \to \Delta^+$ is a τ-decomposable measure with τ being a distributive triangle function. Let us denote by $\Omega^{\mathbb{R}_+}$ the set of all measurable functions $f \colon \Omega \to \mathbb{R}_+$. A *simple* function is a function $f \in \Omega^{\mathbb{R}_+}$ that attains a finite number of values $x_1, x_2, \ldots, x_n \in \mathbb{R}_+$ on pairwise disjoint sets $E_1, E_2, \ldots, E_n \in \Sigma$. Formally, a simple function will be written in the form $\sum_{i=1}^n x_i \mathbb{1}_{E_i}$, where $\mathbb{1}_E$ is the indicator function of a measurable set $E \in \Sigma$. The set of all simple functions will be denoted by $\mathcal{S}$, and for $f \in \mathcal{S}$ we put

$$\int_E f \, d\gamma = \int_E \left(\sum_{i=1}^n x_i \mathbb{1}_{E_i} \right) d\gamma := \bigoplus_{i=1}^n x_i \odot \gamma_{E \cap E_i}, \tag{2}$$

in the sense of $\oplus$-addition and $\odot$-multiplication in Δ^+. Clearly, the γ-integral of a simple function belongs to Δ^+. The important feature of the introduced γ-integral of a simple function is its order reversing property.

Proposition 1. *Let $\tau \in \mathfrak{D}(\Delta^+)$, γ be a τ-decomposable measure on Σ and $f, g \in \mathcal{S}$. If $f \leq g$ then $\int_E f \, d\gamma \geq \int_E g \, d\gamma$ on $E \in \Sigma$.*

The previous property of the integral motivated the following definition of the probabilistic γ-integral of a nonnegative measurable function, see [5]. Here, measurability is understood in the following sense: a function $f \in \Omega^{\mathbb{R}_+}$ is *measurable* with respect to Σ, if there exists a sequence $(f_n)_1^\infty \in \mathcal{S}$ such that $\lim_{n \to \infty} f_n(\omega) = f(\omega)$ for each $\omega \in \Omega$. In what follows $\mathcal{S}_{f,E}$ denotes a set of all simple functions $f \in \mathcal{S}$ such that $f(\omega) \leq f(\omega)$ for each $\omega \in E$.

Definition 1. *Let $\tau \in \mathfrak{D}(\Delta^+)$ and γ be a τ-decomposable measure on Σ. We say that a function $f \in \Omega^{\mathbb{R}_+}$ is γ-integrable on a set $E \in \Sigma$, if there exists $H \in \Delta^+$ such that $\int_E f \, d\gamma \geq H$ for each $f \in \mathcal{S}_{f,E}$. In this case we put*

$$\int_E f \, d\gamma := \inf \left\{ \int_E f \, d\gamma; \, f \in \mathcal{S}_{f,E} \right\}.$$

Remark 1. Infimum in the previous definition is meant in the pointwise sense. It is known that the function $G: \mathbb{R} \to [0,1]$ defined as a pointwise infimum $G(x) = \inf\{G_i(x); i \in I\}$, $x \in \mathbb{R}$, need not be a distance distribution function (it need not be left-continuous on $\mathbb{R}$, e.g. $G(x) = \inf\{\varepsilon_{a+1/n}(x); n = 1, 2, \dots\}$). Fortunately, there exists a "repairing" construction. Taking the left-limit $\mathfrak{G}(x) := \lim_{x' \nearrow x} G(x') = \sup_{x' < x} G(x')$, $x \in \mathbb{R}$, then $\mathfrak{G}(x) \leq G(x)$ for each $x \in \mathbb{R}$, the function $\mathfrak{G}$ belongs to Δ^+ and $\mathfrak{G} = \inf_{i \in I} G_i$ is the infimum of the family $\{G_i\}$ in the ordered set $(\Delta^+, \leq)$, see [3].

3.1 Probabilistic Integral – Properties

Using the fact that each nonnegative (extended) real-valued function f is the pointwise limit of a sequence $(f_n)_1^\infty$ of nondecreasing simple functions, [8, §20, Theorem B], it is possible to define γ-integral via the uniform approximation of a nonnegative function by a sequence of pointwise converging monotone sequence of simple functions. However, we have to assume that a triangle function τ is continuous and a probabilistic-valued measure γ is continuous from below, for the proof we refer to [5, Theorem 5.2]. Recall that a probabilistic-valued set function γ is *continuous from below*, if $\gamma_{E_n} \xrightarrow{w} \gamma_E$ whenever $E_n \nearrow E$, i.e., $E_n \subseteq E_{n+1}$ and $\overset{\infty}{\underset{n=1}{\cup}} E_n = E$.

Proposition 2. *Let $\tau \in \mathfrak{CD}(\Delta^+)$, γ be a τ-decomposable measure on Σ continuous from below, and $f \in \Omega^{\mathbb{R}+}$ be a γ-integrable function on a set $E \in \Sigma$. Then there exists a nondecreasing sequence $(\mathfrak{f}_n)_1^\infty \in \mathcal{S}_{f,E}$ converging pointwisely to f such that*

$$\int_E f \, \mathrm{d}\gamma = \lim_{n \to \infty} \int_E \mathfrak{f}_n \, \mathrm{d}\gamma.$$

A summary of basic properties of the introduced probabilistic-valued integral reads as follows. For the proofs and more details we refer to [5].

Proposition 3. *Let $\tau \in \mathfrak{D}(\Delta^+)$ and γ be a τ-decomposable measure on Σ. Let $f, g \in \Omega^{\mathbb{R}+}$ be γ-integrable functions on the corresponding sets and put $\mathcal{I}: f \mapsto \int f \, \mathrm{d}\gamma$. Then*

(i) γ-integral is an antimonotone operator, i.e., $\mathcal{I}(f) \geq \mathcal{I}(g)$ whenever $f \leq g$;
(ii) γ-integral is positively homogeneous, i.e., for each $c \in \mathbb{R}_+$ it holds $\mathcal{I}(c \cdot f) = c \odot \mathcal{I}(f)$;
(iii) if τ is continuous and γ is continuous from below, then the γ-integral is $\oplus_\tau$-linear, i.e., $\mathcal{I}(f + g) = \mathcal{I}(f) \oplus_\tau \mathcal{I}(g)$ for each $f, g \in \Omega^{\mathbb{R}+}$ if and only if $\tau = \tau_M$. In general, $\mathcal{I}(f + g) \geq \mathcal{I}(f) \oplus_\tau \mathcal{I}(g)$ if and only if $\tau_M \geq \tau$.

Important properties of the probabilistic-valued integral are stated in the following proposition. Recall that a set $E \in \Sigma$ is said to be γ-*null*, if $\gamma_E = \varepsilon_0$. Functions $f, g \in \Omega^{\mathbb{R}+}$ are said to be *equal a.e.* on a set S, we write $f = g$ a.e., if $f(\omega) = g(\omega)$ for all $\omega \in S \setminus E$, where E is a γ-null set.

Proposition 4. *Let $\tau \in \mathfrak{D}(\Delta^+)$ and γ be a τ-decomposable measure on Σ. Let $E, F \in \Sigma$ and $f, g \in \Omega^{\mathbb{R}+}$ be γ-integrable functions on the corresponding sets. Then*

(i) $\int_E f \, d\gamma = \int_\Omega f \mathbb{1}_E \, d\gamma$;
(ii) $\int_E 1 \, d\gamma = \gamma_E$;
(iii) if $E \subseteq F$, then $\int_E f \, d\gamma \geq \int_F f \, d\gamma$;
(iv) if τ is continuous and $E \cap F = \emptyset$, then $\int_{E \cup F} f \, d\gamma = \int_E f \, d\gamma \oplus_\tau \int_F f \, d\gamma$;
(v) $\int_E f \, d\gamma = \varepsilon_0$ *if and only if $f = 0$ a.e. on E.*

Proof. Properties (i) and (ii) follow immediately from Definition 1 and (2). In order to prove (iii) it is enough to realize that $f \cdot \mathbb{1}_E \leq f \cdot \mathbb{1}_F$ and to use antimonotonicity of the integral. The property (iv) we prove as follows. From the proof of [5, Theorem 5.6] the property (iv) is true for simple functions (without any additional conditions), i.e., for each simple function $\mathfrak{f} \in \mathcal{S}_{f,E}$ it holds

$$\int_{E \cup F} \mathfrak{f} \, d\gamma = \int_E \mathfrak{f} \, d\gamma \oplus_\tau \int_F \mathfrak{f} \, d\gamma.$$

Hence, using the monotonicity of triangle function $\oplus_\tau$ we have

$$\int_{E \cup F} f \, d\gamma = \inf \left\{ \int_{E \cup F} \mathfrak{f} \, d\gamma; \, \mathfrak{f} \in \mathcal{S}_{f,E \cup F} \right\} = \inf \left\{ \int_E \mathfrak{f} \, d\gamma \oplus_\tau \int_F \mathfrak{f} \, d\gamma; \, \mathfrak{f} \in \mathcal{S}_{f,E \cup F} \right\}$$
$$\geq \inf \left\{ \int_E \mathfrak{f} \, d\gamma; \, \mathfrak{f} \in \mathcal{S}_{f,E} \right\} \oplus_\tau \inf \left\{ \int_F \mathfrak{f} \, d\gamma; \, \mathfrak{f} \in \mathcal{S}_{f,F} \right\} = \int_E f \, d\gamma \oplus_\tau \int_F f \, d\gamma.$$

For the reverse inequality let us consider the simple functions $\mathfrak{f} \cdot \mathbb{1}_E \in \mathcal{S}_{f,E}$ and $\mathfrak{g} \cdot \mathbb{1}_F \in \mathcal{S}_{f,F}$ and let us construct the simple function $\mathfrak{h} \in \mathcal{S}_{f,E \cup F}$ such that

$$\mathfrak{h}(\omega) = \begin{cases} \mathfrak{f}(\omega), & \omega \in E, \\ \mathfrak{g}(\omega), & \omega \in F, \\ 0, & \text{otherwise.} \end{cases}$$

Then

$$\int_E \mathfrak{f} \, d\gamma \oplus_\tau \int_F \mathfrak{g} \, d\gamma = \int_E \mathfrak{h} \, d\gamma \oplus_\tau \int_F \mathfrak{h} \, d\gamma = \int_{E \cup F} \mathfrak{h} \, d\gamma \geq \int_{E \cup F} f \, d\gamma.$$

On the right hand we have a lower bound which remains valid for all simple functions that lie below f on $E \cup F$. Thus, taking infimum over $\mathfrak{f}$ and $\mathfrak{g}$ separately on the left hand side by continuity of τ we get the inequality $\int_E f \, d\gamma \oplus_\tau \int_F f \, d\gamma \geq \int_{E \cup F} f \, d\gamma$. This completes the proof of (iv). Now it remains to prove the part (v).

"$\Leftarrow$" Let $f = 0$ a.e. on E. Then any simple function $\mathfrak{f} \leq f$ must also be 0 a.e. on E. Since in its formula either $x_i = 0$ or else $\gamma_{E_i} = \varepsilon_0$, its probabilistic-valued integral equals to ε_0. Taking infimum over all $\mathfrak{f} \leq f$ we get the result.

"$\Rightarrow$" By contradiction. Let $\int_E f \, d\gamma = \varepsilon_0$ and let there exist a set $F = \{\omega \in E; \, f(\omega) > 0\} \in \Sigma$ such that $\gamma_F \neq \varepsilon_0$. Then it is easy to check (by contradiction) that there exists some positive integer n such that the τ-decomposable measure of the set $F_n = \{\omega \in E; f(\omega) > \frac{1}{n}\}$ differs from ε_0, i.e., $\gamma_{F_n} \neq \varepsilon_0$. This inequality is equivalent to the fact that there exists $a \in]0, \infty[$ such that $\gamma_{F_n}(x) < \varepsilon_0(x)$ for $x \in]0, a]$. Since $f \geq f \cdot \mathbb{1}_{F_n} \geq \frac{1}{n} \cdot \mathbb{1}_{F_n}$, we have

$$\int_E f \, d\gamma \leq \int_E \frac{1}{n} \cdot \mathbb{1}_{F_n} \, d\gamma = \frac{1}{n} \odot \gamma_{F_n}(x) < \varepsilon_0(x)$$

for $x \in]0, \frac{a}{n}]$. So, $\int_E f \, d\gamma \neq \varepsilon_0$ which contradicts the assumption. $\qquad \square$

Remark 2. Let us underline that the property (iv) of Proposition 4 follows immediately from [5, Theorem 5.6] under the additional assumption γ is continuous from below. In fact, the previous theorem has shown that this assumption is superfluous.

Corollary 1. *Let $\tau \in \mathfrak{CD}(\Delta^+)$ and γ be a τ-decomposable measure on Σ. For each γ-integrable function $f \in \Omega^{\mathbb{R}+}$ the set function $\nu^f : \Sigma \to \Delta^+$ defined by $\nu_E^f := \int_E f \, d\gamma$ for $E \in \Sigma$, is a τ-decomposable measure on Σ.*

3.2 Probabilistic Integral – Convergence Theorems

The crux of integration lies in convergence theory. One convergence theorem has already been mentioned, see Proposition 2. Analyzing the original proof we can immediately formulate the following (generalized) monotone convergence theorem.

Theorem 1 (Monotone convergence theorem I). *Let $\tau \in \mathfrak{CD}(\Delta^+)$, and γ be a τ-decomposable measure on Σ. Then the following assertions are equivalent:*

(i) γ is continuous from below on Σ,
(ii) for each γ-integrable function $f \in \Omega^{\mathbb{R}+}$ on Ω and for each nondecreasing sequence $(f_n)_1^\infty \in \Omega^{\mathbb{R}+}$ converging pointwisely to f it holds

$$\int_\Omega f \, d\gamma = \lim_{n \to \infty} \int_\Omega f_n \, d\gamma. \tag{3}$$

Proof. Since $\left(\int_\Omega f_n \, d\gamma \right)_1^\infty$ is a nonincreasing sequence of distance distribution functions, which is bounded below by distance distribution function $\int_\Omega f \, d\gamma$, the existence of the limit follows.

(i)$\Rightarrow$(ii) By Halmos [8, §20, Theorem B] each nonnegative real-valued function f_n is the pointwise limit of a nondecreasing sequence $(f_{nm})_1^\infty$ of simple functions with respect to m. Then considering $g_n = \max\{f_{1n}, f_{2n}, \ldots, f_{nn}\}$ for each $n \in \mathbb{N}$, we get a nondecreasing sequence of simple functions converging pointwisely to f and $g_n \leq f_n \leq f$ for each $n \in \mathbb{N}$. For this construction, see e.g. [2, Lemma 3.71 c]. Hence, from the antimonotonicity of probabilistic integral we have

$$\int_\Omega g_n \, d\gamma \geq \int_\Omega f_n \, d\gamma \geq \int_\Omega f \, d\gamma$$

for each $n \in \mathbb{N}$. Finally, analyzing the proof [5, Theorem 5.2] (the proposition is true for any nondecreasing sequence of simple functions pointwisely converging to f not only for special one) we have $\int_\Omega f \, \mathrm{d}\gamma = \int_\Omega (\lim_{n\to\infty} \mathfrak{g}_n) \, \mathrm{d}\gamma = \lim_{n\to\infty} \int_\Omega \mathfrak{g}_n \, \mathrm{d}\gamma$, thus we get

$$\int_\Omega f \, \mathrm{d}\gamma = \lim_{n\to\infty} \int_\Omega \mathfrak{g}_n \, \mathrm{d}\gamma \geq \lim_{n\to\infty} \int_\Omega f_n \, \mathrm{d}\gamma \geq \int_\Omega f \, \mathrm{d}\gamma.$$

(ii)$\Rightarrow$(i) Consider $F, (F_n)_1^\infty \in \Sigma$ such that $F_n \nearrow F$. Then $(\mathbb{1}_{F_n})_1^\infty$ is a nondecreasing sequence pointwisely converging to $\mathbb{1}_F$, therefore using the assumptions and Proposition 4 (i), (ii) we get

$$\lim_{n\to\infty} \gamma_{F_n} = \lim_{n\to\infty} \int_\Omega \mathbb{1}_{F_n} \, \mathrm{d}\gamma = \int_\Omega \mathbb{1}_F \, \mathrm{d}\gamma = \gamma_F.$$

Since convergence at every point implies weak convergence, γ is continuous from below on Σ.

$\square$

Theorem 2 (Monotone convergence theorem II). *Let $\tau \in \mathfrak{CD}(\Delta^+)$ and γ be a τ-decomposable measure on Σ continuous from below. Then the following assertions are equivalent:*

(i) γ is continuous from above on Σ,
(ii) for each γ-integrable function $f \in \Omega^{\mathbb{R}_+}$ on Ω and for each nonincreasing sequence $(f_n)_1^\infty \in \Omega^{\mathbb{R}_+}$ converging pointwisely to f such that there exists $M > 0$ with $f_n(\omega) \leq M$ for each n and $\omega \in \Omega$, the equality (3) holds.

Proof. (i)$\Rightarrow$(ii) From antimonotonicity of integral we immediately get the inequality $\lim_{n\to\infty} \int_\Omega f_n \, \mathrm{d}\gamma \leq \int_\Omega f \, \mathrm{d}\gamma$. Thus, it suffices to show the reverse inequality. For any fixed real $t > 1$ put $E_n := \{\omega \in \Omega; f_n(\omega) \leq t \cdot f(\omega)\}$ with $n = 1, 2, \ldots$. Then $E_n \uparrow \Omega$ and $E_n^c = \Omega \setminus E_n \downarrow \emptyset$. Thus,

$$\int_\Omega f_n \, \mathrm{d}\gamma = \left(\int_{E_n} f_n \, \mathrm{d}\gamma \right) \oplus_\tau \left(\int_{E_n^c} f_n \, \mathrm{d}\gamma \right) \geq \left(\int_{E_n} t \cdot f \, \mathrm{d}\gamma \right) \oplus_\tau \left(\int_{E_n^c} M \, \mathrm{d}\gamma \right)$$

$$= \left(t \odot \int_{E_n} f \, \mathrm{d}\gamma \right) \oplus_\tau (M \odot \gamma_{E_n^c}),$$

where Proposition 4 (iv), Corollary 1 and Proposition 3 (ii), have been used. Since τ and γ are continuous, for each $t > 1$ we get

$$\lim_{n\to\infty} \int_\Omega f_n \, \mathrm{d}\gamma \geq \left(t \odot \lim_{n\to\infty} \int_{E_n} f \, \mathrm{d}\gamma \right) \oplus_\tau (M \odot \lim_{n\to\infty} \gamma_{E_n^c}) = \left(t \odot \int_\Omega f \, \mathrm{d}\gamma \right) \oplus_\tau (M \odot \varepsilon_0)$$

$$= t \odot \int_\Omega f \, \mathrm{d}\gamma$$

where [5, Lemma 5.1] is used. Letting $t \to 1$, we get $\lim_{n\to\infty} \int_\Omega f_n \, \mathrm{d}\gamma \geq \lim_{t\to 1} t \odot \int_\Omega f \, \mathrm{d}\gamma = \int_\Omega f \, \mathrm{d}\gamma$, where the last equality follows from the left continuity of distance distribution functions.

(ii)$\Rightarrow$(i) The proof is analogous to the proof of the same implication in Theorem 1.

$\square$

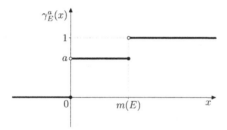

Fig. 1. Graph of probabilistic measure γ_E^a from Example 1

The following example demonstrates that the convergence theorem for a non-monotone sequence of functions also does not work in general. For this reason let us consider an example of probabilistic-valued decomposable measure, see [5, Example 3.6]. For a finite numerical (additive) measure m on a σ-algebra Σ and for $E \in \Sigma$ the mapping $\gamma \colon \Sigma \to \Delta^+$ of the form

$$\gamma_E^a(x) = a \ \mathbb{1}_{]0,m(E)]}(x) + \mathbb{1}_{]m(E),\infty[}(x) = \begin{cases} 0, & x \le 0, \\ a, & x \in]0, m(E)], \\ 1, & x > m(E) \end{cases}$$

with $a \in [0,1]$, is a τ_M-decomposable measure. Graph of γ_E^a for some $a \in [0,1]$ and certain value of $m(E)$ is illustrated in Fig. 1.

Example 1. Let us consider τ_M-decomposable measure γ^a with $a \in [0,1[$ and numerical measure being the Lebesgue measure, set $\Omega = [0,1]$ and put $f_n = n \cdot \mathbb{1}_{]0,\frac{1}{n}]}$. Then $f_n(x) \to 0$ for every $x \in [0,1]$ and $\int_{[0,1]} f_n \, d\gamma^a = a \cdot \mathbb{1}_{]0,1]} + \mathbb{1}_{]1,\infty[}$ for every n. Hence,

$$\int_{[0,1]} f \, d\gamma^a = \varepsilon_0 \neq a \cdot \mathbb{1}_{]0,1]} + \mathbb{1}_{]1,\infty[} = \lim_{n \to \infty} \int_{[0,1]} f_n \, d\gamma^a.$$

As the previous example shows, the convergence theorem need not work if the sequence of functions does not converge monotonically. For the following general convergence theorem let us recall that a probabilistic-valued set function $\gamma \colon \Sigma \to \Delta^+$ is continuous if it is continuous from below and from above.

Theorem 3. *Let $\tau \in \mathfrak{CD}(\Delta^+)$ and γ be a τ-decomposable measure on Σ. Then the following assertions are equivalent:*

(i) γ is continuous on Σ,
(ii) for each γ-integrable function $f \in \Omega^{\mathbb{R}+}$ on Ω and for each sequence $(f_n)_1^\infty \in \Omega^{\mathbb{R}+}$ converging pointwisely to f such that there exists $M > 0$ with $f_n(\omega) \le M$ for each n and $\omega \in \Omega$, the equality (3) holds.

Proof. (i)$\Rightarrow$(ii) Consider the functions $k_n = \sup_{i \ge n} f_i$ and $l_n = \inf_{i \ge n} f_i$. Then for each n the functions k_n, l_n are γ-integrable on Ω and both pointwisely converge to f

such that $k_n \searrow f$ and $l_n \nearrow f$. Since $l_n \leq f_n \leq k_n$ for each n, by antimonotonicity of the integral we obtain $\int_\Omega k_n \, d\gamma \leq \int_\Omega f_n \, d\gamma \leq \int_\Omega l_n \, d\gamma$. Hence, we have

$$\liminf_{n\to\infty} \int_\Omega k_n \, d\gamma \leq \liminf_{n\to\infty} \int_\Omega f_n \, d\gamma \leq \limsup_{n\to\infty} \int_\Omega f_n \, d\gamma \leq \limsup_{n\to\infty} \int_\Omega l_n \, d\gamma.$$

However, by monotone convergence theorem I and II, see Theorem 1 and 2, the sequences $(k_n)_1^\infty$ and $(l_n)_1^\infty$ are γ-integrable with

$$\lim_{n\to\infty} \int_\Omega k_n \, d\gamma = \int_\Omega f \, d\gamma = \lim_{n\to\infty} \int_\Omega l_n \, d\gamma.$$

Thus, from the squeezing inequalities above we have

$$\liminf_{n\to\infty} \int_\Omega f_n \, d\gamma = \limsup_{n\to\infty} \int_\Omega f_n \, d\gamma = \int_\Omega f \, d\gamma.$$

(ii)$\Rightarrow$(i) Consider any nondecreasing sequence $(f_n)_1^\infty \in \Omega^{\mathbb{R}_+}$ converging pointwisely to f for which the equality (3) holds. Then by Theorem 1 the set function γ is continuous from below. Analogously, for any nonincreasing bounded sequence $(f_n)_1^\infty \in \Omega^{\mathbb{R}_+}$ converging pointwisely to f for which the equality (3) holds, Theorem 2 enforces the continuity of γ from above. $\square$

3.3   Probabilistic Integral – Rounding of Reals

In each time a measurement is made, physical quantity is prone to errors, which may be classified into several groups. Among them are errors which are caused by rounding. Indeed, the precision of a measuring instrument is determined by the smallest unit to which it can measure. After all, any calculation may be affected by such errors. For instance, the value of π is approximated very often by $\frac{22}{7}$ or less exactly by 3.14. Similarly, in practice we approximate a lot of partial results. In the context of probabilistic-valued measures, *every rounding of reals corresponds to a distance distribution function* $F_{[a_i,b_i]}$ *which is uniformly distributed over intervals* $[a_i, b_i]$ *with* $a_i, b_i \in \mathbb{R}_+$.

Consider a set $\Omega = \{\omega_1, \omega_2, \ldots, \omega_n\}$ of n physical quantities, whose linear combination states another physical quantity ω. Then ω may be seen as a n-ary mapping $\omega \colon \mathbb{R}_+^n \to \mathbb{R}_+$ of the form

$$\omega = \sum_{i=1}^n c_i \omega_i, \tag{4}$$

with nonnegative constants $c_i \in \mathbb{R}_+$ for $i = 1, 2, \ldots, n$. If the exact values of quantities $\omega_1, \omega_2, \ldots, \omega_n$ are known, then it is easy to compute the exact value of quantity ω. Assuming that original values were rounded, a probabilistic information about their values is expressed by the distribution function $F_{[a_i,b_i]}$ of the uniform distribution on the corresponding nonnegative interval $[a_i, b_i]$. The proper tool for calculating probabilistic values of the quantity ω is the probabilistic integral with respect to the τ-decomposable measure with τ being the convolution.

Remark 3. For two positive real-valued independent random variables X and Y, having distribution functions F and G, respectively, the binary operation $*\colon \Delta^+ \times \Delta^+ \to \Delta^+$ in the form

$$(F * G)(z) = \begin{cases} 0, & z \leq 0, \\ \int_0^z F(z - t)\,\mathrm{d}G(t), & z \in]0, \infty[, \\ 1, & z = \infty, \end{cases} \tag{5}$$

is the convolution of F, G which corresponds to distribution function of the sum of random variables $Z = X + Y$. It is not difficult to show that the convolution is a triangle function. Moreover, it is distributive. Indeed, for $c = 0$ we get the distribution function ε_0 on the both sides of (1). For $c > 0$ it is clear that functions $(c \odot F) * (c \odot G)(z)$ and $c \odot (F * G)(z)$ are both zero on $[-\infty, 0]$, whereas for $z \in]0, \infty[$ it holds

$$(c \odot F) * (c \odot G)(z) = \int_0^z F\left(\frac{z - t}{c}\right) \mathrm{d}G\left(\frac{t}{c}\right) = \int_0^{\frac{z}{c}} F\left(\frac{z}{c} - u\right) \mathrm{d}G(u)$$
$$= c \odot (F * G)(z).$$

Identity between the integrals follows from the substitution $\frac{t}{c} = u$ due to the fact that $\frac{t}{c}$ is strictly increasing and continuous function. The explicit formula of the n-fold convolution of distribution functions corresponding to n mutually independent random variables $X_i \sim U(a_i, b_i)$ with $a_i, b_i \in \mathbb{R}_+$, $a_i < b_i$, for $i = 1, 2, \ldots, n$, can be found e.g. in [11], or [13, p. 238].

For our purpose let us denote by $\Sigma = 2^\Omega$ the power set of Ω, $f\colon \Omega \to \mathbb{R}_+$ the function defined for each $i = 1, 2, \ldots, n$ as follows $f(\omega_i) = c_i$, and γ the $*$-decomposable measure defined for singletons $\{\omega_i\}$, $i = 1, 2, \ldots, n$ as follows $\gamma_{\{\omega_i\}} = F_{[a_i, b_i]}$, with $a_i, b_i \in \mathbb{R}_+$ and $a_i < b_i$. Probabilistic information about the values of quantity ω is now described by the distance distribution function

$$\int_\Omega f\,\mathrm{d}\gamma = \bigoplus_{i=1}^n c_i \odot F_{[a_i, b_i]}, \tag{6}$$

where $\oplus$ is the convolution given by (5). In the special case $c_i = 0$ for each $i = 2, \ldots, n$, i.e., the physical quantity ω is just a multiple of quantity ω_1, the probabilistic information about its values is described by the distribution function $F_{[c_1 \cdot a_1, c_1 \cdot b_1]}$, as expected.

Example 2. Let us consider the formula $f(u, v) = 2 \cdot (u + v)$, where u, v are nonnegative real-valued physical quantities. This expression corresponds to the formula of the perimeter of a rectangle. Let us set the values of rectangle-sides by the measuring and, subsequently, rounding at $u = 1j$ and $v = 1j$ (directed rounding, round up). Then the values of both rectangle sides can be expressed by d.d.f. uniformly distributed over the interval $[0, 1]$. Then the probabilistic information about the values of the perimeter of this rectangle is expressed by the distribution function

$$\int_\Omega f \, d\gamma = (2 \odot F_{[0,1]}) * (2 \odot F_{[0,1]}) = F_{[0,2]} * F_{[0,2]},$$

where $F_{[0,2]}$ is d.d.f. uniformly distributed over the interval $[0,2]$. The explicit form of this convolution is

$$(F_{[0,2]} * F_{[0,2]})(z) = \begin{cases} 0, & z = 0, \\ \frac{z^2}{8}, & z \in]0,2], \\ -\frac{z^2}{8} + z - 1, & z \in]2,4], \\ 1, & \text{otherwise,} \end{cases}$$

see Fig. 2.

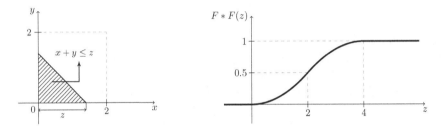

Fig. 2. Convolution of distribution functions related to two independent uniform distributions over $[0,2]$.

4   Concluding Remarks

No measurement of any sort is complete without a consideration of this inherent error. The problem of describing the value of certain measured physical quantities with respect to measurement errors has been studied extensively, see e.g. Moore's interval analysis [14]. The aim of this theory is to find an interval that contains, with absolute certainty, the truth value of a certain computation or measurable physical quantity. What is more, the theory develops the rules for the arithmetic of intervals explicitly. The theory provides just the information that the real value is somewhere in an interval, but it does not state the probabilities with which these values are attained. Thus, every value has the same chance to be acquired. From the statistical point of view, the problem of setting the value of a physical quantity reliably is more extensive. In this text we have considered only the errors of measurement that were caused by rounding. However, there are many other classes of measurement errors. In general, random errors get close to the normal distribution. So, if distribution functions (of a normal distribution) for certain physical quantities $w_1, \ldots, w_n$ are known, it is not difficult to compute the distribution function of the quantity w (which is defined by the form (4)). Then it follows that it is sufficient to use the formula (6) with the convolution of distribution functions which correspond to the normal distribution. It is, however, a well-known fact that the sum of normally distributed random variables is a normally distributed random variable, whereas it does not hold for a general triangle function applied to Gaussian distribution functions.

References

1. Borzová-Molnárová, J., Halčinová, L., Hutník, O.: Probabilistic-valued decomposable set functions with respect to triangle functions. Inf. Sci. **295**, 347–357 (2015)
2. Bukovský, L.: The Structure of the Real Line. Monografie Matematyczne. Springer, Basel (2011). https://doi.org/10.1007/978-3-0348-0006-8
3. Cobzaş, S.: Completeness with respect to the probabilistic Pompeiu-Hausdorff metric. Stud. Univ. Babeş-Bolyai, Math. **52**(3), 43–65 (2007)
4. Dubois, D., Prade, H.: Fuzzy Sets and Systems: Theory and Applications. Academic Press, New York (1980)
5. Halčinová, L., Hutník, O.: An integral with respect to probabilistic-valued decomposable measures. Int. J. Approx. Reason. **55**, 1469–1484 (2014)
6. Halčinová, L., Hutník, O., Mesiar, R.: On distance distribution functions-valued submeasures related to aggregation functions. Fuzzy Sets Syst. **194**(1), 15–30 (2012)
7. Halčinová, L., Hutník, O., Mesiar, R.: On some classes of distance distribution functions-valued submeasures. Nonlinear Anal. **74**(5), 1545–1554 (2011)
8. Halmos, P.: Measure Theory. Van Nostrand, New York (1968)
9. Hutník, O., Mesiar, R.: On a certain class of submeasures based on triangular norms. Int. J. Uncertainty Fuzziness Knowl. Based Syst. **17**, 297–316 (2009)
10. Klement, E.P., Mesiar, R., Pap, E.: Triangular Norms, Trends in Logic, Studia Logica Library, vol. 8. Kluwer Academic Publishers, Dordrecht (2000)
11. Jong, S.K., Sung, L.K., Yang, H.K., Yu, S.J.J.: Generalized convolution of uniform distributions. Appl. Math. Inform. **28**, 1573–1581 (2010)
12. Menger, K.: Statistical metrics. Proc. Nat. Acad. Sci. U.S.A. **28**, 535–537 (1942)
13. Mood, A.M., Graybill, F.A., Boes, D.C.: Introduction to the Theory of Statistics. McGraw-Hill, New York (1974)
14. Moore, R.E., Bierbaum, F.: Methods and Applications of Interval Analysis. Studies in Applied and Numerical Mathematics. SIAM, Philadelphia (1979)
15. Pap, E.: Applications of decomposable measures on nonlinear differential equations. Novi Sad J. Math. **31**, 89–98 (2001)
16. Schweizer, B., Sklar, A.: Probabilistic Metric Spaces. North-Holland Publishing, New York (1983)
17. Šerstnev, A.N.: Random normed spaces: problems of completeness. Kazan. Gos. Univ. Uch. Zap. **122**, 3–20 (1962)
18. Torra, V., Navarro-Arribas, G.: Probabilistic metric spaces for privacy by design machine learning algorithms: modeling database changes. In: Garcia-Alfaro, J., Herrera-Joancomartí, J., Livraga, G., Rios, R. (eds.) DPM/CBT -2018. LNCS, vol. 11025, pp. 422–430. Springer, Cham (2018). https://doi.org/10.1007/978-3-030-00305-0_30
19. Weber, S.: ⊥-decomposable measures and integrals for Archimedean t-conorm. J. Math. Anal. Appl. **101**, 114–138 (1984)

Distorted Probabilities and Bayesian Confirmation Measures

Andrea Ellero[1]([✉])[iD] and Paola Ferretti[2]([✉])[iD]

[1] Department of Management, Ca' Foscari University of Venice,
Fondamenta San Giobbe, Cannaregio 873, 30121 Venezia, Italy
`ellero@unive.it`
[2] Department of Economics, Ca' Foscari University of Venice,
Fondamenta San Giobbe, Cannaregio 873, 30121 Venezia, Italy
`ferretti@unive.it`

Abstract. Bayesian Confirmation Measures (BCMs) assess the impact of the occurrence of one event on the credibility of another. Many measures of this kind have been defined in literature. We want to analyze how these measures change when the probabilities involved in their computation are distorted. Composing distortions and BCMs we define a set of Distorted Bayesian Confirmation Measures (DBCMs); we study the properties that DBCMs may inherit from BCMs, and propose a way to measure the degree of distortion of a DBCM with respect to a corresponding BCM.

Keywords: Bayesian Confirmation Measures · Distorted probabilities · Distorted Bayesian Confirmation Measures

1 Introduction

The degree to which an event E supports or contradicts the occurrence of another event H can be assessed comparing prior probability $P(H)$ with posterior probability $P(H|E)$. An accustomed way to carry out the comparison is by using a Bayesian Confirmation Measure (BCM) usually denoted as a function of the two events, $C(H, E)$.

More precisely, a BCM $C(H, E)$ is always expressed (or can be rewritten) as a function of prior probability, posterior probability and the probability of evidence $P(E)$; quite often only $P(H)$ and $P(H|E)$ are necessary (in this case the BCM is also called Initial Final Probability Dependent (IFPD) measure, see e.g. [6]).

The main subject of this paper concerns the fact that the evaluation of probabilities involved in the definition of the BCM could be subject to distortions when dealing with practical situations. Biased probabilities evaluations influence,

---

Authors are listed in alphabetical order. All authors contributed equally to this work, they discussed the results and implications and commented on the manuscript at all stages.

© Springer Nature Switzerland AG 2020
V. Torra et al. (Eds.): MDAI 2020, LNAI 12256, pp. 92–103, 2020.
https://doi.org/10.1007/978-3-030-57524-3_8

in turn, the values of Bayesian Confirmation Measures. To go into some detail, let us consider a widely adopted definition of Bayesian Confirmation Measure (BCM). Observe that, as soon as an evidence E occurs, the knowledge changes and conclusion H may be confirmed, when $P(H|E) > P(H)$, or disconfirmed, when $P(H|E) < P(H)$. For these reasons, a Bayesian Confirmation Measure $C(H, E)$ is required to satisfy the following sign properties:

$$\begin{cases} C(H, E) > 0 & \text{if } P(H|E) > P(H) & \text{(confirmation case)} \\ C(H, E) = 0 & \text{if } P(H|E) = P(H) & \text{(neutrality case)} \\ C(H, E) < 0 & \text{if } P(H|E) < P(H) & \text{(disconfirmation case).} \end{cases}$$

Confirmation measures have been explored from different perspectives (see e.g. [6,10,11]), we will recall the definitions of some well known BCMs in Sect. 2.

The second concept we need to recall concerns distorted probabilities.

Humans often fail to evaluate an expected gain or loss and instead of evaluating the true probability P of an event they perceive its distorted probability. Systematic (undetected) errors may occur when probabilities are estimated via frequency measurements, also leading to distorted measurements. The assessed value of probability P becomes then a distorted value $g(P)$ where the distortion g is modeled as a function $g : [0, 1] \to [0, 1]$, non-decreasing and such that $g(0) = 0$ and $g(1) = 1$.

Section 3 recalls some possible distortion functions coupling them with BCMs, which will lead us to define the Distorted Bayesian Confirmation Measures (DBCMs), the main subject of this paper. The study of the DBCMs proposed in Sect. 3 provides some insights on the differences with respect to the corresponding BCMs, focusing then on their symmetry properties. Concluding remarks are presented in Sect. 4.

2   Bayesian Confirmation Measures

As mentioned, a BCM evaluates how much the knowledge of an evidence E supports or contradicts the occurrence of a conclusion H. We need now some *guinea pig* BCMs that will serve hereinafter to build examples to illustrate our proposal of coupling the concepts of distortion and confirmation measure. The definitions of a set of well known BCMs is reported in Table 1[1].

Besides d, K, F and R, which are already written in terms of only prior probability $P(H)$ and posterior probability $P(H|E)$, some algebraic manipulation allow to express also $\mathcal{L}$ making use of $P(H)$ and $P(H|E)$ only:

$$\mathcal{L}(H, E) = \log \left[\frac{P(H|E)[1 - P(H)]}{P(H)[1 - P(H|E)]} \right].$$

---

[1] Throughout the paper, the formulas are assumed to be well defined, e.g. denominators do not vanish, as implicitly done in the literature.

Table 1. A set of known Bayesian Confirmation Measures

BCM	Definition
d (Carnap [1])	$d(H, E) = P(H|E) - P(H)$
K (Keynes [16])	$K(H, E) = \log[P(H|E)/P(H)]$
F (Finch [9])	$F(H, E) = [P(H|E) - P(H)]/P(H)$
$\mathcal{L}$ (Good [13])	$\mathcal{L}(H, E) = \log[P(E|H)/P(E|\neg H)]$
R (Rips [18])	$R(H, E) = [P(H|E) - P(H)]/[1 - P(H)]$
M (Mortimer [17])	$M(H, E) = P(E|H) - P(E)$

The BCMs that can be written as functions of $P(H|E)$ and $P(H)$ only, constitute the special class of IFPD (Initial and Final Probability Dependence) confirmation measures. For not-IFPD BCMs a third variable must be included in the definition, for example the probability of the evidence $P(E)$. This is the case for measure M [17] which can be rewritten using only $P(H|E)$, $P(H)$ and $P(E)$ (but not less than three variables) as

$$M(H, E) = P(E) \frac{P(H|E) - P(H)}{P(H)}.$$

We observe that, once BCMs are rewritten in terms of the three variables $x = P(H|E)$, $y = P(H)$ and $z = P(E)$, several considerations in terms of monotonicity and symmetry, become much easier to be made, borrowing simple calculus ideas (see, e.g., [2,3,19]). Nevertheless the rewriting of BCMs requires close attention when the probabilities involved in the definition of the BCM are subject to distortion (see Sect. 3).

We will give here some examples concerning symmetry properties of BCMs. Until recently, the symmetry properties of BCM have been the subject of debate in the literature, especially for what concerns desirability or undesirability of certain properties (see, e.g., [14] and [12]). We focus on three of the symmetry definitions that have been recently considered for a confirmation measure C (see, e.g., [5]):

Evidence Symmetry (ES) if $C(H, E) = -C(H, \neg E)$ for all (H, E);
Hypothesis Symmetry (HS) if $C(H, E) = -C(\neg H, E)$ for all (H, E);
Inversion Symmetry (IS) if $C(H, E) = C(E, H)$ for all (H, E).

Note that the considered symmetries are defined by means of different combinations of H, E and of their negations $\neg H$ and $\neg E$, and it is possible to reformulate the definitions in terms of the probabilities $x = P(H|E)$, $y = P(H)$ and $z = P(E)$. For example, if we consider symmetry HS, its definition can be reformulated as

$$C(P(H|E), P(H), P(E)) = -C(P(\neg H|E), P(\neg H), P(E))$$
$$= -C(1 - P(H|E), 1 - P(H), P(E))$$

that is, $C(x, y, z) = -C(1 - x, 1 - y, z)$ for all feasible (x, y, z). As a further example consider symmetry IS; in this case the definition can be rewritten as

$$C(P(H|E), P(H), P(E)) = C(P(E|H), P(E), P(H))$$
$$= C(P(H|E)P(E)/P(H), P(E), P(H))$$

or $C(x, y, z) = C(xz/y, z, y)$ for all feasible (x, y, z).

In a similar way we obtain that a confirmation measure C satisfies:

(ES) if $C(x, y, z) = -C((y - xz)/(1 - z), y, 1 - z)$ for all feasible $(x, y, z)^2$;
(HS) if $C(x, y, z) = -C(1 - x, 1 - y, z)$ for all feasible (x, y, z);
(IS) if $C(x, y, z) = C(xz/y, z, y))$ for all feasible (x, y, z).

Most confirmation measures do not satisfy all the above defined symmetry properties[3]. And neither are those properties always desirable: for what concerns eligibility of symmetry properties, Eells and Fitelson [8] and Glass [12] argue that an acceptable BCM should not exhibit symmetries ES, and IS, while consider HS as the only suitable symmetry property for a BCM (for another point of view see e.g. [14]). Therefore it becomes also interesting to measure how far a BCM is from the considered symmetry, is it far from the desirable HS? Is it close to the unwanted ES? This issue has been faced in [4] suggesting a way to measure asymmetry, considering the mean distance of measure $C(x, y, z)$ from its symmetric counterparts.

3 Distorted Probabilities and Bayesian Confirmation Measures

In a recent paper, Vassend [21] studied the role assumed by the *interval measurement hypothesis* in the right choice of a BCM function: he proved that under the previous assumption only the Good BCM [13], $G(H, E)$ that is, the log-likelihood measure, works well. He made the choice of considering only IFPD confirmation measures (see in [21] the discussion regarding this choice) and in order to take into account the uniform lack of dependence of a BCM on small variations on $P(H|E)$ and $P(H)$, the Author introduced a function in order to capture the notion of a variation in the probability, given the size of variation. More precisely, a small variation of size ϵ in the value p of the probability, is modeled by a function $v(p, \epsilon)$, so that, for example, the variation in C resulting from a variation of size ϵ with respect to $p = P(H|E)$ is given by

$$C(v(P(H|E), \epsilon), P(H)) - C(P(H|E), P(H))$$

---

[2] Feasibility of (x, y, z) requires that probabilities x, y and z satisfy in particular the Total Probability Theorem.

[3] A BCM that satisfies all the above considered symmetries is Carnap's b [1] which is defined as $b(H, E) = P(E \cap H) - P(E) P(H)$.

where $C(H, E)$ is a function of $P(H|E)$ and $P(H)$ only, i.e., Vassend considers C as an IFPD measure. Similarly, if the variation is computed with respect to $p = P(H)$ the variation to be considered becomes

$$C(P(H|E), v(P(H), \epsilon)) - C(P(H|E), P(H)).$$

Arguing on properties of BCMs, Vassend [21] discussed the minimal requirements they should satisfy: among them the main one concerns the uniform insensitivity of a BCM to small variations in $P(H|E)$ and $P(H)$. From those requirements, he deduced the most plausible form for v the function $v(x, \epsilon) = x + x(1 - x)\epsilon$, where $x \in [0, 1]$. Note that, in the particular case of $\epsilon = 1$, the variation function v corresponds to the Dual Power Transform $g(x) = 1 - (1 - x)^k$ with $k = 2$ (see [23] for a discussion of risk measure as expectation with probability distortions). Nevertheless, in [21] no reference is made about the possible link between the variation function $v(x, \epsilon)$ and the general class of distortion functions $g(x)$.

Distortion functions have been extensively investigated in Literature, under different perspectives and with diverse aims. Following the seminal work of Denneberg [7], given a probability measure P on a σ-algebra $\mathcal{A} \subset 2^\Omega$ and a distortion function g, then the composed function $\mu = g(P)$ is a monotone set function called distorted probability with corresponding distortion g. Moreover, if g is a concave (convex) function then μ is a submodular (supermodular) set function. In literature various distortion functions have been proposed, between them the most cited are the power distortion

$$g(u) = u^r \qquad 0 < r \leq 1$$

corresponding to the PH transform studied in [22] mainly in the context of a premium principle ($1/r$ is called parameter of risk aversion when distortions are considered in the field of risk measures), and the dual power transform

$$g(u) = 1 - (1 - u)^k, \qquad k \geq 1.$$

Both are examples of beta transforms (see [23]) in which the distortion is expressed in the form of an incomplete beta function. Moreover, in the context of risk measures, the Conditional Tail Expectation CTE and the Value at Risk VaR$_\alpha$ can respectively be expressed in terms of the distortion functions ($0 \leq \alpha < 1$)

$$g(u) = \begin{cases} u/(1 - \alpha), & 0 \leq u < 1 - \alpha; \\ 1, & 1 - \alpha \leq u \leq 1, \end{cases} \qquad g(u) = \begin{cases} 0, & 0 \leq u < 1 - \alpha; \\ 1, & 1 - \alpha \leq u \leq 1. \end{cases}$$

so that the standard risk measures can be classified according to their associated distortion functions.

When probabilities are distorted, regardless of the reason that induces the distortion, in the framework of BCMs it is natural to expect that the corresponding functions may be modified, in fact if probabilities are distorted, this may lead to a change in the assessments provided by a BCM.

If this can be easily accepted, a new question arises: what probabilities are distorted, namely, how to take into account a distorted probability in the definition of a BCM? Let us consider, for example, the Bayesian Confirmation Measure R proposed by Rips [18] and the power distortion $g(u) = \sqrt{u}$.

Referring to the definition of Rips as in [18]

$$R(H, E) = \frac{P(H|E) - P(H)}{1 - P(H)}$$

and by assuming that all probabilities considered in the definition are distorted by g, then it is possible to define the new function R_g as follows

$$R_g(H, E) = \frac{g(P(H|E)) - g(P(H))}{1 - g(P(H))} = \frac{\sqrt{P(H|E)} - \sqrt{P(H)}}{1 - \sqrt{P(H)}}.$$

On the other hand though, the BCM R can be expressed not only as a function of prior and posterior probabilities $P(H)$ and $P(H|E)$ that, on their own, can be distorted by g, but it can be easily rewritten as a function of $P(\neg H|E)$ and $P(\neg H)^4$ and so a new function $\tilde{R}_g$ can be defined through a distortion of all involved probabilities

$$\tilde{R}_g(H, E) = 1 - \frac{g(P(\neg H|E))}{g(P(\neg H))} = 1 - \frac{\sqrt{1 - P(H|E)}}{\sqrt{1 - P(H)}}.$$

It turns out that extremely important in the study of distortion impact on BCM functions is the choice of variables with respect to which the BCM functions are defined and which will be subsequently subject to distortion. Here we assume that all considered BCM functions $C(H, E)$ are expressed in terms of $P(H|E)$, $P(H)$ and $P(E)$; this assumption together with monotonicity of each distortion function g ensure that the new functions

$$C_g(H, E) = C(g(P(H|E)), g(P(H)), g(P(E)))$$

are new Bayesian Confirmation Measures:

Definition 1. *Given a BCM $C(H, E)$ and a distortion $g(u)$, the function C_g defined by $C_g(H, E) = C(g(P(H|E)), g(P(H)), g(P(E)))$ is called Distorted Bayesian Confirmation Measure (DBCM).*

In the following we will also use variables $x = P(H|E)$, $y = P(H)$ and $z = P(E)$ so that a DBCM will be expressed as

$$C_g(H, E) = C_g(x, y, z) = C(g(x), g(y), g(z)).$$

Note that the new function can be properly defined (Distorted) Bayesian Confirmation Measure, in fact it is a new interestingness measure aimed at evaluating

---

[4] Most of the literature reports in fact the measure as $R(H, E) = 1 - [P(\neg H|E)/P(\neg H)]$, see, e.g., [5].

the degree to which the antecedent E supports or contradicts the conclusion H, using the prior probability $P(H)$, posterior probability $P(H|E)$ and the probability $P(E)$ of antecedent E that are considered in $C(H,E)$; moreover, by monotonicity of g, the confirmation/neutrality/disconfirmation cases rely on comparisons that involve just the original probabilites $P(H|E)$ and $P(H)$:

$$C_g(H,E) > 0 \text{ if } P(H|E) > P(H) \qquad \text{(confirmation case)}$$
$$C_g(H,E) = 0 \text{ if } P(H|E) = P(H) \qquad \text{(neutrality case)}$$
$$C_g(H,E) < 0 \text{ if } P(H|E) < P(H) \qquad \text{(disconfirmation case).}$$

Table 2 reports some examples of DBCMs where the distortion g is represented by the power distortion $g(u) = u^r$ or the dual power transform $g(u) = 1 - (1-u)^r$.

Table 2. A set of Distorted Bayesian Confirmation Measures

	BCM	DBCM: $g(u) = u^r$	DBCM: $g(u) = 1 - (1-u)^r$
d	$P(H|E) - P(H)$	$[P(H|E)]^r - [P(H)]^r$	$[1 - P(H)]^r - [1 - P(H|E)]^r$
K	$\log\left[\frac{P(H|E)}{P(H)}\right]$	$r\log\left[\frac{P(H|E)}{P(H)}\right]$	$\log\left[\frac{1-[1-P(H|E)]^r}{1-[1-P(H)]^r}\right]$
F	$\frac{P(H|E)-P(H)}{P(H)}$	$\left[\frac{P(H|E)}{P(H)}\right]^r - 1$	$\frac{[1-P(H)]^r-[1-P(H|E)]^r}{1-[1-P(H)]^r}$
L	$\log\left[\frac{P(E|H)}{P(E|\neg H)}\right]$	$r\log\left[\frac{P(H|E)}{P(H)}\right] + \log\left[\frac{1-[P(H)]^r}{1-[P(H|E)]^r}\right]$	$r\log\left[\frac{1-P(H)}{1-P(H|E)}\right] + \log\left[\frac{1-[1-P(H|E)]^r}{1-[1-P(H)]^r}\right]$
R	$\frac{P(H|E)-P(H)}{1-P(H)}$	$\frac{[P(H|E)]^r-[P(H)]^r}{1-[P(H)]^r}$	$1 - \left[\frac{1-P(H|E)}{1-P(H)}\right]^r$
M	$P(E|H) - P(E)$	$\frac{[P(H|E)]^r-[P(H)]^r}{[P(H)]^r}[P(E)]^r$	$\frac{[1-P(H)]^r-[1-P(H|E)]^r}{1-[1-P(H)]^r}[1-[1-P(E)]^r]$

It is worth to observe that in general a BCM and the corresponding DBCM are not ordinally equivalent, in fact, depending on how we fix prior and posterior probabilities, we may obtain different rankings of the couples $(P(H|E), P(H))$. Examples illustrating the order reversion on BCM and DBCM values are reported in Table 3[5] for the case of Carnap measure d, distortions d_{g_1} ($g_1(u) = \sqrt[2]{u}$) and $d_{g_2}(g_2(u) = \sqrt[3]{u})$. The ranking of the four points P_1, P_2, P_3 and P_4, in the space $(P(H|E), P(H))$ (rankings are reported in the last three columns of the table) allow to conclude that the three measures are not ordinally equivalent.

To investigate the possible influences exerted by a distortion function $g(u)$ on the definition of a new Bayesian confirmation measure DBCM, we focus our attention on the function $C(x,y,z) - C_g(x,y,z)$ and on the relative strength that the function g has in setting $C_g(x,y,z)$. To this goal, the relative distortion force of g on C_g, denoted by $\mu^p(C,g)$, is computed as

$$\mu^p(C,g) = \frac{\|C - C_g\|_p}{\|C\|_p} \qquad (1)$$

---

[5] Computations in this paper were performed with Wolfram's software *Mathematica* (version 11.0.1.0).

Table 3. Carnap's $d(H, E) = P(H|E) - P(H)$ and the distorted measures d_{g_1} with $g_1(u) = \sqrt{u}$ and d_{g_2} with $g_2(u) = \sqrt[3]{u}$ are NOT ordinally equivalent.

CASE	$P(H|E)$	$P(H)$	$d(H,E)$	d_{g_1}	d_{g_2}	Rank d	Rank d_{g_1}	Rank d_{g_2}
P_1	0.1	0.03	0.07	0.143023	0.153436	1	2	2
P_2	0.2	0.1	0.1	0.130986	0.120645	2	1	1
P_3	0.2	0.02	0.18	0.305792	0.313362	3	3	4
P_4	0.5	0.15	0.35	0.319808	0.262371	4	4	3

where $\| \cdot \|_p$ denotes the L^p-norm ($p \in [1, \infty]$) (see, e.g., [7]). We assume that functions C and C_g are continuous in their domains (as all BCMs defined in the literature are). The computation of the integrals involved in the L^p-norm must take into account restrictions on probabilities, i.e. x, y and z belong to $[0, 1]$ and $xz \leq y \leq 1 - z + zx$ (i.e. $P(H \cap E) \leq P(H) \leq P(\neg E) + P(H \cap E)$ which corresponds to $P(E) \leq P(H \cup E) \leq 1)^6$.

Table 4 presents the computations when $p = 2$ and the considered distortion functions are $g(u) = u^r$ ($r = 1/2$ and $r = 1/3$) and $g(u) = 1 - (1 - u)^r$ ($r = 2$, and $r = 3$) respectively, and the analyzed BCMs are those in Table 1. Remarkable are the rankings implicitly induced by those values. Let us consider for example DBCMs $\mathcal{L}_g$ and F_g: the relative distance of $\mathcal{L}_g$ from $\mathcal{L}$ is greater with distortion $g_3(u) = 1 - (1 - u)^2$ and $g_4(u) = 1 - (1 - u)^3$ than with $g_1(u) = u^{1/2}$ and $g_2(u) = u^{1/3}$ while the order is reversed for F_g: the two BCMs react in a different way to different distortions. In particular, since the relative distortion induced on $\mathcal{L}$ by g_1 is lower than on F_g implies that $\mathcal{L}$ is more robust with respect to this kind of distortion than $F_g{}^7$.

Table 4. Relative distortion force of g on C_g (L_2 norm)

DBCM	$g_1(u) = u^{1/2}$	$g_2(u) = u^{1/3}$	$g_3(u) = 1 - (1 - u)^2$	$g_4(u) = 1 - (1 - u)^3$
d_g	0.28193601	0.43067074	0.3055055	0.47888368
K_g	1/2	2/3	0.20071986	0.32832845
F_g	0.95982788	0.97915816	0.35675502	0.52461323
$\mathcal{L}_g$	0.20222796	0.26186300	0.46914148	0.98068926
R_g (confirmation case)	0.09870233	0.13389174	0.34887014	0.53803290
R_g (disconfirmation case)	0.39514893	0.61095485	∞	∞
M_g	0.30918101	0.44992216	0.34396375	0.52515358

---

[6] Moreover, depending on the chosen DBCM, some of the above recalled inequalities may be required to be strict.

[7] Note that in the case of R_g, confirmation and disconfirmation cases are considered separately. We decided to split the computation of the L_2 norm in the latter case, due to the fact that in some cases (distortions g_3 and g_4) the value of $\mu^p(C, g)$ diverges on its whole domain, but converges in the subset where $P(H|E) > P(H)$, i.e., the confirmation case.

The definition of DBCMs allows to study how to extend typical properties of a BCM, such as, for example, symmetry properties, on which we focus hereinafter. In fact, if a BCM function C exhibits a symmetry behavior, the same is expected to be showed by C_g for each distortion function g. In the same way, it is expected that each DBCM C_g results to be asymmetric if the related function C is, but with different levels of strength, when the related distortion function g varies.

Let us consider for example the following definitions of ES, HS and IS symmetries which are inspired by their definitions for a (non-distorted) BCM of Sect. 2, expressed in terms of the probabilities $x = P(H|E)$, $y = P(H)$ and $z = P(E)$.

Definition 2. *A DBCM* $C_g(H, E) = C(g(x), g(y), g(z))$ *satisfies*

Evidence Symmetry(ES) if
$$C(g(x), g(y), g(z)) = -C((g(y) - g(x)g(z))/(1 - g(z)), g(y), 1 - g(z))$$
Hypothesis Symmetry(HS) if
$$C(g(x), g(y), g(z)) = -C(1 - g(x), 1 - g(y), g(z)) \ \forall \ (x, y, z)$$
Inversion Symmetry(IS) if
$$C(g(x), g(y), g(z)) = C(g(x)g(z)/g(y), g(z), g(y)) \ \forall \ (x, y, z)$$

where the equalities must hold for all feasible (x, y, z).

Definition 2 extends the definitions of symmetries of a BCM, and coincide with them when $g(x) = x$, i.e., when there is no distortion.

With reference to a symmetry $\sigma \in \{ES, HS, IS\}$, by setting

$$C_g^{ES}(x, y, z) = -C((g(y) - g(x)g(z))/(1 - g(z)), g(y), 1 - g(z))$$
$$C_g^{HS}(x, y, z) = -C(1 - g(x), 1 - g(y), g(z))$$
$$C_g^{IS}(x, y, z) = C(g(x)g(z)/g(y), g(z), g(y))$$

we can say that a DBCM C_g satisfies symmetry σ if

$$C_g(x, y, z) = C_g^{\sigma}(x, y, z) . \tag{2}$$

Observe that C_g^{σ} depends on the chosen symmetry σ, the original BCM C and the distortion g under consideration.

Considering the L^p-norm of the function $C_g(x, y, z) - C_g^{\sigma}(x, y, z)$ as a measure of the distance among C_g and C_g^{σ}, we can propose a set of σ-asymmetry measures for C_g, defined as:

$$\mu_{\sigma a}^{p}(C_g) = \frac{\|C_g - C_g^{\sigma}\|_p}{\|C_g\|_p} \tag{3}$$

where $\| \cdot \|_p$ denotes the L^p-norm, $p \in [1, \infty]$, and $\|C_g\|_p$ is used to allow comparisons of the asymmetry measures $\mu_{\sigma a}$ among different DBCM functions.

As it is well-known, Carnap $d(H, E)$ satisfies only HS, Mortimer $M(H, E)$ only ES, while for Finch $F(H, E)$ only IS is verified. The same conclusions can be deduced if we consider a distortion g and the corresponding DBCMs $d_g(H, E)$, $M_g(H, E)$ and $F_g(H, E)$. On the other hand, if the attention is focused on the symmetry properties not satisfied by the BCM, they will not be satisfied even by the corresponding DBCM, but the degree of asymmetry will change according to the considered distortion. Tables 5 and 6 present the σ-asymmetry measures of the DBCM $d_g(H, E)$ and $M_g(H, E)$ with respect to different distortions: g_1, g_2, g_3 and g_4 (the degenerate distortion $g_0(u) = u$ is presented for the sake of completeness: in this case, in fact, $d_g(H, E)$ and $M_g(H, E)$ coincide with the BCMs $d(H, E)$ and $M(H, E)$ presented in Table 1). The two Tables highlight how different the asymmetry can become depending on the distortion that operates on probabilities. In particular we can observe that a distortion can even reduce the asymmetry: for example, in Table 5 the distorted measure d_g appears to be less IS-symmetric under all distortions g than undistorted (i.e., under distortion g_0). The comparison of the two Tables, on the other hand, allows to compare asymmetries of different BCMs under different distortions, for example considering distortion g_1, d_g is less asymmetric (degree of asymmetry equal to 0.92724449) than M_g (degree 2.4730445).

Table 5. Degree of asymmetry of $d_g(H, E)$ (L_2 norm)

Distortion	Distorted BCM	ES	HS	IS
$g_0(u) = u$	$x - y$	0.81651086	0	0.69176312
$g_1(u) = u^{1/2}$	$x^{1/2} - y^{1/2}$	0.92724449	0	0.39614199
$g_2(u) = u^{1/3}$	$x^{1/3} - y^{1/3}$	1.3614209	0	0.25029642
$g_3(u) = 1 - (1 - u)^2$	$-(1 - x)^2 + (1 - y)^2$	11.225865	0	0.67806788
$g_4(u) = 1 - (1 - u)^3$	$-(1 - x)^3 + (1 - y)^3$	∞	0	0.60935907

Table 6. Degree of asymmetry of $M_g(H, E)$ (L_2 norm)

Distortion	Distorted BCM	ES	HS	IS
$g_0(u) = u$	$(x - y)z/y$	0	0.88138727	0.95774881
$g_1(u) = u^{1/2}$	$(x^{1/2} - y^{1/2})z^{1/2}/y^{1/2}$	0	2.4730445	0.54846018
$g_2(u) = u^{1/3}$	$(x^{1/3} - y^{1/3})z^{1/3}/y^{1/3}$	0	4.6039914	0.3465364
$g_3(u) = 1 - (1 - u)^2$	$[-(1 - x)^2 + (1 - y)^2][1 - (1 - z)^2]/[1 - (1 - y)^2]$	0	25.488363	0.9387877
$g_4(u) = 1 - (1 - u)^3$	$[-(1 - x)^3 + (1 - y)^3][1 - (1 - z)^3]/[1 - (1 - y)^3]$	0	∞	0.84366008

4 Conclusions

In this paper we investigated the effects of probability distortion on Bayesian Confirmation Measures (BCMs). While different expressions of the BCM lead to the same measure when probabilities are not distorted, distortion leads to different measures depending on the approach, e.g. likelihoodist or bayesian, adopted in expressing the BCM. Even the sign condition on $C(H, E)$ which is related to the sign of $P(H|E) - P(H)$ in the Bayesian approach, when probabilities are not distorted allows equivalent formulations that represent different perspectives on confirmation, e.g. $P(H|E) - P(H) > 0$ can be rewritten as $P(E|H) - P(E) > 0$ in a likelihoodist approach (as discussed by Greco et al. in [15]). This is not the case when probabilities are distorted. As a starting point for our research, we considered distorted probabilities in computing a BCM expressed in terms of prior probability $P(H)$, posterior probability $P(H|E)$ and probability of evidence $P(E)$ which led us to define Distorted Bayesian Confirmation Measures (DBCMs). Those measures, in general, are not ordinally equivalent to the original BCMs also if the same distortion affects all the involved probabilities. The properties of the DBCMs can be studied as for usual BCMs. In this paper the focus was put on symmetry properties and asymmetry evaluations, observing the effects of distortion on those properties for some well known BCMs.

If distortion acts on a different formulation of the confirmation measure or when fuzzy measures are considered in place of distortions (see e.g., [20]), clearly the results will differ, this will be the subject of further investigations. In particular we will study the effects of distortion on BCMs expressed in their original formulation and to BCMs expressed in terms of contingency tables frequencies. Also the cases in which only some of the probability involved in the definition of a BCM are distorted seems to be interesting; in fact the evaluation of one of those probabilities can be more affected by distortion than another one. At this aim, the use of the distance between a BCM and a corresponding DBCM suggested in Sect. 3, may be useful: in fact it allows to compare the behavior of different BCM under the same distortion or the effect of different distortions on the same BCM.

References

1. Carnap, R.: Logical Foundations of Probability. University of Chicago Press, Chicago (1950)
2. Celotto, E.: Visualizing the behavior and some symmetry properties of Bayesian confirmation Measures. Data Min. Knowl. Disc. **31**(3), 739–773 (2016)
3. Celotto, E., Ellero, A., Ferretti, P.: Monotonicity and symmetry of IFPD Bayesian confirmation measures. In: Torra, V., Narukawa, Y., Navarro-Arribas, G., Yañez, C. (eds.) MDAI 2016. LNCS (LNAI), vol. 9880, pp. 114–125. Springer, Cham (2016). https://doi.org/10.1007/978-3-319-45656-0_10
4. Celotto, E., Ellero, A., Ferretti, P.: Asymmetry degree as a tool for comparing interestingness measures in decision making: the case of Bayesian confirmation measures. In: Esposito, A., Faundez-Zanuy, M., Morabito, F.C., Pasero, E. (eds.) WIRN 2017 2017. SIST, vol. 102, pp. 289–298. Springer, Cham (2019). https://doi.org/10.1007/978-3-319-95098-3_26
5. Crupi, V., Tentori, K., Gonzalez, M.: On Bayesian measures of evidential support: theoretical and empirical issues. Philos. Sci. **74**(2), 229–252 (2007)

6. Crupi, V., Festa, R., Buttasi, C.: Towards a grammar of Bayesian confirmation. In: Suárez, M., Dorato, M., Rédei, M. (eds.) Epistemology and Methodology of Science, pp. 73–93. Springer, Dordrecht (2010). https://doi.org/10.1007/978-90-481-3263-8_7

7. Denneberg, D.: Non-Additive Measure and Integral. Kluwer, Dordrecht (1994)

8. Eells, E., Fitelson, B.: Symmetries and asymmetries in evidential support. Philos. Stud. **107**(2), 129–142 (2002)

9. Finch, H.A.: Confirming power of observations metricized for decisions among hypotheses. Philos. Sci. **27**(4), 293–307 (1960)

10. Fitelson, B.: The plurality of Bayesian measures of confirmation and the problem of measure sensitivity. Philos. Sci. **66**, 362–378 (1999)

11. Geng, L., Hamilton, H.J.: Interestingness measures for data mining: a survey. ACM Comput. Surv. **38**(3), 1–32 (2006)

12. Glass, D.H.: Entailment and symmetry in confirmation measures of interestingness. Inform. Sci. **279**, 552–559 (2014)

13. Good, I.J.: Probability and the Weighing of Evidence. Charles Griffin, London (1950)

14. Greco, S., Słowiński, R., Szczęch, I.: Analysis of symmetry properties for Bayesian confirmation measures. In: Li, T., et al. (eds.) RSKT 2012. LNCS (LNAI), vol. 7414, pp. 207–214. Springer, Heidelberg (2012). https://doi.org/10.1007/978-3-642-31900-6_27

15. Greco, S., Słowiński, R., Szczęch, I.: Properties of rule interestingness measures and alternative approaches to normalization of measures. Inf. Sci. **216**, 1–16 (2012)

16. Keynes, J.: A Treatise on Probability. Macmillan, London (1921)

17. Mortimer, H.: The Logic of Induction. Prentice Hall, Paramus (1988)

18. Rips, L.: Two kinds of reasoning. Psychol. Sci. **12**(2), 129–134 (2001)

19. Susmaga, R., Szczęch, I.: Can interestingness measures be usefully visualized? Int. J. Appl. Math. Comput. Sci. **25**, 323–336 (2015)

20. Torra, V., Guillen, M., Santolino, M.: Continuous m-dimensional distorted probabilities. Inf. Fusion **44**, 97–102 (2018)

21. Vassend, O.: Confirmation measures and sensitivity. Philos. Sci. **82**(5), 892–904 (2015)

22. Wang, S.S.: Insurance pricing and increased limits ratemaking by proportional hazards transforms. Insur. Math. Econ. **17**(1), 43–54 (1995)

23. Wirch, J.L., Hardy, M.R.: A synthesis of risk measures for capital adequacy. Insur. Math. Econ. **25**(3), 337–347 (1999)

Constructive k-Additive Measure and Decreasing Convergence Theorems

Ryoji Fukuda[1](✉), Aoi Honda[2], and Yoshiaki Okazaki[3]

[1] Oita University, Dannoharu, 700, Oita City 870-1192, Japan
`rfukuda@oita-u.ac.jp`
[2] Kyushu Institute of Technology, 680-4, Kawazu Izuka City 820-8502, Japan
[3] Fuzzy Logic Systems Institute, 680-41, Kawazu Izuka City 820-0067, Japan

Abstract. Pan and concave integrals are division-type nonlinear integrals defined on a monotone measure space. These satisfy the monotone increasing convergence theorem under reasonable conditions. However, the monotone decreasing convergence theorem for these integrals was proved only under limited conditions. In this study we consider the k-additive monotone measure on a non-discrete measurable space, and argue this property of Pan and concave integrals on a k-additive monotone measure space.

Keywords: Monotone measure · Nonlinear integral · k-additivity

1 Introduction

In the past several decades, various studies have been conducted on monotone measures (fuzzy measures or capacities) and corresponding nonlinear integrals for both practical and theoretical purpose. In several fields, such as game theory [1], economics [2], and decision theory [3], these concepts are used as strong tools to analyze the targets. In most of these analysis, monotone measures are defined on a finite set and the number of degrees of freedom for a monotone measure represents its exponential growth with respect to the cardinal number of data. Some theoretical studies were motivated by the problem of these combinatorial explosions. In this study, we selected the two perspectives for examining this problem, i.e., "k-additivity" and the "convergence theorem." The former is a concept for the reduction of parameters to determine a monotone measure and the latter is used for the approximation of the integral values of a measurable function with respect to a monotone measure.

k-additivity was defined using the Möbius inversion formula for a non-additive set function defined on a finite set [4,5]. It is utilized to express the set function using the correlation values among subset elements of the space. k-additivity is one typical reduction method for the number of these correlations. Consider a monotone measure on a set with cardinality n; thus, there are $2^n - 1$ parameters to determine the set function. However, for a two-additive measure

© Springer Nature Switzerland AG 2020
V. Torra et al. (Eds.): MDAI 2020, LNAI 12256, pp. 104–116, 2020.
https://doi.org/10.1007/978-3-030-57524-3_9

$(k = 2)$, we only need $\dfrac{n(n+1)}{2}$ parameters to determine the monotone measure. The definition of k-additivity for general measurable space was introduced by R. Mesiar [6]. For our arguments and calculations, we reconstruct the concept of k-additivity in the next section, which is essentially same with the Mesiar's definition.

In this paper, $(X, \mathcal{B})$ denotes a general measurable space. A set function $\mu : \mathcal{B} \to \mathbb{R}$ is a monotone measure if $\mu(\emptyset) = 0$ and $A \subset B \Rightarrow \mu(A) \le \mu(B)$. Thus $(X, \mathcal{B}, \mu)$ is called a monotone measure space. Several integrals are defined on this space. In this study, we focus on the Pan integral [7] and the concave integral [8]. These integrals, defined using the basic sum of simple functions, are called "division-type integrals." For a simple function $\sum_{j=1}^{n} a_j 1_{A_j}$, the basic sum is defined by $\sum_{j=1}^{n} a_j \mu(A_j)$. Several complex situations can be derived from the non-additivity of μ. For example, two basic sums are not the same even if their corresponding simple functions are the same. In this study, we consider some convergence theorems for these integrals when the corresponding monotone measure satisfies "k-additivity."

Let $(X, \mathcal{B}, \mu)$ be a monotone measure space, and f be a measurable function on X. We consider a function $\rho(r) = \mu(f \ge r)$. This function is used to define the Choquet, Sugeno, and Shilkret integrals (for example, the Choquet integral $\int^{ch} f d\mu = \int_0^\infty \rho(r) dr$). Monotone increasing/decreasing convergence theorems are valid for them under some reasonable conditions. Details regarding these integrals and further theoretical investigations were given by J. Kawabe [10]. These integrals were also dealt in [11] (by Klement et al.), they call the function ρ a survival function.

Our target integrals, Pan and concave, satisfy the monotone increasing convergence theorem under certain conditions [12]. The monotone decreasing convergence theorem is very sensitive toward these integrals, this property was proved in some limited conditions [12]. There are two conditions for the monotone decreasing convergence theorem for the Pan integral. The first condition is the sub-additivity. When a monotone measure μ is sub-additive, its Pan integral is a linear function with non-negative coefficients [13]. Using this property, the monotone decreasing convergence theorem for the Pan integral was demonstrated under the condition of continuity of μ from below [12]. When μ is sub-additive, its Pan integral is identical to its concave integral. Thus the monotone decreasing theorem is also valid for the concave integral under the same conditions. The second condition entails that "the function sequence converges to 0." In this case, the monotone decreasing convergence theorem is valid if μ is continuous at $\emptyset$" [12]. In this study, we prove the monotone decreasing convergence theorem for the Pan integral under the condition that "the function sequence is uniformly bounded and μ is k-additive." The idea for this proof is not valid for the concave integral; however, if "the function sequence converges to 0," the monotone decreasing convergence theorem becomes valid for the concave integral, under the above conditions (uniformly bounded, and k-additive).

2 Constructive k-Additive Measure

k-additivity was originally defined for a set-function on a finite set [4,5] and extended for general measure space by R. Mesiar [6]. In this section we define a k-additive measure constructively, which concept is almost same with Mesiar's definition.

First, we give an example of a three-points set.

Example 1. Let $X = \{a, b, c\}$ and μ be a set function with

$$\mu(\emptyset) = 0, \quad \mu(\{a\}) = \mu_a, \quad \mu(\{b\}) = \mu_b, \quad \mu(\{c\}) = \mu_c,$$

$$\mu(\{a, b\}) = \mu_{ab}, \quad \mu(\{b, c\}) = \mu_{bc}, \quad \mu(\{a, c\}) = \mu_{ac}, \quad \mu(\{a, b, c\}) = \mu_{abc}.$$

We identify ab with $\{a, b\}$, abc with $\{a, b, c\}$, and so on. Then, its Möbius transform $\{\nu_B\}_{B \subset X}$ is given by

$$\nu_\emptyset = 0, \quad \nu_a = \mu_a, \quad \nu_b = \mu_b, \quad \nu_c = \mu_c,$$

$$\nu_{ab} = \mu_{ab} - \mu_a - \mu_b, \quad \nu_{bc} = \mu_{bc} - \mu_b - \mu_c, \quad \nu_{ca} = \mu_{ac} - \mu_c - \mu_a,$$

$$\nu_{abc} = \mu_{abc} - \mu_{ab} - \mu_{bc} - \mu_{ac} + \mu_a + \mu_b + \mu_c.$$

For any $A \subset X$, we have

$$\mu(A) = \sum_{B \subset A} \nu_B.$$

Let us consider the set spaces

$$X^{(1)} = \{a, b, c\}, \quad X^{(2)} = \{ab, bc, ac\}, \quad X^{(3)} = \{abc\},$$

and define

$$\mu_1 = \nu_a \delta_a + \nu_b \delta_b + \nu_c \delta_c, \quad \mu_2 = \nu_{ab} \delta_{ab} + \nu_{bc} \delta_{bc} + \nu_{ac} \delta_{ac}, \quad \mu_3 = \nu_{abc} \delta_{abc},$$

where δ_p is the Dirac measure with respect to the point p. Then $\mu_1, \mu_2,$ and μ_3 are signed measures satisfying

$$\mu(A) = \sum_{j=1}^{3} \mu_j(\{B \in X^{(j)} : B \subset A\}).$$

The set function μ is expressed using three measures.

In general, a set function μ of a finite set X with $\mu(\emptyset) = 0$ can be expressed by the Möbius transform $\{\nu_B\}_{B \subset X}$ as follows.

$$\mu(A) = \sum_{B \subset A} \nu_B.$$

When μ is a k-additive measure ($k \leq |X|$), that is, $\nu_B = 0$ if $|B| > k$, μ can be expressed using k signed measures on finite set spaces using the same method with the above example.

In this paper, $(X, \mathcal{B})$ denotes a general measurable space, that is, X is a non-discrete set and $\mathcal{B}$ is a σ-algebra of X. For the precise definitions of σ-algebra, measurable function, Lebesgue integral, and other terms used in the measure theory, can be found in [14]. To define k-additivity on a non-discrete monotone measure space, we define the "finite set space" of a general measurable space.

Definition 1. *Let $(X, \mathcal{B})$ be a measurable space and $j \in \mathbb{N}$. The j-set space $X^{(j)}$ of X is defined by*

$$X^{(j)} = \{(x_i)_{i=1}^{j} : i \neq i' \ \Rightarrow \ x_i \neq x_{i'}\}.$$

We identify $(x_i)_{i=1}^{j}$ with $(x_i')_{i=1}^{j}$ if $(x_i)_{i=1}^{j}$ is a permutation of $(x_i')_{i=1}^{j}$, and the equivalence relation is denoted by $\sim$. For example, when $X = \{a, b, c\}$, $X^{(2)} = \{\{(a, b), (b, a)\}, \{(b, c), (c, b)\}, \{(c, a), (a, c)\}\}$. Compare with the corresponding set space given in Example 1.

On $X^{(j)}$, we consider the natural σ-algebra $\mathcal{B}^{(j)}$ determined by the direct product and the equivalence relation $\sim$.

For $A \in \mathcal{B}$, we define

$$A^{(j)} = \{(x_i)_{i=1}^{j} \in X^{(j)}, \ x_i \in A \ (\forall i \leq j) \}.$$

Then $A^{(j)} \in \mathcal{B}^{(j)}$. We constructively define the non-discrete k-additivity as follows.

Definition 2. *Let $(X, \mathcal{B})$ be a measurable space, $k \in \mathbb{N}$ and μ is a monotone measure on X. μ is a constructive k-additive measure (in short k-additive measure) if there exists a finite signed measure μ_j on each $X^{(j)}$, $1 \leq j \leq k$, such that*

$$\mu(A) = \sum_{j=1}^{k} \mu_j(A^{(j)}).$$

Remark 1. A finite signed measure ν can be expressed by two non-negative finite measures ν^+ and ν^- as follows:

$$\nu = \nu^+ - \nu^-, \ \exists S \in \mathcal{B}, \ \nu^+(A) = \nu(A \cap S), \ \nu^-(A) = -\nu(A \cap S^c), \ \forall A \in \mathcal{B}.$$

(We define $|\nu| = \nu^+ + \nu^-$.) This expression is called "Hahn decomposition" [15], and these notations are used in the sequel.

3 Set Operations

In this section, we establish some properties of set operations. First, we add a notation of a set operation on a set space.

Definition 3. *For $j > 1$, $A \in \mathcal{B}$, and $B \in \mathcal{B}^{(j-1)}$, we define*

$$A(\times)B = \{(a_i)_{i=0}^{j} : \ a_{i_0} \in A, (a_i)_{i \neq i_0} \in B, \ \exists i_0 \leq j\}.$$

Next, we show the following set inclusions.

Proposition 1. *Fix $j \in \mathbb{N}$, let $\{A_\ell\}_{\ell=1}^{L}$ $(L \in \mathbb{N})$ be a disjoint subfamily of $\mathcal{B}$, and $B \in \mathcal{B}$ be a measurable set. Then, the following inclusions hold.*

(a)

$$\bigcup_{\ell=1}^{L} \left\{ A_\ell^{(j)} \setminus (A_\ell \cap B)^{(j)} \right\} \subset (B^{(j)})^c.$$

(b)

$$\left(\bigcup_{\ell=1}^{L} A_\ell \right)^{(j)} \setminus \left(\bigcup_{\ell=1}^{L} (A_\ell \setminus B) \right)^{(j)} \subset \left(\bigcup_{\ell=1}^{L} (A_\ell \cap B^c) \right) (\times) X^{(j-1)}.$$

Proof. (a)   Consider an element

$$(a_i)_{i=1}^{j} \in \bigcup_{\ell=1}^{L} \left\{ A_\ell^{(j)} \setminus (A_\ell \cap B)^{(j)} \right\}.$$

Then,

$$(a_i)_{i=1}^{j} \in A_{\ell_0}^{(j)} \setminus (A_{\ell_0} \cap B)^{(j)}$$

for some $\ell_0 \leq L$. We have $a_i \in A_{\ell_0}$ for any $i \leq j$ because $(a_i)_{i=1}^{j} \in A_{\ell_0}^{(j)}$. Since

$$(a_i)_{i=1}^{j} \notin (A_{\ell_0} \cap B)^{(j)},$$

we have

$$(a_i)_{i=1}^{j} \in (B^{(j)})^c.$$

(b)   Consider an element

$$(a_i)_{i=1}^{j} \in \left(\bigcup_{\ell=1}^{L} A_\ell \right)^{(j)} \setminus \left(\bigcup_{\ell=1}^{L} (A_\ell \cap B) \right)^{(j)}.$$

Then, $a_i \in A_{\ell_i}$ $(\ell_i \leq L)$ for any $i \leq j$ because $(a_i)_{i=1}^{j} \in \left(\bigcup_{\ell=1}^{L} A_\ell \right)^{(j)}$. Since $(a_i)_{i=1}^{j} \notin \left(\bigcup_{\ell=1}^{L} (A_\ell \setminus B) \right)^{(j)}$, there exists $i_0 \leq j$ such that $a_{i_0} \notin A_\ell \cap B$ for any $\ell \leq L$. This implies that $a_{i_0} \in \bigcup_{\ell=1}^{L} A_\ell \setminus B$ and

$$(a_i)_{i=1}^{j} \in (\bigcup_{\ell=1}^{L} A_\ell \setminus B)(\times) X^{(j)}. \qquad \square$$

Proposition 2. *Let* $A_1, A_2, \ldots, A_n \in \mathcal{B}$ *be measurable sets. Set* $C = \bigcup_{i=1}^n A_i$, *then there exists a partition* $\{B_1, \ldots, B_L\} \subset \mathcal{B}$ *of* C *satisfying*

$$A_i = \bigcup_{\ell : B_\ell \subset A_i} B_\ell.$$

Additionally, for any set of coefficients $\{a_i\}_{i=1}^n$,

$$\sum_{i=1}^n a_i 1_{A_i} = \sum_{\ell=1}^L \left(\sum_{i : B_\ell \subset A_i} a_i \right) 1_{B_\ell}.$$

Proof. For a measurable set $A \in \mathcal{B}$, define

$$A^{[s]} = \begin{cases} A, & s = 1, \\ A^c, & s = 0. \end{cases}.$$

For $(s_1, s_2, \ldots, s_n) \in \{0, 1\}^n$, set

$$D(s_1, s_2, \cdots, s_n) = \bigcap_{i=1}^n A_i^{[s_i]}.$$

Then, $\{D(s_1, \cdots, s_n)\}_{(s_1, \cdots, s_n) \in \{0,1\}^n}$ is a finite disjoint family of sets. Set $\{B_\ell\}_{\ell=1}^L = \{D(s_1, \cdots, s_n)\}_{(s_1, \cdots, s_n) \in \{0,1\}^n}$. Then, $\{B_\ell\}_{\ell=1}^L$ satisfies the proposition. $\qquad\qquad\Box$

4 Monotone Decreasing Convergence Theorems

In this section, we state the monotone decreasing convergence theorems, for Pan and concave integrals on a k-additive monotone measure space. First, we define these integrals.

Definition 4. *We define two families* $\mathcal{S}^p$ *and* $\mathcal{S}^c$ *as follows.*

$$\mathcal{S}^p = \{(a_i, A_i)_{i=1}^n, a_i \in [0, \infty), A_i \in \mathcal{B}, n \in \mathbb{N}, \{A_i\}_{i=1}^n \text{ is a partition of } X\},$$
$$\mathcal{S}^c = \{(a_i, A_i)_{i=1}^n, a_i \in [0, \infty), A_i \in \mathcal{B}, n \in \mathbb{N}, \{A_i\}_{i=1}^n \text{ is a covering of } X\}.$$

We regard the families $\mathcal{S}^p$ and $\mathcal{S}^c$ as the corresponding families of simple functions using the identification between $(a_i, A_i)_{i=1}^n$ and $\sum_{i=1}^n a_i 1_{A_i}(x)$. For a monotone measure μ and $\varphi = (a_i, A_i)_{i=1}^n \in \mathcal{S}^p$ or $\mathcal{S}^c$, the basic sum $\mu(\varphi)$ is defined by

$$\mu(\varphi) = \sum_{i=1}^n a_i \mu(A_i).$$

Let $\varphi_1 = (a_i, A_i)_{i=1}^{n_1}, \varphi_2 = (b_i, B_i)_{i=1}^{n_2}$ be two elements of $\mathcal{S}^p$ or $\mathcal{S}^c$.

$$\sum_{i=1}^{n_1} a_i 1_{A_i}(x) = \sum_{i=1}^{n_2} b_i 1_{B_i}(x), \quad \forall x \in X$$

does not imply that $\mu(\varphi_1) = \mu(\varphi_2)$. Then, we describe a simple function as a finite sequence of pairs of a coefficient and a measurable set. However, we often consider $\varphi = (a_i, A_i)_{i=1}^n$ as the function

$$\varphi(x) = \sum_{i=1}^n a_i 1_{A_i}(x),$$

when there is no confusion.

Definition 5. *For a non-negative measurable function f on X, The Pan integral $\int^{\text{pan}}$ and the concave integral $\int^{\text{cav}}$ are defined by*

$$\int^{\text{pan}} f d\mu = \sup\{\mu(\varphi) : \varphi \in \mathcal{S}^p, \varphi \le f\},$$

$$\int^{\text{cav}} f d\mu = \sup\{\mu(\varphi) : \varphi \in \mathcal{S}^c, \varphi \le f\}.$$

These definitions and basic properties are described in [7–9].

Remark 2. By the above definitions, we have

$$f \le g \Rightarrow \int^{\text{pan}} f d\mu \le \int^{\text{pan}} g d\mu, \int^{\text{cav}} f d\mu \le \int^{\text{cav}} g d\mu.$$

The following proposition can be easily obtained from the uniform convergence theorem [12]. The referred article was originally written in Japanese; therefore, we have provided the direct proof.

Proposition 3. *Assume that $\int^{\text{pan}} 1_X d\mu < \infty$ and μ is continuous from below $(A_n \nearrow A \Rightarrow \mu(A_n) \nearrow \mu(A))$. Let f be a non-negative measurable function satisfying $\int^{\text{pan}} f d\mu < \infty$. Then,*

$$\int^{\text{pan}} (f + \delta) d\mu \searrow \int^{\text{pan}} f d\mu \quad as\ \delta \searrow 0.$$

Proof. Fix $\delta > 0$ and let $\varphi \in \mathcal{S}^p$ be an arbitrary simple function satisfying $\varphi(x) \le f(x) + \delta$.

Define $\varphi_\delta(x) = \sum_{i=0}^n ((a_i - \delta) \vee 0) \, 1_{A_i}(x)$. Then, we have $\phi_\delta \le f$ and

$$\mu(\phi) \le \mu(\phi_\delta) + \sum_{i=1}^n \delta\mu(A_i) \le \int^{\text{pan}} f d\mu + \delta \int^{\text{pan}} 1_X d\mu \searrow \int^{\text{pan}} f d\mu.$$

Therefore,

$$\lim_{\delta \searrow 0} \int^{\text{pan}} (f + \delta) d\mu \le \int^{\text{pan}} f d\mu.$$

Using the clear reverse inequality, we have the required equality. $\square$

A k-additive measure satisfies the condition of the above proposition.

Proposition 4. *If μ is k-additive ($k \in \mathbb{N}$),*

$$\int^{\mathrm{pan}} 1_X d\mu \leq \int^{\mathrm{cav}} 1_X d\mu < \infty.$$

Proof. Let $\varphi = (a_i, A_i)_{i=1}^n \in \mathcal{S}^a$ be a simple function satisfying $\varphi(x) = \sum a_i 1_{A_i}(x) \leq 1$. By Proposition 2, there exists $\{B_1, \ldots, B_L\} \subset \mathcal{B}$ satisfying

$$A_i = \bigcup_{\ell : B_\ell \subset A_i} B_\ell, \ i \leq n$$

and

$$\sum_{i=1}^n a_i 1_{A_i} = \sum_{\ell=1}^L \left(\sum_{i : B_\ell \subset A_i} a_i \right) 1_{B_\ell}.$$

Set $I_\ell = \{i : B_\ell \subset A_i\}$, then the condition $\varphi \leq 1$ implies that

$$\sum_{i \in I_\ell} a_i \leq 1$$

for any $\ell \leq L$. Then,

$$
\begin{aligned}
\sum_{i=1}^n a_i \mu_{A_i} &= \sum_{i=1}^n a_i \sum_{j=1}^k \mu_j \left(\left(\bigcup_{\ell : B_\ell \subset A_i} B_\ell \right)^{(j)} \right) \\
&\leq \sum_{i=1}^n a_i \sum_{j=1}^k |\mu_j| \left(\left(\bigcup_{\ell : B_\ell \subset A_i} B_\ell \right)^{(j)} \right) \\
&\leq \sum_{i=1}^n a_i \sum_{j=1}^k |\mu_j| \left(\left(\bigcup_{\ell : B_\ell \subset A_i} B_\ell \right) (\times) X^{(j-1)} \right) \\
&\leq \sum_{j=1}^k \sum_{\ell=1}^L \left(\sum_{i \in I_\ell} a_i \right) |\mu_j| (B_\ell (\times) X^{(j-1)}) \\
&\leq \sum_{j=1}^k \sum_{\ell=1}^L |\mu_j| (B_\ell (\times) X^{(j-1)}) \\
&\leq \sum_{j=1}^k |\mu_j| (X^{(j)}).
\end{aligned}
$$

The right hand side is finite and does not depend on φ. Thus we have

$$\left(\int^{\mathrm{pan}} 1_X d\mu \leq \right) \int^{\mathrm{cav}} 1_X d\mu \leq \sum_{j=1}^k |\mu_j| (X^{(j)}) < \infty. \qquad \square$$

Theorem 1. *Let $\{f_n\}$ be a decreasing non-negative measurable function on X. Assume that a monotone measure μ on $(X, \mathcal{B})$ is k-additive ($k \in \mathbb{N}$), and $f_n(x) \leq M < \infty$ for any $n \in \mathbb{N}$ and $x \in X$. Set $f = \lim\limits_{n \to \infty} f_n$, then we have*

$$\int^{\mathrm{pan}} f d\mu = \lim_{n \to \infty} \int^{\mathrm{pan}} f_n d\mu.$$

Proof. For an arbitrary $\delta > 0$ and $n \in \mathbb{N}$, set

$$B_n^{(\delta)} = \{x | f_n(x) \leq f(x) + \delta\}.$$

Because the monotone measure μ is k-additive, there exists a signed measure μ_j on $X^{(j)}$ for each $j \leq k$ such that

$$\mu(A) = \sum_{j=1}^{k} \mu_j(A^{(j)}).$$

By the continuity of signed measures $\{\mu_j\}_{j=1}^{k}$, $B_n^{(\delta)} \nearrow X$ as $n \to \infty$ implies that $\mu(B_n^{(\delta)c}) \searrow 0$ as $n \to \infty$.

Fix an arbitrary $\varepsilon > 0$. Then, there exists $\delta > 0$ satisfying:

$$\int^{\mathrm{pan}} (f + \delta) d\mu \leq \int^{\mathrm{pan}} f d\mu + \varepsilon.$$

For each $n \in \mathbb{N}$, there exists $\varphi_n = (a_{n,i}, A_{n,i})_{i=1}^{N_n}$ satisfying

$$\mu(\varphi_n) \geq \int^{\mathrm{pan}} f_n d\mu - \varepsilon.$$

Set

$$\widetilde{\varphi_n} = \sum a_{n,i} 1_{A_{n,i} \cap B_n^{(\delta)}} \in \mathcal{S}^p,$$

then $\widetilde{\varphi_n} \leq f + \delta$. By the definition of Pan integral, we have

$$\mu(\widetilde{\varphi_n}) \leq \int^{\mathrm{pan}} (f + \delta) d\mu \leq \int^{\mathrm{pan}} f d\mu + \varepsilon.$$

Using Proposition 1,

$$\mu(\varphi_n) - \mu(\widetilde{\varphi}) \leq \sum_i a_{n,i} \{\mu(A_{n,i}) - \mu(A_{n,i} \cap B_n^{(\delta)})\}$$

$$= \sum_j \sum_i a_{n,i} \{\mu_j(A_{n,i}^{(j)}) - \mu_j((A_{n,i} \cap B_n^{(\delta)})^{(j)})\}$$

$$= \sum_j \{\mu_j(A_i^{(j)} \setminus (A_i \cap B_n^{(\delta)})^{(j)})\}$$

$$\leq \sum_j (\sum_i a_i) |\mu_j| (A_i^{(j)} \setminus (A_i \cap B_n^{(\delta)})^{(j)})$$

$$\leq M \sum_j |\mu_j| (B_n^{(\delta)})^{(j)c}).$$

The right hand side tends to 0 as $n \to \infty$, and depends only on n. There exists $n_0 \in \mathbb{N}$ such that

$$\mu(\varphi_n) - \mu(\widetilde{\varphi_n}) < \varepsilon$$

for any $n \geq n_0$. Therefore, we have

$$\mu(\varphi_n) \leq \mu(\widetilde{\varphi_n}) + \varepsilon \leq \int^{pan} f d\mu + 2\varepsilon.$$

This implies that

$$\lim_{n\to\infty} \int^{Pan} f_n d\mu \leq \int^{pan} f_n d\mu \leq \int^{pan} f d\mu + 3\varepsilon \searrow \int^{pan} f d\mu \quad (\varepsilon \searrow 0).$$

Because the reverse inequality is evident, we have the required conclusion.   □

We cannot prove the monotone decreasing convergence theorem for concave integral in similar way. However, in the case where $f_n \searrow 0$, we can prove this property.

Theorem 2. *Let $\{f_n\}$ be a decreasing non-negative measurable function on X satisfying $f_n \searrow 0$ as $n \to \infty$. Assume that a monotone measure on $(X, \mathcal{B})$ is k-additive $(k \in \mathbb{N})$ and $f_n(x) \leq M < \infty$ for any $n \in \mathbb{N}$ and $x \in X$. Then,*

$$\lim_{n\to\infty} \int^{cav} f_n d\mu = 0.$$

Proof. For $\delta > 0$ and $n \in \mathbb{N}$, set:

$$B_n^{(\delta)} = \{x | f_n(x) \leq \delta\} \nearrow X, \quad \text{as } n \to \infty.$$

Because μ is continuous from above, $(A_n \searrow A \Rightarrow \mu(A_n) \searrow \mu(A))$, $\mu\left(\left(B_n^{(\delta)}\right)^c\right) \searrow 0$ as $n \to \infty$.

By Proposition 4 we have $\int^{cav} 1_X d\mu < \infty$. Then, we have

$$\int^{cav} \delta 1_X d\mu = \delta \int^{cav} 1_X d\mu \searrow 0, \quad \delta \searrow 0.$$

Let $\varphi_n = \sum_{i=1}^{N} a_i 1_{A_i} \in \mathcal{S}^c$ be a simple function satisfying $\varphi_n \leq f_n$. Set

$$\widetilde{\varphi_n} = \sum_i a_i 1_{A_i \cap B_n^{(\delta)}}.$$

Then, we have

$$\mu(\widetilde{\varphi_n}) \leq \delta \int^{cav} 1_X d\mu.$$

In contrast,

$$\mu(\varphi_n) - \mu(\widetilde{\varphi_n}) = \sum_i a_i \left(\mu(A_i) - \mu(A_i \cap B^{(\delta)})\right). \tag{1}$$

Using some signed measure μ_j on $X^{(j)}$ for each $j \leq k$, μ can be expressed as follows:

$$\mu(A) = \sum_{j=1}^{k} \mu_j(A^{(j)}), \quad (A \in \mathcal{B}).$$

Therefore,

$$(1) = \sum_i a_i \sum_j \left(\mu_j(A_i^{(j)}) - \mu_j((A_i \cap B_n^{(\delta)})^{(j)}) \right)$$

$$= \sum_i a_i \sum_j \mu_j \left(A_i^{(j)} \setminus (A_i \cap B_n^{(\delta)})^{(j)} \right)$$

$$\leq \sum_i a_i \sum_j |\mu_j| \left((A_i^{(j)}) \setminus (A_i \cap B_n^{(\delta)})^{(j)} \right). \tag{2}$$

By Proposition 2, there exists a partition $\{D_\ell\}_{\ell=1}^L$ such that

$$\sum_i a_i 1_{A_i} = \sum_\ell \left(\sum_{i \in I_\ell} a_i \right) 1_{D_\ell}, \quad A_i = \bigcup_{\ell \in \widetilde{I}_i} D_\ell,$$

where $I_\ell = \{i : D_\ell \subset A_i\}$, $\widetilde{I}_i = \{\ell : D_\ell \subset A_i\}$. Using Proposition 1, we have

$$A_i^{(j)} \setminus (A_i \cap B_n^{(\delta)})^{(j)} = \left(\bigcup_{\ell \in I_i} D_\ell \right)^{(j)} \setminus \left(\bigcup_{\ell \in I_i} (D_\ell \cap B_n^{(\delta)}) \right)^{(j)}$$

$$\subset \bigcup_{\ell \in I_i} (D_\ell \cap B_n^{(\delta)^c})(\times) X^{(j-1)}$$

for any $i \leq N$. Then, we have

$$(2) \leq \sum_i a_i \sum_j \sum_{\ell \in \widetilde{I}_i} |\mu_j| \left((D_\ell \cap B_n^{(\delta)^c})(\times) X^{(j-1)} \right)$$

$$= \sum_j \sum_\ell \left(\sum_{i \in I_\ell} a_i \right) |\mu_j| \left((D_\ell \cap B_n^{(\delta)^c})(\times) X^{(j-1)} \right)$$

$$\leq M \sum_j |\mu_j| \left(B_n^{(\delta)^c}(\times) X^{(j-1)} \right) \searrow 0.$$

By the above arguments, for any $\varepsilon > 0$, there exists $\delta > 0$ such that

$$\mu(\widetilde{\varphi_n}) < \frac{\varepsilon}{2}$$

for any $n \in \mathbb{N}$, where $\widetilde{\varphi_n}$ depends on δ for each $n \in \mathbb{N}$. By fixing $\delta > 0$, we can select $n_0 \in \mathbb{N}$ that satisfies

$$\mu(\varphi_n) - \mu(\widetilde{\varphi_n}) < \frac{\varepsilon}{2},$$

for any $n \geq n_0$. Thus, we found $\varphi_n \in \mathcal{S}^c$ that satisfies

$$\mu(\varphi_n) = \mu(\widetilde{\varphi_n}) + \mu(\varphi_n) - \mu(\widetilde{\varphi_n}) < \varepsilon$$

for any $n \geq n_0$. This implies that

$$\lim_{n \to \infty} \int^{cav} f_n d\mu = 0. \qquad \Box$$

5   Conclusion

In this study, we constructively defined k-additivity for a non-discrete monotone measure space. For this space, we demonstrated the monotone decreasing convergence theorem for Pan integrals under the condition that the function sequence is uniformly bounded. Furthermore, we demonstrated the monotone decreasing convergence theorem for concave integrals under an additional condition that "the function sequence converges to zero." We believe that these properties can play important role in some arguments of functional analyses where these non-linear integrals are used.

References

1. Fujimoto, K.: Cooperative game as non-additive measure. Stud. Fuzziness Soft Comput. **310**, 131–172 (2014)
2. Ozaki, H.: Integral with respect to non-additive measure in economics. Stud. Fuzziness Soft Comput. **310**, 97–130 (2014)
3. Grabisch, M.: Alternative representations of discrete fuzzy measures for decision making. Int. J. Uncert. Fuzziness Knowl.-Based Syst. **5**(5), 587–607 (1997)
4. Grabisch, M.: k-order additive discreet fuzzy measures and their representation. Fuzzy Sets Syst. **92**, 167–189 (1997)
5. Combarro, E.F., Miranda, P.: On the structure of the k-additive fuzzy measures. Fuzzy Sets Syst. **161**(17), 2314–2327 (2010)
6. Mesiar, R.: Generalizations of k-order additive discrete fuzzy measures. Fuzzy Sets Syst. **102**, 423–428 (1999)
7. Yang, Q.: The pan-integral on the fuzzy measure space. Fuzzy Mathematica **3**, 107–114 (1985). (in Chinese)
8. Lehrer, E., Teper, R.: The concave integral over large spaces. Fuzzy Sets Syst. **159**, 2130–2144 (2008)
9. Even, Y., Lehrer, E.: Decomposition-integral: unifying Choquet and the concave integrals. Econ. Theor. **56**(1), 33–58 (2013)
10. Kawabe, J.: A unified approach to the monotone convergence theorem for nonlinear integrals. Fuzzy Sets Syst. **304**, 1–19 (2016)
11. Klement, E.P., Li, J., Mesiar, M., Pap, E.: Integrals based on monotone set functions. Fuzzy Sets Syst. **281**, 88–102 (2015)
12. Fukuda, R., Honda, A., Okazaki, Y.: Comparison of decomposition type nonlinear integrals based on the convergence theorem. J. Japan Soc. Fuzzy Theory Intell. Informat. (2020). (Accepted, in Japanese)

13. Ouyang, Y., Li, J., Mesiar, R.: On linearity of pan-integral and pan-integrable functions space. Int. J. Approximate Reasoning Arch. **90**(8), 307–318 (2017)
14. Tao, T.: An Introduction to Measure Theory, Graduate Studies in Mathematics, vol. 126. American Mathematical Society (2011)
15. Billingsley, P.: Probability and Measure. Wiley, Hoboken (1995)

Data science and Data Mining

Generalization Property of Fuzzy Classification Function for Tsallis Entropy-Regularization of Bezdek-Type Fuzzy C-Means Clustering

Yuchi Kanzawa[(✉)]

Shibaura Institute of Technology, Tokyo, Japan
kanzawa@sic.shibaura-it.ac.jp

Abstract. In this study, Tsallis entropy-regularized Bezdek-type fuzzy c-means clustering method is proposed. Because the proposed method reduces to four conventional fuzzy clustering methods by appropriately controlling fuzzification parameters, the proposed method is considered to be their generalization. Through numerical experiments, this generalization property is confirmed; in addition, it is observed that the fuzzy classification function of the proposed method approaches a value equal to the reciprocal of the cluster number.

1 Introduction

Fuzzy clustering is a framework of clustering methods wherein data points can belong to more than one cluster. The most representative fuzzy clustering algorithm is the fuzzy c-means (FCM) algorithm proposed by Bezdek [1], wherein linear membership weights in the objective function of the hard c-means (HCM) [2] algorithm are replaced with power of membership as weights; to distinguish it from other variants discussed herein, this algorithm is referred to as the Bezdek-type fuzzified FCM (BFCM) algorithm.

Another approach to fuzzifying the HCM algorithm involves regularization of the HCM objective function. For example, Miyamoto and Mukaidono introduced a regularization term for Shannon's entropy in the HCM objective function; the resulting method is referred to as entropy-regularized FCM (EFCM) algorithm [3]. Furthermore, Kanzawa proposed the power-regularized FCM (PFCM) algorithm wherein the power of membership is adopted as a regularizer [4].

Menard et al. discussed the application of nonextensive thermodynamics in fuzzy clustering, and proposed the Tsallis entropy-based FCM (TFCM) method [5]. Because the TFCM method can be reduced to either BFCM or EFCM method by controlling the value of fuzzification parameters, the TFCM method is a generalization of the BFCM and EFCM method; consequently, the TFCM method can be used to obtain flexible clustering results. In contrast, there is no association between the TFCM and PFCM methods. However, a generalization of both TFCM and PFCM methods would yield more flexible clustering results than the TFCM method.

© Springer Nature Switzerland AG 2020
V. Torra et al. (Eds.): MDAI 2020, LNAI 12256, pp. 119–131, 2020.
https://doi.org/10.1007/978-3-030-57524-3_10

Thus, in this study, we proposed a novel fuzzy clustering approach, referred to as Tsallis entropy-regularized Bezdek-type FCM (TBFCM) method, wherein the BFCM objective function was regularized using Tsallis entropy. It should be noted that the TFCM method could also be considered as a regularization of the BFCM method. However, in the case of the TFCM method, the value of fuzzification parameter for BFCM as well as that of Tsallis entropy must be same. In contrast, in the case of the proposed TBFCM method, these values can be different. Because the proposed TBFCM method can be reduced to the TFCM, PFCM, BFCM and EFCM method by controlling the values of its fuzzification parameters, the TBFCM method is a generalization of all these methods; therefore, the TBFCM method has the potential to yield more flexible clustering results than these conventional methods. As an initial step before evaluating the clustering accuracy of the proposed TBFCM method, in this study, the generalization property of the proposed method is investigated comparing its fuzzy classification function (FCF) with those of conventional methods. Such the investigation is according to the literature as follows. For example, in [6], it was theoretically shown that the FCFs of the BFCM and EFCM methods yield an allocation rule to classify a new object into a Voronoi cell with the Voronoi seeds as the cluster centers; in particular, the FCF of in the BFCM method approaches the reciprocal of the cluster number at infinity, while that in the EFCM method approaches 1 at infinity. Furthermore, in [4], it was numerically shown that the value of the FCF in the PFCM method is 1 at points that are sufficiently far from cluster centers. In this study, through numerical experiments using an artificial dataset, we observed that the FCF of the TBFCM method reduces to that of the TFCM, PFCM, BFCM and EFCM methods by appropriately controlling the fuzzification parameter values; in addition, we observed that the FCF in the of the TBFCM method approaches the reciprocal of the cluster number at infinity even if the fuzzification parameter setting of the proposed method is the case similar to the PFCM and EFCM methods.

The remainder of this paper is organized as follows. Section 2 presents the notations in the paper, and introduces the convectional fuzzy clustering methods as well as their FCFs. In Sect. 3, basic concepts used in the proposed method are discussed and the method is described. Section 4 explains the generalization property of the proposed method based on FCFs. Section 5 provides a summary of the work performed in this study.

2   Preliminaries

2.1   Entropy

For a probability distribution P, Shannon's entropy $H_{\mathsf{Shannon}}(P)$ is defined as follows:

$$H_{\mathsf{Shannon}}(P) = -\sum_{k} P(k)\ln(P(k)). \tag{1}$$

In a previous work, Shannon's entropy was introduced in the EFCM method [3]. Furthermore, Shannon's entropy has also been extended by using the q-logarithmic function with a real number q

$$\ln_q(x) = \frac{1}{1-q}(x^{1-q} - 1) \quad \text{(for } x > 0\text{)} \tag{2}$$

as

$$H_q(P) = \frac{1}{q-1}\left(\sum_k P(k)^q - 1\right), \tag{3}$$

which is referred to as Tsallis entropy [7]. Using the limit, $q \to 1$, Shannon's entropy can be recovered. As the name suggests, Tsallis entropy has been used in the TFCM method [5].

2.2 Some Conventional Fuzzy Clustering Methods

In this subsection, we introduce four optimization problems from which four representative fuzzy clustering methods are derived. These optimization problems and their relationships with each other form the basis of the method proposed in this study.

Let $X = \{x_k \in \mathbb{R}^M \mid k \in \{1, \cdots, N\}\}$ be a dataset of M-dimensional points. Consider classifying the objects in X into C disjoint subsets $\{G_i\}_{i=1}^C$, which are referred to as clusters. The membership of x_k that belongs to the i-th cluster is denoted by $u_{i,k}$ ($i \in \{1, \cdots, C\}, k \in \{1, \cdots, N\}$) and the set $u_{i,k}$ denoted by u obeys the following constraint:

$$\sum_{i=1}^C u_{i,k} = 1, \quad u_{i,k} \in [0, 1]. \tag{4}$$

Then, the set of cluster centers is denoted by $v = \{v_i \mid v_i \in \mathbb{R}^M, i \in \{1, \cdots, C\}\}$.

The BFCM, EFCM, PFCM, and TFCM methods are derived by solving the following optimization problems:

$$\underset{u,v}{\text{minimize}} \sum_{i=1}^C \sum_{k=1}^N (u_{i,k})^m \|x_k - v_i\|_2^2, \tag{5}$$

$$\underset{u,v}{\text{minimize}} \sum_{i=1}^C \sum_{k=1}^N u_{i,k}\|x_k - v_i\|_2^2 + \lambda^{-1} \sum_{i=1}^C \sum_{k=1}^N u_{i,k} \ln(u_{i,k}), \tag{6}$$

$$\underset{u,v}{\text{minimize}} \sum_{i=1}^C \sum_{k=1}^N u_{i,k}\|x_k - v_i\|_2^2 + \frac{\lambda^{-1}}{m-1} \sum_{i=1}^C \sum_{k=1}^N (u_{i,k})^m, \tag{7}$$

and

$$\underset{u,v}{\text{minimize}} \sum_{i=1}^C \sum_{k=1}^N (u_{i,k})^m \|x_k - v_i\|_2^2 + \frac{\lambda^{-1}}{m-1} \sum_{k=1}^N \left(\sum_{i=1}^C (u_{i,k})^m - 1\right), \tag{8}$$

respectively, subject to Eqs. (4), where $m > 1$ and $\lambda > 0$ are the fuzzification parameters.

Because the PFCM method reduces to the EFCM method as $m \searrow 1$, the PFCM method is a generalization of the EFCM method. Furthermore, because the TFCM method reduces to the BFCM method as $\lambda \to +\infty$, and to the EFCM method as $m \searrow 1$, the TFCM method is a generalization of both the BFCM and EFCM methods. Therefore, the TFCM method can be used to obtain flexible clustering results by modifying two fuzzification parameters. On the contrary, there is no connection between the TFCM and PFCM methods; a generalization for both these methods would yield even more flexible clustering results than the TFCM method.

2.3 Fuzzy Classification Function (FCF)

An FCF for FCMs indicates the degree to which an arbitrary point in a data space is prototypical to each cluster obtained from the FCMs. The FCF can be considered as an extension of the membership $u_{i,k}$ to the entire data space. In particular, an FCF $u_i(x)$ for the BFCM, EFCM, PFCM, and TFCM methods with respect to a new datum $x \in \mathbb{R}^M$ is defined as the solution of the following optimization problems

$$\underset{u}{\text{minimize}} \sum_{i=1}^{C} (u_i(x))^m \|x - v_i\|_2^2, \tag{9}$$

$$\underset{u}{\text{minimize}} \sum_{i=1}^{C} u_i(x)\|x - v_i\|_2^2 + \lambda^{-1} \sum_{i=1}^{C} u_i(x)\ln(u_{i,k}), \tag{10}$$

$$\underset{u}{\text{minimize}} \sum_{i=1}^{C} u_i(x)\|x - v_i\|_2^2 + \lambda^{-1} \sum_{i=1}^{C} (u_{i,k})^m, \tag{11}$$

$$\underset{u}{\text{minimize}} \sum_{i=1}^{C} (u_i(x))^m \|x - v_i\|_2^2 + \frac{\lambda^{-1}}{m-1} \left(\sum_{i=1}^{C} (u_{i,k})^m - 1 \right), \tag{12}$$

respectively, subject to

$$\sum_{i=1}^{C} u_i(x) = 1, \tag{13}$$

where $\{v_i\}_{i=1}^{C}$ are the cluster centers obtained by the corresponding fuzzy clustering algorithms. Here, based on [8], we define a crisp allocation rule to classify $\mathbb{R}^M$ using the FCF as follows:

$$x \in G_i \overset{\text{def}}{\equiv} u_i(x) > u_j(x) \text{ for } j \neq i. \tag{14}$$

Then, the subsets $\{G_i\}_{i=1}^{C}$ produced from all methods, including the BFCM, EFCM, PFCM, and TFCM methods, yield Voronoi sets because

$$u_i(x) > u_j(x) \text{ for } j \neq i \Leftrightarrow \|x - v_i\| < \|x - v_j\| \text{ for } j \neq i \tag{15}$$

for all FCFs in the BFCM, EFCM, PFCM, and TFCM methods. Figure 1 depicts the Voronoi sets produced from four cluster centers in a two-dimensional space. The FCFs of the EFCM and PFCM methods have the feature that $u_i(x)$ approaches 1 as $\|x\| \to +\infty$ under $x^\mathsf{T}(v_i - v_j) > 0$ for all $j \neq i$, whereas the FCFs of the BFCM and TFCM methods have the feature that $u_i(x)$ approaches $1/C$ as $\|x\| \to +\infty$.

3 Proposed Method

3.1 Basic Concepts

The objective of this study was to construct a clustering method that is a generalization of both the TFCM and PFCM methods. First, we considered the relationship between the TFCM and PFCM methods. In the TFCM optimization problem given by Eq. (8), both the first and second term have the factor $(u_{i,k})^m$. In contrast, in the PFCM optimization problem given by Eq. (7), the second term has the factor $(u_{i,k})^m$, while the first term has the factor $u_{i,k}$. Therefore, the TFCM method cannot be reduced to the PFCM method by controlling the fuzzification parameter m. In order to reduce the TFCM method to the PFCM method, the fuzzification parameter value m only in the first term of the TFCM objective function has to approach one, whereas that in the second term should remain larger than one.

Thus, we considered the fuzzification parameter m in the first and second term in the TFCM objective function as different, namely m_1 and m_2, respectively, and proposed a novel optimization problem, which is as follows:

$$\underset{u,v}{\text{minimize}} \sum_{i=1}^{C} \sum_{k=1}^{N} (u_{i,k})^{m_1} \|x_k - v_i\|_2^2 + \frac{\lambda^{-1}}{m_2 - 1} \sum_{k=1}^{N} \left(\sum_{i=1}^{C} (u_{i,k})^{m_2} - 1 \right) \quad (16)$$

subject to Eq. (4), where $m_1 > 1$, $m_2 > 1$ and $\lambda > 0$ are the fuzzification parameters. The clustering method obtained by solving this optimization problem is referred to as the Tsallis entropy-regularized Bezdek-type fuzzy c-means (TBFCM) method because its objective function is obtained by Tsallis-entropy regularization of the BFCM method with different values for the Tsallis entropy parameter and fuzzification parameter of the BFCM method. Thus, the TBFCM method reduces to the TFCM method with $m_1 = m_2$, to the BFCM method with $\lambda \to +\infty$ or $m_2 \searrow 1$, to the EFCM method with $m_1 = m_2 \searrow 1$, and to the PFCM method with $m_1 \searrow 1$. Therefore, the proposed TBFCM method is a generalization of the TFCM, BFCM, EFCM and PFCM methods. Hence, the proposed method could potentially yield more flexible clustering results than the other conventional clustering algorithms in this paper by controlling three fuzzification parameters.

3.2   Algorithm

The TBFCM method is obtained by solving the optimization problem given by Eqs.(16) and (4), where the Lagrangian $L(u, v)$ is defined as

$$
L(u, v) = \sum_{i=1}^{C} \sum_{k=1}^{N} (u_{i,k})^{m_1} \|x_k - v_i\|_2^2 + \frac{\lambda^{-1}}{m_2 - 1} \sum_{k=1}^{N} \left(\sum_{i=1}^{C} (u_{i,k})^{m_2} - 1 \right)
$$
$$
+ \sum_{k=1}^{N} \gamma_k \left(1 - \sum_{i=1}^{C} u_{i,k} \right) \tag{17}
$$

with Lagrangian multipliers $(\gamma_1, \cdots, \gamma_N)$. The necessary conditions for optimality are given as follows:

$$
\frac{\partial L(u, v)}{\partial u_{i,k}} = 0, \tag{18}
$$

$$
\frac{\partial L(u, v)}{\partial v_i} = 0, \tag{19}
$$

$$
\frac{\partial L(u, v)}{\partial \gamma_k} = 0, \tag{20}
$$

The optimal cluster center is obtained from Eq. (19) in a manner similar to that in the cases of the BFCM and TFCM methods:

$$
v_i = \frac{\sum_{k=1}^{N} (u_{i,k})^{m_1} x_k}{\sum_{k=1}^{N} (u_{i,k})^{m_1}}. \tag{21}
$$

Based on Eq. (18), the optimal membership conditions are

$$
m_1 (u_{i,k})^{m_1 - 1} \|x_k - v_i\|_2^2 + \frac{m_2 \lambda^{-1}}{m_2 - 1} (u_{i,k})^{m_2 - 1} = \gamma_k \tag{22}
$$

and that given by Eq. (4), where γ_k is the Lagrange multiplier. However, because it is difficult to explicitly obtain its optimal membership, we adopt the bisection method, which is described as follows. If the γ_k value is given, we obtain the optimal membership using the following algorithm:

Algorithm 1

STEP 1. Let the lower bound of $u_{i,k}$, $\underline{u_{i,k}}$ be 0. Let the upper bound of $u_{i,k}$, $\overline{u_{i,k}}$ be 1.

STEP 2. Set $\hat{u}_{i,k} = (\underline{u_{i,k}} + \overline{u_{i,k}})/2$. If $\left| \overline{u_{i,k}} - \underline{u_{i,k}} \right|$ is sufficiently small, terminate the algorithm and let the optimal $u_{i,k}$ be $\hat{u}_{i,k}$.

STEP 3. If $m_1 (u_{i,k})^{m_1 - 1} \|x_k - v_i\|_2^2 + \frac{m_2 \lambda^{-1}}{m_2 - 1} (u_{i,k})^{m_2 - 1} > \gamma_k$, let $\overline{u_{i,k}} = \hat{u}_{i,k}$; otherwise, let $\underline{u_{i,k}} = \hat{u}_{i,k}$. Go to STEP 2.

The optimal γ_k value is obtained using the following algorithm.

Algorithm 2

STEP 1. Let the lower bound of γ_k, $\underline{\gamma_k}$ be $m_1(1/C)^{m_1-1}\min_{1\leq i'\leq C}\{\|x_k-v_{i'}\|_2^2\}+\frac{m_2\lambda^{-1}}{m_2-1}(1/C)^{m_2-1}$. Let the upper bound of γ_k, $\overline{\gamma_k}$ be $m_1(1/C)^{m_1-1}\max_{1\leq i'\leq C}\{\|x_k-v_{i'}\|_2^2\}+\frac{m_2\lambda^{-1}}{m_2-1}(1/C)^{m_2-1}$.

STEP 2. Set $\hat{\gamma}_k=(\underline{\gamma_k}+\overline{\gamma_k})/2$. If $\left|\overline{\gamma_k}-\underline{\gamma_k}\right|$ is sufficiently small, terminate the algorithm and let the optimal γ_k be $\hat{\gamma}_k$.

STEP 3. Calculate $u_{i,k}$ using Algorithm 1.

STEP 4. If $\sum_{i=1}^{C}u_{i,k}>1$, let $\overline{\gamma_k}=\hat{\gamma}_k$; otherwise, let $\underline{\gamma_k}=\hat{\gamma}_k$. Go to STEP 2.

Based on the above discussion, we propose the algorithm for the proposed TBFCM algorithm clustering method as follows:

Algorithm 3 (TBFCM)

STEP 1. Given the number of clusters C and fuzzification parameter (m_1, m_2, λ), where $m_1 > 1$, $m_2 > 1$ and $\lambda > 0$, let the set of the initial cluster centers be v.

STEP 2. Calculate γ_k using Algorithm 2, and obtain the membership using Algorithm 1.

STEP 3. Obtain v using Eq. (21).

STEP 4. Check the stopping criterion for (u, v). If the criterion is not satisfied, go to STEP 2.

4   Numerical Experiment

In this section, we present some numerical examples to investigate the generalization property of the proposed method using an artificial dataset obtained from four clusters, wherein each cluster was comprised of 66 points in a two-dimensional space, as shown in Fig. 2. We observe that, for all combinations of fuzzification parameter values, appropriate clustering results are obtained using our proposed method, as shown in Fig. 3, where the triangles denote the data belonging to the first cluster, circles denote the data belonging to the second cluster, inverse triangles denote the data belonging to the third cluster, and squares denote the data belonging to the fourth cluster.

Figures 4, 5, 6 and 7 show the FCFs for the first cluster obtained using the proposed method with $(m_1, m_2, \lambda) = (1 + 10^{-6}, 1 + 10^{-6}, 0.5)$, $(m_1, m_2, \lambda) = (1 + 10^{-6}, 2, 0.5)$, $(m_1, m_2, \lambda) = (2, 2, 0.5)$, and $(m_1, m_2, \lambda) = (2, 2, 1000)$, respectively. These results were compared to the FCFs obtained using the conventional methods; in particular, Fig. 8 shows the FCF results for the EFCM method with $\lambda = 0.5$, Fig. 9 shows those for the PFCM with $(m, \lambda) = (2, 0.5)$, Fig. 10 shows those for the TFCM with $(m, \lambda) = (2.0.5)$, and Fig. 11 shows those for the BFCM method with $m = 2$.

Figure 4 and Fig. 8 are quite similar; thus, we can deduce that the proposed method with $(m_1, m_2) \searrow (1,1)$ reduces to the EFCM method. Similarly, we observe that Fig. 5 and Fig. 9 are quite similar; thus, we can deduce that the proposed method with $m_1 \searrow 1$ reduces to the PFCM method. Furthermore, Fig. 6 and Fig. 10 are quite similar; therefore, we can deduce that the proposed method with $m_1 = m_2$ reduces to the TFCM method. Finally, wee also note that Fig. 7 and Fig. 11 are quite similar; thus, it can be considered that the proposed method with $\lambda \to +\infty$ reduces to the BFCM method. From these results, we numerically confirmed that the proposed method generalizes all conventional methods discussed in this paper.

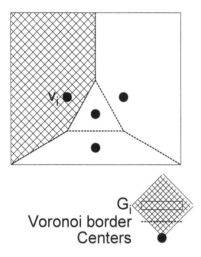

Fig. 1. Example of a Voronoi diagram generated from four cluster centers for $M = 2$ and $C = 4$

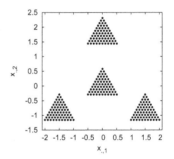

Fig. 2. Artificial dataset used for numerical verification of proposed method.

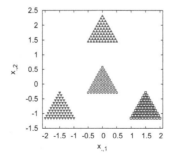

Fig. 3. Clustering result obtained using the proposed method.

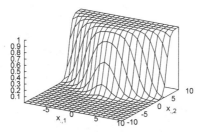

Fig. 4. FCF of the proposed method with $(m_1, m_2, \lambda) = (1 + 10^{-6}, 1 + 10^{-6}, 0.5)$.

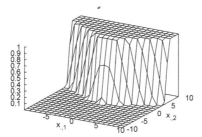

Fig. 5. FCF of the proposed method with $(m_1, m_2, \lambda) = (1 + 10^{-6}, 2, 0.5)$.

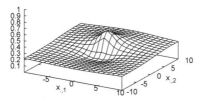

Fig. 6. FCF of the proposed method with $(m_1, m_2, \lambda) = (2, 2, 0.5)$.

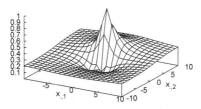

Fig. 7. FCF of the proposed method with $(m_1, m_2, \lambda) = (2, 2, 1000)$.

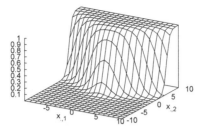

Fig. 8. FCF of the EFCM method with $\lambda = 0.5$

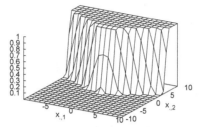

Fig. 9. FCF of the PFCM method with $(m, \lambda) = (2, 0.5)$.

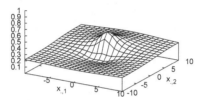

Fig. 10. FCF of the TFCM method with $(m, \lambda) = (2, 0.5)$.

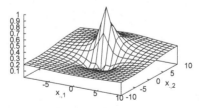

Fig. 11. FCF of the BFCM method with $m = 2$.

Next, we investigated the behavior of the FCF of the proposed method at infinity. From Figs. 7 and 10, we observed that the the FCF value at infinity converges to 0.25, which is the reciprocal of the cluster number, i.e., 4. In contrast, from Figs 5 and 4, we observed the FCF value at infinity seems to converges to one or zero. However, when the FCF value is obtained in the range of $[-10^7, 10^7]^2$ while keeping the same fuzzification parameter values as those in Figs. 4 and 5, we observed that the FCF value at infinity converges to 0.25,

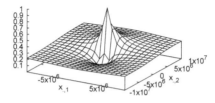

Fig. 12. FCF of the proposed method with $(m_1, m_2, \lambda) = (1 + 10^{-6}, 1 + 10^{-6}, 0.5)$ on $x \in [-10^{-7}, 10^7]^2$.

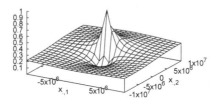

Fig. 13. FCF of the proposed method with $(m_1, m_2, \lambda) = (1 + 10^{-6}, 2, 0.5)$ on $x \in [-10^{-7}, 10^7]^2$.

which is the reciprocal of the cluster number, i.e., 4, as shown in Figs. 12 and 13, This implies that the fuzzification parameter m_1 with a value even slightly higher than 1 ensures that the FCF value at infinity is fuzzy, and that the power of fuzzification parameter m_1 is stronger than the other parameters, namely m_2 and λ.

5 Summary

In this study, we proposed the TBFCM clustering method. Through numerical experiments using an artificial dataset, it was confirmed that the proposed method reduces to the EFCM, PFCM, TFCM, and BFCM methods by appropriately controlling the fuzzification parameters; in addition, the FCF of the proposed method approaches the reciprocal of the cluster number at infinity, which implies that the power of fuzzification parameter m_1 is stronger than those of other fuzzification parameters.

However, this generalization feature was not proved theoretically, but only observed numerically; in contrast, similar features of conventional methods have been investigated theoretically [9,10]. Additionally, because the distinct meanings of the three fuzzification parameters (m_1, m_2, λ) are unknown, a flexibility of the proposed method is still vague.

Thus, in a future research,

- the generalization feature of the proposed method will be proved theoretically in a manner similar to [9,10],
- The proposed method will be applied to various datasets, and the distinct meanings of the three fuzzification parameters will be clear along with showing the flexibility of the proposed method.

- The proposed method will be applied to real datasets, and the results will be compared with those obtained using conventional methods in terms of clustering accuracy.
- A variable to control cluster size [11,12] will be introduced for the proposed method,
- The proposed fuzzification technique will applied to other types of data, such as spherical data [13–15] and categorical multivariate data [16–18].

References

1. Bezdek, J.: Pattern Recognition with Fuzzy Objective Function Algorithms. Plenum Press, New York (1981)
2. MacQueen, J.B.: Some methods of classification and analysis of multivariate observations. In: Proceedings 5th Berkeley Symposium on Mathematical Statistics and Probability, pp. 281–297 (1967)
3. Miyamoto, S., Mukaidono, M.: Fuzzy c-means as a regularization and maximum entropy approach. In: Proceedings 7th International Fuzzy Systems Association World Congress (IFSA 1997), vol. 2, pp. 86–92 (1997)
4. Kanzawa, Y.: Generalization of quadratic regularized and standard fuzzy c-means clustering with respect to regularization of hard c-means. In: Torra, V., Narukawa, Y., Navarro-Arribas, G., Megías, D. (eds.) MDAI 2013. LNCS (LNAI), vol. 8234, pp. 152–165. Springer, Heidelberg (2013). https://doi.org/10.1007/978-3-642-41550-0_14
5. Menard, M., Courboulay, V., Dardignac, P.: Possibilistic and probabilistic fuzzy clustering: unification within the framework of the non- extensive thermostatistics. Pattern Recogn. **36**, 1325–1342 (2003)
6. Miyamoto, S., Umayahara, K.: Methods in hard and fuzzy clustering. In: Liu, Z.Q., Miyamoto, S. (eds.) Soft Computing and Human-Centered Machines. Springer, Tokyo (2000). https://doi.org/10.1007/978-4-431-67907-3_5
7. Tsallis, C.: Possible generalization of Boltzmann-Gibbs statistics. J. Statist. Phys. **52**, 479–487 (1988)
8. Miyamoto, S., Ichihashi, H., Honda, K.: Algorithms for Fuzzy Clustering. Springer (2008). https://doi.org/10.1007/978-3-540-78737-2
9. Kanzawa, Y., Miyamoto, S.: Regularized fuzzy c-means clustering and its behavior at point of infinity. JACIII **23**(3), 485–492 (2019)
10. Kanzawa, Y., Miyamoto, S.: Generalized fuzzy c-means clustering and its theoretical properties. In: Torra, V., Narukawa, Y., Aguiló, I., González-Hidalgo, M. (eds.) MDAI 2018. LNCS (LNAI), vol. 11144, pp. 243–254. Springer, Cham (2018). https://doi.org/10.1007/978-3-030-00202-2_20
11. Miyamoto, S., Kurosawa, N.: Controlling cluster volume sizes in fuzzy c-means clustering. In: Proceedings SCIS&ISIS2004, pp. 1–4 (2004)
12. Ichihashi, H., Honda, K., Tani, N.: Gaussian mixture PDF approximation and fuzzy c-means clustering with entropy regularization. In: Proceedings 4th Asian Fuzzy System Symposium, pp. 217–221 (2000)
13. Kanzawa, Y.: On kernelization for a maximizing model of Bezdek-like spherical fuzzy c-means clustering. In: Torra, V., Narukawa, Y., Endo, Y. (eds.) MDAI 2014. LNCS (LNAI), vol. 8825, pp. 108–121. Springer, Cham (2014). https://doi.org/10.1007/978-3-319-12054-6_10

14. Kanzawa, Y.: A maximizing model of bezdek-like spherical fuzzy c-means. J. Adv. Comput. Intell. Intell. Informat. **19**(5), 662–669 (2015)
15. Kanzawa, Y.: A maximizing model of spherical bezdek-type fuzzy multi-medoids clustering. J. Adv. Comput. Intell. Intell. Informat. **19**(6), 738–746 (2015)
16. Kanzawa, Y.: Fuzzy co-clustering algorithms based on fuzzy relational clustering and TIBA imputation. J. Adv. Comput. Intell. Intell. Informat. **18**(2), 182–189 (2014)
17. Kanzawa, Y.: On possibilistic clustering methods based on Shannon/Tsallis-entropy for spherical data and categorical multivariate data. In: Torra, V., Narukawa, Y. (eds.) MDAI 2015. LNCS (LNAI), vol. 9321, pp. 115–128. Springer, Cham (2015). https://doi.org/10.1007/978-3-319-23240-9_10
18. Kanzawa, Y.: Bezdek-type fuzzified co-clustering algorithm. J. Adv. Comput. Intell. Intell. Informat. **19**(6), 852–860 (2015)

Nonparametric Bayesian Nonnegative Matrix Factorization

Hong-Bo Xie[1]($\boxtimes$) ⓘ, Caoyuan Li[2,3] ⓘ, Kerrie Mengersen[1] ⓘ, Shuliang Wang[2] ⓘ, and Richard Yi Da Xu[3] ⓘ

[1] The ARC Centre of Excellence for Mathematical and Statistical Frontiers, Queensland University of Technology, Brisbane, QLD 4001, Australia
`hongbo.xie@qut.edu.au`
[2] School of Computer Science and Technology, Beijing Institute of Technology (BIT), Beijing 100081, China
[3] Faculty of Engineering and Information Technology, University of Technology Sydney (UTS), Ultimo, NSW 2007, Australia

Abstract. Nonnegative Matrix Factorization (NMF) is an important tool in machine learning for blind source separation and latent factor extraction. Most of existing NMF algorithms assume a specific noise kernel, which is insufficient to deal with complex noise in real scenarios. In this study, we present a hierarchical nonparametric nonnegative matrix factorization (NPNMF) model in which the Gaussian mixture model is used to approximate the complex noise distribution. The model is cast in the nonparametric Bayesian framework by using Dirichlet process mixture to infer the necessary number of Gaussian components. We derive a mean-field variational inference algorithm for the proposed nonparametric Bayesian model. Experimental results on both synthetic data and electroencephalogram (EEG) demonstrate that NPNMF performs better in extracting the latent nonnegative factors in comparison with state-of-the-art methods.

Keywords: Dirichlet process · Nonnegative matrix factorization · Nonparametric Bayesian methods · Gaussian mixture model · Variational Bayes

1 Introduction

Nonnegative matrix factorization (NMF) plays an important role to solve challenging problems in machine learning, for example, blind source separation, hyperspectral unmixing, audio spectra analysis, text mining, image restoration, spectral clustering, and sources localization in neuroscience [4,8,13]. It aims to factorise a given matrix $\mathbf{Y}$ into two nonnegative low rank latent factor matrices $\mathbf{U}$ and $\mathbf{V}$, so that their product reconstructs the original matrix [4,8,13]. NMF was originally posited via optimizing a suitable cost function subject to non-negativity constraints. It is well-known that most popular NMF cost function estimates can be interpreted as the maximum likelihood (ML) estimator

© Springer Nature Switzerland AG 2020
V. Torra et al. (Eds.): MDAI 2020, LNAI 12256, pp. 132–141, 2020.
https://doi.org/10.1007/978-3-030-57524-3_11

of a particular statistical model. For instance, the ℓ_2-norm distance measure is related to Gaussian error statistics, while KL- or IS-divergence can be approximated by alternative error statistics given by Poisson or Gamma noise model. Hence, constrained optimization of proper cost functions can be achieved within a statistical framework in terms of maximum likelihood estimation [12]. This results in the development of more conceptually principled approaches based on Bayesian probabilistic interpretations of NMF.

One of the most popular and straightforward Bayesian NMF models under nonnegative constraint is to use the Poisson likelihood (noise) function with a conjugate Gamma prior for $\mathbf{U}$ and $\mathbf{V}$ [3]. Vincent and Hugo [11] replaced the Gamma prior with an exponential distribution to be coupled with the Poisson likelihood. However, the Poisson distribution is formally defined only for integers, which impairs the statistical interpretation of KL-NMF on uncountable data such as real-valued signals or images. There are also some other variant algorithms which employ an exponential or truncated Gaussian prior for the purpose of nonnegativity. However, a common shortcoming of these existing methods is that they consider only a single noise kernel or noise distribution. In practice, it is well known that most observations contain complex noise components or distributions. To address this issue, this study aims to develop a hierarchical Bayesian non-negative matrix factorization model. A Gaussian mixture model (GMM), a universal approximator for any continuous distribution [10], is employed to approximate the distribution of the complex noise components. For GMMs, the choice of the number of mixture components is an important issue. Insufficient components result in underfitting, while an excessive number of components leads to over-fitting. In this paper, we leverage the power of a nonparametric Bayesian technique, i.e., Dirichlet process mixtures, to determine the required number of Gaussian components. The model is thus termed nonparametric nonnegative matrix factorization (NPNMF).

Experimental results on both synthetic data sets and a real-world signal (electroencephalogram, EEG) demonstrate the proposed NPNMF model achieves improved performance compared to representative baseline methods, including Lee and Seung's seminal NMF algorithm and three other sparse or robust counterparts [9].

2   Nonparametric Nonnegative Matrix Factorization

In this section, we first elaborate on the model specification of the nonparametric nonnegative matrix factorization. We then present the variational Bayesian method to infer all parameters and latent variables of this model in detail.

2.1   Model Specification of NPNMF

For an observation matrix $\mathbf{Y} \in \mathbb{R}^{m \times n}$, nonnegative matrix factorization can be formulated as decomposing $\mathbf{Y}$ into two latent matrices $\mathbf{U} \in \mathbb{R}_+^{m \times r}$ and $\mathbf{V} \in \mathbb{R}_+^{n \times r}$, whose values are constrained to be positive. In other words, the task is to solve

$$\mathbf{Y} = \mathbf{U}\mathbf{V}^\top + \mathbf{E}, \tag{1}$$

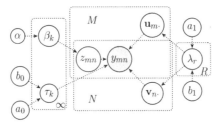

Fig. 1. Directed graphical representation of the NPNMF model

where $\mathbf{E} \in \mathbb{R}^{m \times n}$ represents a noise term. We take the variational Bayesian approach to this problem. Figure 1 shows the graphical model of the proposed hierarchical nonparametric Bayesian nonnegative matrix factorization with latent variables and their corresponding priors. In particular, we utilize the mixture Gaussian distributions to model the noise:

$$p(e_{mn}) = \sum_{k=1}^{\infty} \theta_k \mathcal{N}(0, \tau_k), \tag{2}$$

where $\theta_k = \beta_k \prod_{l=1}^{k-1}(1 - \beta_l)$ and β_k is drawn independently from a Beta distribution $\mathcal{B}(1, \alpha)$ according to the stick-breaking construction [2], and τ_k is the precision of the k-th component. In the model, a_0, b_0, α, a_1 and b_1 are hyperparameters. Let z_{mn} be a latent variable that assigns the index of the parameter associated with the entry e_{mn}. The distribution of z_{mn} can be regarded as a multinomial distribution with parameters $\{\theta_1, \cdots, \theta_\infty\}$. In practice, we set a relatively large value, K, as the initial number of Gaussian components to model the noise term.

For the precision τ_k we use a Gamma distribution with shape $a_0 > 0$ and rate $b_0 > 0$,

$$p(\tau_k) \sim \mathcal{G}(a_0, b_0). \tag{3}$$

We choose an exponential prior over $\mathbf{U}$ and $\mathbf{V}$. In order to automatically prune the rank of $\mathbf{U}$ and $\mathbf{V}$, and we assign rate parameters λ_r to the exponential prior for $\mathbf{u}_{\cdot r}$ and $\mathbf{v}_{\cdot r}$, the r-th columns of $\mathbf{U}$ and $\mathbf{V}$.

$$u_{mr} \sim f(u_{mr}|\lambda_r), \tag{4}$$

$$v_{nr} \sim f(v_{nr}|\lambda_r), \tag{5}$$

where $f(x|\lambda) = \lambda \exp(-\lambda x) s(x)$ is the density of the exponential distribution, and $s(x)$ is the unit step function, which means when $x < 0$, $s(x) = 0$, otherwise $s(x) = 1$. With the constraint of the same rate parameters λ_r across $\mathbf{u}_{\cdot r}$ and $\mathbf{v}_{\cdot r}$, most of the rate parameters λ_r will be iteratively updated to very large values. The corresponding columns of $\mathbf{U}$ and $\mathbf{V}$ are removed since they make little contribution to the approximation of $\mathbf{Y}$, and hence the rank of the latent factors $\mathbf{U}$ and $\mathbf{V}$ are automatically determined.

Combining the likelihood and the priors, we can formulate the joint distribution as

$$p(\mathbf{Y}, \mathbf{U}, \mathbf{V}, \mathbf{z}, \boldsymbol{\tau}, \boldsymbol{\lambda}, \boldsymbol{\beta}|a_0, b_0, a_1, b_1, \alpha)$$
$$= p(\mathbf{Y}|\mathbf{U}, \mathbf{V}, \mathbf{z}, \boldsymbol{\tau})p(\mathbf{U}|\boldsymbol{\lambda})p(\mathbf{V}|\boldsymbol{\lambda})p(\boldsymbol{\lambda}|a_1, b_1) \qquad (6)$$
$$\cdot p(\boldsymbol{\tau}|a_0, b_0)p(\mathbf{z}|\boldsymbol{\beta})p(\boldsymbol{\beta}|\alpha),$$

where $\boldsymbol{\beta} = (\beta_1, \ldots, \beta_K)$ and $\boldsymbol{\lambda} = (\lambda_1, \ldots, \lambda_R)$.

In order to approximate the posterior distribution, we consider the mean-field variational approximation:

$$q(\mathbf{U}, \mathbf{V}, \mathbf{z}, \boldsymbol{\tau}, \boldsymbol{\lambda}, \boldsymbol{\beta}) = \prod_{m=1}^{M} \prod_{r=1}^{R} q_{u_{mr}}(\mu_{mr}^{U}, \tau_{mr}^{U})$$

$$\prod_{n=1}^{N} \prod_{r=1}^{R} q_{v_{nr}}(\mu_{nr}^{V}, \tau_{nr}^{V}) \prod_{r=1}^{R} q_{\lambda_r}(a_r^*, b_r^*) \prod_{k=1}^{K} q_{\beta_k}(\gamma_{k,1}, \gamma_{k,2}) \qquad (7)$$

$$\prod_{k=1}^{K} q_{\tau_k}(\rho_{k,1}, \rho_{k,2}) \prod_{m=1}^{M} \prod_{n=1}^{N} q_{z_{mn}}(\phi_{mn}),$$

where the approximation of each entry of $\mathbf{U}$ ($q_{u_{mr}}$) follows a truncated normal distribution with mean μ_{mr}^{U} and covariance τ_{mr}^{U} restricted to positive values, and similarly for $\mathbf{V}$, τ_k and λ_r follow gamma distribution parametrized by $\rho_{k,1}$, $\rho_{k,2}$ and a_r^*, b_r^*, respectively, $q_{\beta_k}(\gamma_{k,1}, \gamma_{k,2})$ is a beta distribution, and $q_{z_{mn}}(\phi_{mn})$ is a multinomial distribution (see next section for details).

2.2   Model Inference of NPNMF

By applying the variational Bayesian inference principle, one can derive the mean-field posterior representation of each latent variable and parameter from the observed data and the assumed priors. We summarize the major update rules in this section while the details of posterior inference are omitted due to the space restriction.

$q(\beta_k)$ follows a Bata distribution,

$$q(\beta_k) = \text{Beta}(\beta_k|\gamma_{k,1}, \gamma_{k,2}), \qquad (8)$$

with

$$\gamma_{k,1} = 1 + \sum_{m=1}^{M} \sum_{n=1}^{N} \phi_{mnk}, \qquad (9)$$

$$\gamma_{k,2} = \alpha + \sum_{m=1}^{M} \sum_{n=1}^{N} \sum_{t=k+1}^{K} \phi_{mnt}, \qquad (10)$$

and

$$\mathbb{E}_q(\ln \beta_k) = \psi(\gamma_{k,1}) - \psi(\gamma_{k,1} + \gamma_{k,2}), \qquad (11)$$

$$\mathbb{E}_q(\ln(1 - \beta_k)) = \psi(\gamma_{k,2}) - \psi(\gamma_{k,1} + \gamma_{k,2}), \qquad (12)$$

where ψ denotes the digamma function.

$q(z_{mn})$ obeys a multinomial distribution with parameters ϕ_{mnk} for $k = 1, 2, \cdots K$ given by

$$
\begin{aligned}
\phi_{mnk} \propto \exp\{&\frac{1}{2}(\psi(\rho_{k,1}) - \ln \rho_{k,2}) + \mathbb{E}_q \ln \beta_k \\
&- \frac{1}{2}\frac{\rho_{k,1}}{\rho_{k,2}}\mathbb{E}_q\{(y_{mn} - \mathbf{u}_{m\cdot}\mathbf{v}_{n\cdot}^\top)^2\} + \sum_{t=1}^{k-1}\mathbb{E}_q \ln(1 - \beta_t)\}.
\end{aligned}
\tag{13}
$$

$q(u_{mr})$ follows a truncated normal distribution (defined on positive values), with mean and precision

$$
\mu_{mr}^U = \frac{1}{\tau_{mr}^U}(-\langle\lambda_r\rangle + \sum_{n\in\Omega_m}\sum_{k=1}^{K}\phi_{mnk}\langle\tau_k\rangle(y_{mn} - \sum_{r'\neq r}\langle u_{mr'}\rangle\langle v_{nr'}\rangle)\langle v_{nr}\rangle),
\tag{14}
$$

$$
\tau_{mr}^U = \sum_{n\in\Omega_m}\sum_{k=1}^{K}\phi_{mnk}\langle\tau_k\rangle\langle v_{nr}^2\rangle,
\tag{15}
$$

where $\Omega = \{(m,n) : m = 1,\ldots,M, n = 1,\ldots,N\}$ and $\Omega_m = \{m|(m,n)\in\Omega\}$.

Similarly, the element v_{nr} of factor matrix $\mathbf{V}$ follows truncated normal distribution with mean and precision

$$
\mu_{nr}^V = \frac{1}{\tau_{nr}^V}(-\langle\lambda_r\rangle + \sum_{m\in\Omega_n}\sum_{k=1}^{K}\phi_{mnk}\langle\tau_k\rangle(y_{mn} - \sum_{r'\neq r}\langle u_{mr'}\rangle\langle v_{nr'}\rangle)\langle u_{mr}\rangle),
\tag{16}
$$

$$
\tau_{nr}^V = \sum_{m\in\Omega_n}\sum_{k=1}^{K}\phi_{mnk}\langle\tau_k\rangle\langle u_{mr}^2\rangle.
\tag{17}
$$

where $\Omega_n = \{n|(m,n)\in\Omega\}$.

The precision parameter τ_k follows a Gamma distribution

$$
q(\tau_k) = G(\tau_k|\rho_{k,1},\rho_{k,2}),
\tag{18}
$$

where

$$
\begin{aligned}
\rho_{k,1} &= a_0 + \frac{1}{2}\sum_{m=1}^{M}\sum_{n=1}^{N}\phi_{mnk}, \\
\rho_{k,2} &= b_0 + \frac{1}{2}\sum_{m=1}^{M}\sum_{n=1}^{N}\phi_{mnk}\mathbb{E}_q[(y_{mn} - \mathbf{u}_{m\cdot}\mathbf{v}_{n\cdot}^\top)^2].
\end{aligned}
\tag{19}
$$

Here $\mathbb{E}_q[(y_{mn} - \mathbf{u}_m\mathbf{v}_n^\top)^2]$ is given by

$$
\begin{aligned}
\mathbb{E}_q[(y_{mn} - \mathbf{u}_m\mathbf{v}_n^\top)^2] &= (y_{mn} - \sum_{r=1}^{R}\langle u_{mr}\rangle\langle v_{nr}\rangle)^2 \\
&+ \sum_{r=1}^{R}(\langle u_{mr}^2\rangle\langle v_{nr}^2\rangle - \langle u_{mr}\rangle^2\langle v_{nr}\rangle^2).
\end{aligned}
\tag{20}
$$

After updating the parameters ρ via Eq. (19), the weight coefficients θ of the K clusters are also updated. A θ_k smaller than a pre-defined threshold means that the probability of some entries to be represented by this cluster is very rare. We therefore prune those clusters, and the noise is represented by a limited number of Gaussians.

λ_r follows a Gamma distribution with shape parameter

$$a_r^* = a_1 + M + N, \tag{21}$$

and inverse scale parameter

$$b_r^* = b_1 + \sum_{m=1}^{M} \langle u_{mr} \rangle + \sum_{n=1}^{N} \langle v_{nr} \rangle. \tag{22}$$

We update each parameter in turn while holding others fixed. Convergence of the NPNMF algorithm to a local minimum can be guaranteed after a finite number of iterations following standard variational Bayes arguments [2].

3   Results

In this Section, we empirically compare the proposed NPNMF model with several state-of-the-art methods including MUNMF in [9], a sparseness-constrained NMF (SCNMF) [7], a Bayesian NMF (PSNMF) [6], and the outlier-robust Mah-NMF [5].

3.1   Results on Synthetic Data

We first compare the performance of each NMF method using synthetic data sets. The elements of the two latent matrices $\mathbf{U}, \mathbf{V}$ are generated from unit mean exponential distributions, and the matrices have three different ranks $r = 5$, 10 and 15. The ground-truth $\mathbf{Y}_0$ is the product of $\mathbf{U} \in \mathbb{R}_+^{500 \times r}$ and $\mathbf{V} \in \mathbb{R}_+^{500 \times r}$. Three types of noise are considered for $\mathbf{E}$, i.e. Gaussian, sparse, and mixed (see Table 1). The initial number of Gaussian components K is set to 300 for NPNMF, which is large enough to model a wide range of noise structures.

Because some methods for comparison are unable to automatically infer the rank r, we set the ground-truth rank r as a known model input parameter. We use the relative error of the Frobenius norm with respect to the ground truth, defined by

$$Error = \frac{|\mathbf{Y}_0 - \overline{\mathbf{U}}\,\overline{\mathbf{V}}^\top|_F}{|\mathbf{Y}_0|_F}, \tag{23}$$

as the metric to quantify the performance of each algorithm, where $\overline{\mathbf{U}}$ and $\overline{\mathbf{V}}$ are the recovered latent matrices. For each noise setting, we run each method 20 times with different random input of $\mathbf{Y}_0$. The resulting mean relative errors are shown in Table 2.

It can be seen that NPNMF yields the lowest relative error under most noise and rank settings. For Gaussian noise, PSNMF achieves comparable performance to NPNMF, however the relative errors for the other methods are much higher than NPNMF and PSNMF, even given the correct rank input, which is difficult to estimate in practice. The relative error of these methods increases significantly when the initial rank deviates from the ground-truth rank. In the case of sparse noise, it is evident that the performance of MUNMF and MahNMF methods degrade. This is reasonable since only Gaussian noise is considered in these two models. The NPNMF algorithm performs better than PSNMF for rank 5, and is slightly inferior to NPNMF for rank 10 and 15. Finally, for the mixture noise, it is not surprising that NPNMF outperforms all methods with significantly lower errors. When the noise type is simple, PSNMF achieves comparable and even slightly better results than NPNMF. However, the performance of PSNMF is significantly inferior to NPNMF when the noise is more complicated, due to the ability of NPNMF to fit unknown complex noise, as well as automatically tune the rank.

Table 1. Noise parameter for the synthetic data sets. $\mathbf{U}$ denotes the uniformly distributed noise followed by its range.

	$\mathcal{N}(0, 0.5^2)$	$\mathcal{N}(0, 0.1^2)$	$U[-5, 5]$
Gaussian noise	100%	0	0
Sparse noise	0	0	30%
Mixture noise	60%	20%	20%

We now empirically evaluate the stability of DPNMF in terms of the varying initial number of Gaussians components K and rank r under mixture noise. Figure 2(a) shows the mean relative error in terms of the different initial rank for (true) $r = 5$, 10, and 15, respectively. In each case, the mean relative error has tiny fluctuation as the initial rank varies. However the relative error remains stable at low values as the initial rank increases. Figure 2(b) shows the effect of the initial number of Gaussian components on the mean relative error for $r = 5$, 10, 15. Similar to the effect of fixing K in Fig. 2(a), the mean relative error is almost flat as the initial number of components varying from 10 to 150. Figure 2(c) shows the estimated number of Gaussian components versus the initial number, for $r = 5$. As the initial number of components increases, the estimated number of components stabilise at around 25 components, which is adequate to model the mixture noise. Overall we may conclude from Fig. 2 that DPNMF is fairly robust to initialization and input parameters. The computational complexity of the factor matrices $\mathbf{U}$ and $\mathbf{V}$ in Eqs. (14)–(17) is $O(Kmnr)$ in each iteration, where K is the number of initial mixture components, mn represents the data size, and r the rank. Since this is the major computational cost of the model, DPNMF hence has linear complexity w.r.t. the data size. It should be noted that because latent factors are pruned out in the first few iterations,

r reduces rapidly in practice. With a recent 3.6 GHz CPU, it costs around 10 min to fit the model with $K = 300$ mixture components.

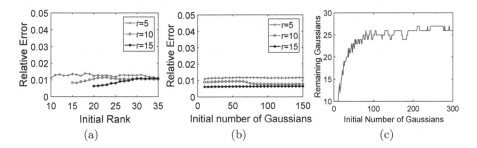

Fig. 2. Effect of initial number of Gaussians and rank, r on the relative error performance of DPNMF for mixture noise.

Table 2. Mean relative errors of five algorithms under three noise types (Table 1) with three different initial ranks r. Best results are shown in bold.

Noise	r	DPNMF	MUNMF	SCNMF	MahNMF	PSNMF
Gaussian	5	**0.0746**	0.0897	0.6360	0.0934	0.0766
	10	**0.1009**	0.1372	0.4675	0.1329	0.1049
	15	**0.1207**	0.1767	0.2853	0.1688	0.1249
Sparse	5	**0.0212**	0.0535	0.6232	0.0572	0.0350
	10	0.0468	0.0815	0.4585	0.1348	**0.0275**
	15	0.0330	0.0975	0.2542	0.1766	**0.0233**
Mixture	5	**0.0109**	0.0551	0.6229	0.0929	0.0449
	10	**0.0089**	0.0856	0.4578	0.1623	0.0355
	15	**0.0061**	0.0965	0.2585	0.2004	0.0300

3.2   Classification of Motor Imagery EEG

Classification of motor imagery EEG is an important research area of machine learning for brain-machine interface and neuroscience. We investigate the performance of NPNMF in a motor imagery EEG classification problem, which heavily relies on nonnegative matrix factorization for feature extraction. A public available BCI competition data set is used [1]. Each data segment has a duration of 9 s with the first 3 s as preparation period. We therefore only analyze the EEG in the final 6 s of the task. The aim is to dynamically identify the actual class with high probability as soon as possible during the trail, rather than to provide the class label using the entire data segment. We follow [1] to construct the data matrix and perform training and test for left and right movement identification. Figure 3 illustrates the time course of five methods to classify the single-trial imagery movements on test data continuously. Accuracy is initially low at the beginning of the EEG, peaks twice at around 3.2 and 3.8 s before levelling out from around 5.5 s. Of all methods, NPNMF and MUNMF perform the best, with one of these models providing the best classification accuracy at any time point. NPNMF provides the highest classification accuracy at the two peaks, with a maximum accuracy of 83% compared to the 80%–81% maximum of the other methods. Overall NPNMF provides a highly competitive classification accuracy through the duration of the time course.

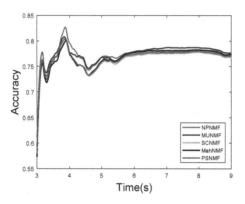

Fig. 3. Time course of the classification accuracy using the encoding variable matrix extracted from each NMF model.

4   Conclusions

In this paper, we have proposed a novel full nonparametric Bayesian model for nonnegative matrix factorization to fit the complex noise distribution. Using both synthetic and real dataset, experimental results show that the proposed method outperforms other four state-of-the-art nonnegative matrix factorization approaches. In particular, the proposed method can effectively extract the latent factors under complex noise distribution. A major limitation of the current research is the size of both synthetic and experimental data sets is relatively small. In addition, the model is trained on the full data set. NPBMF can be further improved to accommodate streaming data for applications such as background/foreground separation in videos.

References

1. Blankertz, B., et al.: The BCI competition 2003: progress and perspectives in detection and discrimination of EEG single trials. IEEE Trans. Biomed. Eng. **51**(6), 1044–1051 (2004). https://doi.org/10.1109/TBME.2004.826692
2. Blei, D.M., Jordan, M.I., et al.: Variational inference for Dirichlet process mixtures. Bayesian Anal. **1**(1), 121–143 (2006)
3. Cemgil, A.T.: Bayesian inference for nonnegative matrix factorisation models. Computational Intelligence and Neuroscience **2009** (2009)
4. Fu, X., Huang, K., Sidiropoulos, N.D., Ma, W.K.: Nonnegative matrix factorization for signal and data analytics: Identifiability, algorithms, and applications. IEEE Signal Process. Mag. **36**, 59–80 (2019)
5. Guan, N., Tao, D., Luo, Z., Shawe-Taylor, J.: MahNMF: Manhattan non-negative matrix factorization. ArXiv abs/1207.3438 (2012)
6. Hinrich, J.L., Mørup, M.: Probabilistic sparse non-negative matrix factorization. In: Deville, Y., Gannot, S., Mason, R., Plumbley, M.D., Ward, D. (eds.) LVA/ICA 2018. LNCS, vol. 10891, pp. 488–498. Springer, Cham (2018). https://doi.org/10.1007/978-3-319-93764-9_45
7. Hoyer, P.O.: Non-negative matrix factorization with sparseness constraints. J. Mach. Learn. Res. **5**(Nov), 1457–1469 (2004)
8. Huang, K., Sidiropoulos, N.D.: Putting nonnegative matrix factorization to the test: a tutorial derivation of pertinent Cramer-Rao bounds and performance benchmarking. IEEE Signal Process. Mag. **31**(3), 76–86 (2014)
9. Lee, D.D., Seung, H.S.: Learning the parts of objects by non-negative matrix factorization. Nature **401**(6755), 788 (1999)
10. Maz'ya, V., Schmidt, G.: On approximate approximations using Gaussian kernels. IMA J. Numer. Anal. **16**(1), 13–29 (1996)
11. Renkens, V., et al.: Automatic relevance determination for nonnegative dictionary learning in the Gamma-Poisson model. Sig. Process. **132**, 121–133 (2017)
12. Schachtner, R., Po, G., Tomé, A.M., Puntonet, C.G., Lang, E.W., et al.: A new Bayesian approach to nonnegative matrix factorization: Uniqueness and model order selection. Neurocomputing **138**, 142–156 (2014)
13. Wang, Y.X., Zhang, Y.J.: Nonnegative matrix factorization: a comprehensive review. IEEE Trans. Knowl. Data Eng. **25**(6), 1336–1353 (2012)

SentiRank: A System to Integrate Aspect-Based Sentiment Analysis and Multi-criteria Decision Support

Mohammed Jabreel[1,2], Najlaa Maaroof[1(✉)], Aida Valls[1], and Antonio Moreno[1]

[1] Dept. Enginyeria Informàtica i Matemàtiques, Universitat Rovira i Virgili,
43007 Tarragona, Spain
{mohammed.jabreel,najlaa.maaroof,aida.valls,antonio.moreno}@urv.cat
[2] Department of Computer Science, Hodeidah University, Hodeidah 1821, Yemen

Abstract. This study proposes a novel ranking method, called Senti-Rank, that integrates data coming from aspect-level sentiment analysis into a multi-criteria decision aiding procedure. The novelty of the work is the combination of the evaluation of the features of a set of products (based on user preferences) with the reviews in social networks about the same products. The contribution of this work is twofold: theoretical and practical. From the theoretical side, we propose an automatic method to extract the aspect categories and their evaluations from the online opinions of users. Next, we describe how to merge that information with the decision maker's preferences about the features of a set of products. The ELECTRE method is used afterwards to rank the products with both types of input data. From the practical side, we have implemented a tool that can be used to rank a set ofs restaurants in Tarragona, using available data and reviews on the Web. The tests show that the ranking is effectively modified when online reviews are included in the analysis.

Keywords: Multiple criteria decision aid · ELECTRE · Sentiment analysis · Aspect-based sentiment analysis

1 Introduction

The massive spread of online platforms, such as social networks and online shopping sites, which allow sharing reviews and opinions, has changed the way in which users make decisions, especially the purchase processes. For many of our decision-making tasks, "what other people think" has become an essential piece of information, as much as what we prefer. It is well known that price and other quality criteria have a significant influence on people' purchases. However, nowadays, online consumers' reviews have also become a common source of influence and an essential input in decision making [14, 16]. Some studies have shown that 90% of consumers read online reviews before making a purchase decision [13].

M. Jabreel and N. Maaroof—Equal contribution.

© Springer Nature Switzerland AG 2020
V. Torra et al. (Eds.): MDAI 2020, LNAI 12256, pp. 142–153, 2020.
https://doi.org/10.1007/978-3-030-57524-3_12

Service and products providers try to attract customers by publishing the best description of the features of their services, making it easy to find the alternatives that fit the preferences of decision-makers. However, it is also essential to consider the quality of the advertised features. For instance, people usually prefer to visit restaurants with free WiFi, which is information that we can find in the description of most restaurants, but, without customers' reviews, it is not possible to figure out the quality of such service.

Thus, in order to obtain better recommendations for each user, we need to consider two sources of information: the decision-maker preferences and online reviews. We can find in the literature of Multi-Citeria Decision Aid (MCDA) numerous systems proposed for modelling the decision makers preferences and ranking alternatives to support the decision-making process [1,3,4]. These systems only focus on the user preferences. Moreover, there are different systems for analysing reviews on products or services [9]. However, there are not any works on exploiting user reviews alongside decision-maker preferences to support the decision-making process. The main aim of this work is to improve the current multi-criteria decision support methods by including the information contained in online reviews about a set of alternatives. Hence, our intention is to understand how we can exploit Sentiment Analysis (SA) to integrate the users' reviews in a multi-criteria ranking system to improve the quality of the decision-making process. Towards this objective, we identify the following research questions: **Q1**: Which properties of user reviews can be used to improve the quality of the decision-making process? **Q2**: How can we extract those properties? **Q3**: How can we integrate the extracted properties in the decision-making problem formulation? **Q4**: How can we solve automatically the resulting decision-making problem?

In response to the objectives described above, this paper proposes the SentiRank system. It is based on a multi-criteria decision aid system that gives to the decision maker the ability to exploit online reviews alongside his/her preferences in the process of alternatives ranking. The main contributions of this paper are the following:

1. We develop an aspect-based SA (ABSA) system to extract the set of aspect categories mentioned in reviews and identify the expressed sentiments towards the extracted aspects. We use a data preparation unit in the decision support system to transform the information into a set of utility criteria.
2. We propose a domain-independent methodology to use ELECTRE to rank a set of alternatives taking into account criteria relate to the decision-makers' preferences and criteria generated from online reviews.
3. We evaluate the proposed system empirically by collecting information on a set of restaurants in Tarragona and the reviews about them.

The layout of this paper is as follows. In Sect. 2, a review of related works is presented. In Sect. 3, the proposed system, SentiRank, is explained step by step. Section 4 presents a case study and discusses the results. Finally, in Sect. 5, we conclude the work and outline some planned future research lines.

2   Related Work

In the literature, there are a limited number of works for products ranking based on customers' reviews. Moreover, most of these works apply MCDA tools only to online reviews in the ranking procedure, ignoring the personal preferences of the decision maker about the products' features. In this section, we outline the most relevant MCDA works based on customers' reviews.

The work presented in [18] proposed a method for ranking products through online reviews. In their method, a dynamic programming technique was proposed to identify subjective sentences and comparative sentences in the reviews. Afterwards, they used the extracted sentences to build a weighted, directed product graph and used it to find the top-ranked products.

The authors in [11] extracted aspects from online reviews available on a Chinese website for a set of mobile phones. After that, they ranked the phones using PROMETHEE. The work presented in [17] proposed a system for products ranking based only on the customers' reviews. Given the reviews of a product, they first identified their aspects and then they analysed the opinions of the consumers on the aspects using a sentiment classifier. Then, they proposed a probabilistic aspect ranking algorithm to indicate the influence of the aspects by simultaneously taking into account the frequency of each aspect and the importance given to each aspect in the reviews. The authors in [7] proposed a system that combines SA and VIKOR methods to estimate customers' satisfaction on mobile application services and rank the alternatives. The work in [10] proposed a ranking method combining SA and intuitionistic fuzzy sets. A lexicon-based system was used to identify the sentiment on each product considering the product features that appeared in each review. Given the identified sentiments, an intuitionistic fuzzy number was constructed to represent the performance of each product. The ranking of products was determined by an intuitionistic fuzzy weighted averaging operator and PROMETHEE-II.

The authors in [20] developed a methodology to rank the alternatives considering customer comments about the features of the products. First, they transformed the online customers' comments into degrees of satisfaction in terms of intuitionistic fuzzy sets using an ABSA system. After that, they used IF-ELECTRE to rank the products based on the obtained satisfaction degrees.

3   Methodology

This section explains the proposed system, SentiRank, and the tools and resources used to implement it. SentiRank receives as inputs a set of alternatives (e.g. restaurants, hotels or laptops) to be ranked and their corresponding consumers' reviews. Figure 1 shows the overall architecture of the system. It is composed by three primary units, which are explained in the next subsections: the customers' reviews unit, the domain analysis unit and the ranking unit.

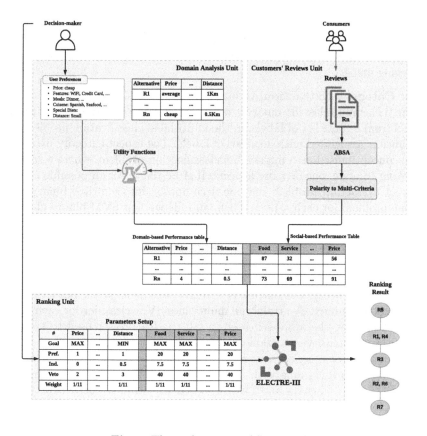

Fig. 1. The architecture of SentiRank.

3.1 Customers' Reviews Unit

The input to this unit is a collection of customers' reviews about a set of alternatives, and the output is a numerical matrix called social-based performance table. Each column in the matrix represents a specific criterion. The main component of the customers' review unit is the ABSA model, which contains two submodels. The first one extracts the set of aspects commented in a given review, and the second one identifies the sentiment expressed in the review towards each extracted aspect. We follow four steps. First, we pre-process the text in the review. After that, we split it into a list of sentences and tokenize them. The next step is to use the aspect extraction sub-model to extract the aspects from each sentence. Finally, we feed the sentence and the extracted aspects to the sentiment classifier.

Text Preprocessing: The text in reviews and social posts (like tweets) is usually noisier than the one in articles and blogs. People tend to write their posts, messages and reviews in an informal way. Thus, in our system, text cleaning and processing is essential. Concretely, the Text is transformed to lower case, and elements such as HTML tags, date, time, URLs and emails are removed. Once we have the normalized text, we use Stanford's tokenizer and Part-of-Speech (POS) tagger to split the text into a set of sentences and perform the word tokenization and tagging steps.

Aspect Category Extraction: Unlike the traditional single-label classification problem (i.e., multi-class or binary), where an instance is associated with only one label from a finite set of labels, in the multi-label classification problem [19], an instance is associated with a subset of labels. The aspect category extraction may be conceptualised as a multi-label classification problem, where a sentence may belong to none, one or more aspects. If A be the set of all possible aspects, the aspect extraction problem may be decomposed into multiple binary problems, one problem for each aspect. Then, an independent SVM binary classifier, h_j is trained for each aspect $a_j \in A$. During the inference, given a new and unseen sentence s, the assigned aspects will be $Aspects(s) = \{j|h_j(s) = 1\}$.

Aspect Polarity Detection: The goal of this sub-model is to detect the sentiment expressed towards a given aspect in a given sentence. For each input pair (sentence, aspect), the output is a single sentiment label: positive, negative, neutral or conflict. We train one multi-class SVM classifier for each aspect category. Following the approach used in [8], the feature set used to train the aspect-category polarity detection model is extended to incorporate the information about a given aspect category. In summary, the following features are used: General Word-Ngrams with $N = 3$, Category-Based Word-Ngrams with $N = 3$, POS tags, Clusters and Lexicon-based features. We used two types of lexicons (global lexicons and target-specific lexicons).

From Polarity to Multi-criteria: Our main goal is to integrate consumers' reviews about a set of alternatives with the decision-maker's preferences in the decision-making process. Hence, we consider each aspect in the social domain as a criterion. We call these set of aspects *social-criteria*. The inputs to this step are a set of alternatives $X = \{x_1, ..., x_r\}$ (e.g. restaurants), a set of aspects $A = \{a_1, ..., a_m\}$ (from the domain) and a collection of reviews about each alternative. We construct a matrix $X \times A$, where the value for (x_i, a_j) corresponds to the polarity index that represents the degree of consumers' satisfaction about the aspect a_j for alternative x_i. The polarity index can be calculated as defined in [6]:

$$Pol(x_i, a_j) = \begin{cases} 1 - \frac{N_{ij}}{P_{ij}} & \text{if } P_{ij} > N_{ij} \\ \frac{P_{ij}}{N_{ij}} - 1 & \text{if } P_{ij} < N_{ij} \\ 0 & otherwise \end{cases} \quad (1)$$

Here, P_{ij} is the number of positive opinions about aspect a_j for alternative x_i, and N_{ij} is the number of negative opinions. The range of *Pol* is $[-1, 1]$. For ease of interpretation, we transform it into the range $[0, 100]$, by adding 1 and multiplying the result by 50.

3.2   Domain Analysis Unit

As shown in Fig. 1, this unit receives as input the decision-maker preferences and the description of the alternatives. After that, it feeds them to the utility functions to transform the preferences into numerical values. Both the description of the alternatives and the user preferences values are given for specific criteria, which we call the domain criteria in this work. Decision-makers can express their preferences in different ways, depending on the type of each criterion. For example, users can specify their preferences on the price criterion as a range of amounts or as a linguistic value (e.g., very cheap, cheap, average or expensive). The goal of this unit is to convert the expression of user's preferences on the domain criteria into a numerical value that reflects the satisfaction degree of the user. Hence, we define in this work three different utility functions that measure the performance of a feature of the alternative.

Identity: This function is used in numerical features. In that case, as ELECTRE can manage heterogeneous scales, the utility function does not need to perform any processing. We simply express this function as $g(v) = v$, being v a value in the numerical scale.

Categorical: It is used in features that take multiple values from an unordered set of categories. The range of this kind of utility function is always $[0, 1]$, and the goal is maximization. The value 0 means the lowest degree of satisfaction, whereas 1 means the highest. The utility function is defined as the following: $g(U, F) = \frac{|U \cap F|}{|U|}$. In this expression, U refers to the items the user prefers and F is the set of available items in this criterion; it is also called the domain of the criterion.

Linguistic: It is used in variables that take one value v from an ordered set of linguistic terms. The range of this kind of function is $[1, |L|]$, where L is the set of linguistic terms in the domain of the function and the goal is maximization. The value 1 means the lowest degree of satisfaction and $|L|$ means the highest. The utility function is defined as the following: $g(v, p) = |L| - abs\,(pos(v) - pos(p))$. In this expression $v, p\ \in L$ are the alternative value and the preferred value respectively, *pos* is the position of the linguistic value in L and *abs* is the absolute value.

3.3   Ranking Unit

The inputs to this unit are the performance table, that contains both the domain and the social-based matrices, and the criteria properties setup. Each (domain-based or social-based) criterion has the following properties: minimum value, maximum value, the goal (MAX or MIN), the preference threshold, the indifference threshold, the veto threshold and the weight. The decision-maker is responsible for setting the values of these properties. The outranking procedure used in SentiRank is ELECTRE with Net-Flow Score (NFS) ranking. ELECTRE has been widely acknowledged as an efficient multi-criteria decision aiding tool with successful applications in many domains [3]. It is based on constructing and exploiting a valued outranking relation $S(x, y) \in [0, 1]$, meaning that "option x is at least as good as option y". There are different versions of ELECTRE. We will focus on the one that uses pseudo-criteria, known as ELECTRE-III. Given the performance table and the criteria' properties, we first apply the construction procedure to get the outranking relation. Afterwards, we apply the NFS procedure to perform the exploitation step to get the final ranking of the alternatives. These two steps work as follows:

1. The construction of the outranking binary relation is made from the combination of two tests that make use of some parameters (i.e., thresholds) to establish the agreement or disagreement level for each criterion. The values of these parameters model the user preferences and should be selected appropriately [2,15]. The tests are the following:

   – Concordance test (known as "majority opinion"): it measures the strength of the criteria coalition in favour of the agreement with the outranking relation $S(x, y)$. In each criterion, two discrimination thresholds are used to model the uncertainty: *Indifference threshold*, below which the user is indifferent to two alternatives in terms of their performances on one criterion, and *Preference threshold*, above which the user shows an ostensible strict preference of one alternative over the other in terms of their performances on one criterion.

   – Discordance test (known as "the respect of minorities"): it measures the rejection against the assertion $S(x, y)$. It gives to some criteria the right to veto the majority opinion if there are essential reasons to refute it. The criteria with this right must have a *Veto threshold*, so that considerable differences in favour of y in a criterion will eliminate the possibility that option x outranks y.

2. The exploitation of the outranking relation permits to solve the decision problem. Different procedures can be used to obtain a ranking of alternatives. In this paper we propose to use the Net Flow Score of each option $NFS(x)$, which is calculated as $NFS(x) = strength(x) - weakness(x)$, being the difference between the number of alternatives that are outranked by x, and the number of alternatives that outrank x.

4   Experiments and Results

We have evaluated and tested the proposed system in the restaurant domain. To this end, we built an aspect extraction model and an aspect polarity detection model for this domain. After that, we collected the details and reviews of a set of restaurants from Tarragona as alternatives. Finally, we tested the system with different users' profiles (i.e., user preferences). The following subsections explain the data used in these experiments, show the performance of SentiRank on one user profile and analyse the obtained results.

4.1   Data

Table 1 shows the description of the domain criteria and the social-based criteria. We chose the domain criteria from the details of the restaurants on TripAdvisor. The social criteria are the aspects defined in SemEval-2014 Task 4: Aspect Based Sentiment Analysis [12].

Table 1. The description of the criteria.

Criterion	Utility function	Domain	Goal
A-Domain			
Price	Linguistic	{Very Cheap, Cheap, Average, Expensive}	MAX
Cuisines	Categorical	Spanish, Seafood, Bar, ...	MAX
Special Diets	Categorical	Vegetarian Friendly, ...	MAX
Meals	Categorical	Lunch, Drinks, ...	MAX
Features	Categorical	Private Dining, Seating, WiFi, ...	MAX
Distance	Identity	$[0, \infty]$	MIN
B-Social			
Food	Identity	$[0, 100]$	MAX
Price	Identity	$[0, 100]$	MAX
Service	Identity	$[0, 100]$	MAX
Ambience	Identity	$[0, 100]$	MAX
Anecdotes	Identity	$[0, 100]$	MAX

We trained our aspect extraction and aspect polarity detection models on the publicly available dataset provided by the organisers of SemEval-2014 Task 4: Aspect Based Sentiment Analysis [12]. The training data was composed by 3041 English sentences, whereas the test data were 800 English sentences. Each sentence is labelled with none, one or more aspects with their corresponding polarities. Our aspect extraction and aspect polarity detection models achieved state-of-the-art systems' performance by obtaining an F1 score of 87.45% for the aspect extraction model and accuracy of 82.44% for the aspect-category polarity detection model.

4.2   Case Study

As a case study, we collected the information and the users' feedback of 23 restaurants in Tarragona from TripAdvisor. The details of each restaurant are the domain aspects used in this study. In addition to the users' reviews, TripAdvisor provides the overall rating and the rating of each aspect.

We analysed the correlation between the output of our sentiment analysis system and the ratings provided by TripAdvisor and found that it is quite strong. The correlation score obtained was 80% which reveals the robustness of our sentiment analysis system and indicates that it is applicable to the restaurant domain. After that, we tested our system, SentiRank, on the following user profile.

User Profile. In this profile, we assume the decision-maker has the preferences shown in the last column of Table 2.

We also assumed that the decision-maker is strict with the Special Diets and Meals criteria and flexible or tolerant with the Cuisine and Features criteria. Based on that, we defined the values of the indifference, preference and veto thresholds, as shown in Table 2. All the social-based criteria have the same values of the thresholds. Moreover, all the criteria are given the same importance (i.e., the same weight).

Table 2. The values of the criteria thresholds and user profile. Features shortcuts, PD: Private Dining, S: Seating, WA: Wheelchair Accessible, CC: Accepts Credit Cards, TS: Table Service, PA: Parking Available, HA: Highchairs Available, and FB: Full Bar.

Criterion	Thresholds			User preferred values
	Indifference	Preference	Veto	
A - Domain				
Price	0	1	2	Cheap
Cuisine	0	0.4	No	Spanish, Seafood, Bar
Special Diets	0	0	0.5	Vegetarian Friendly, Gluten Free
Meals	0	0	0.5	Lunch, Drinks, Brunch
Features	0	0.4	No	PD, S, WA, Free Wifi, CC, TS, PA, HA, FB
Distance	0.5	1	3	Close to town's center
B - Social				
Food	7.5	20	40	
Price	7.5	20	40	
Service	7.5	20	40	
Ambience	7.5	20	40	
Anecdotes	7.5	20	40	

4.3   Results

We analysed and compared two situations. In the first case, we only use the user's preferences (i.e., the domain criteria) as inputs to the ranking procedure. The second case combines the user's preferences with the consumers' feedback. Figure 2 shows the ranking results. The left side of the figure, i.e., (a), is the ranking when we only consider the domain criteria. The results in the right side, i.e., (b), are the ranking of the same alternatives by adding the social-based criteria to the process. The alternatives are ranked based on the NFS value. The first indication from the results is the effect of the social-based criteria. There are notable differences in the ranking results.

Alternative	(a) Domain Criteria		(b) All Criteria	
	NFS	Rank	NFS	Rank
Barquet	18	1	11	4 ↓
El Taller	16	2	17	1 ↑
Sadoll	13	3	16	2 ↑
ELIAN Cafe	10	4	15	3 ↓
Octopussy	9	5	-1	12 ↓
La Xarxa	5	6	5	5 ↑
Arcs	1	7	4	7
The Cotton Club & Cocktails	0	8	-3	14 ↓
AQ	0	9	-3	15 ↓
Les Coques	0	10	-5	18 ↓
La Capital	0	11	-1	11
La Caleta	0	12	5	6 ↑
Les Fonts de Can Sala	-2	13	2	8 ↑
Lizarran Parc Central	-3	14	-8	20 ↓
El Encuentro	-4	15	-4	16 ↓
El Trull	-4	16	-1	10 ↑
Club Nautico Salou	-5	17	-5	17
Ca L Eulalia	-5	18	-8	19 ↓
Palermo Scp	-7	19	-9	21 ↓
Mas Rosello	-7	20	-2	13 ↑
Indian Mirchi	-8	21	-11	22 ↓
Tarakon	-8	22	0	9 ↑
Buffalo Grill	-15	23	-12	23

Fig. 2. The ranking result.

The disagreement between the two cases can be attributed to the fact that most of the alternatives try to attract customers as much as possible by advertising features that match the customers' preferences. On the other hand, consumers give their feedback and experiences with each alternative. This can be confirmed by checking the utility values of the domain criteria. We found that most of the alternatives have similar values in most of the criteria, especially with the special diets and meals criteria on which the user is strict. Moreover, we found that the most discriminative criteria is the distance to the city centre followed by the features.

Hence, by considering the social-based criteria we can make the following observations:

- 'El Taller' is in position 2 in the first case (a) and in position 1 in the second one (b). Inspecting the utility values of the alternatives in the social performance table reveals that the restaurant 'El Taller' has more positive opinions than 'Barquet', which was the top ranked restaurant in the first case.
- The alternatives 'Sadoll' and 'ELIAN Cafe' are promoted to positions 2 and 3, respectively, also due to the positive social opinions.
- The position of the alternatives 'La Caleta' and 'Les Fonts de Can Sala' are changed from 12 and 13 to 6 and 8 respectively. By inspecting the values of the domain criteria of these two restaurants, we found that they have very similar performance to the best two (i.e., Barquet and El Taller) except with the

distance and the price criteria. So, by checking the values of the social-based criteria (i.e., the polarity index), we found that they have very high scores, which cause the large increase. This can indicate that price and distance are no longer critical criteria in people's decisions. The same conclusion can be derived by analysing the alternative 'Tarakon'.
– The alternatives in the worst positions have small differences.

Generally speaking, the reported results show that the integration of the social-based criteria impacts the ranking process.

5   Conclusion and Future Work

In this paper we have proposed an outranking system, called SentiRank, that combines the decision-maker preferences with the users' reviews about alternatives that the decision-maker needs to rank.

We have given answers to the main research questions about integrating the users' feedback with the decision-maker preferences in the ranking process. We have defined a methodology to use the aspect categories, which can be extracted from the users' reviews, as additional attributes in the decision-making process. As a case study, we have applied the proposed system to a dataset of restaurants in Tarragona. The findings of this case study showed the impact of integrating the users' reviews in the decision-making process.

The work done in this paper opens an exciting and promising research line that enables the use of people' feedback and reviews for decision support. For future work, we would like to work on different levels of sentiment analysis. For example, in this work, we have performed the analysis on the aspect category level, hence, in the future, we plan to perform the analysis on the aspect term level [5]. Moreover, in this work, we have only considered the reviews written in English. Thus, we would like to work in a multi-lingual system.

Acknowledgements. This work is supported by URV grant 2018PFR-URV-B2-61. N. Maaroof is funded by a URV doctoral grant 2019PMF-PIPF-17.

References

1. Bouyssou, D., Marchant, T., Pirlot, M., Tsoukias, A., Vincke, P.: Evaluation and Decision Models with Multiple Criteria: Stepping Stones for the Analyst, vol. 86. Springer, Boston (2006). https://doi.org/10.1007/0-387-31099-1
2. Doumpos, M., Grigoroudis, E.: Multicriteria Decision Aid and Artificial Intelligence: Links, Theory and Applications. Wiley, Hoboken (2013)
3. Greco, S., Figueira, J., Ehrgott, M.: Multiple Criteria Decision Analysis. Springer, New York (2016). https://doi.org/10.1007/978-1-4939-3094-4
4. Ishizaka, A., Nemery, P.: Multi-Criteria Decision Analysis: Methods and Software. Wiley, Hoboken (2013)

5. Jabreel, M., Hassan, F., Moreno, A.: Target-dependent sentiment analysis of tweets using bidirectional gated recurrent neural networks. In: Hatzilygeroudis, I., Palade, V. (eds.) Advances in Hybridization of Intelligent Methods. SIST, vol. 85, pp. 39–55. Springer, Cham (2018). https://doi.org/10.1007/978-3-319-66790-4_3

6. Jabreel, M., Moreno, A.: SentiRich: sentiment analysis of tweets based on a rich set of features. In: CCIA, pp. 137–146 (2016)

7. Kang, D., Park, Y.: Review-based measurement of customer satisfaction in mobile service: sentiment analysis and VIKOR approach. Expert Syst. Appl. **41**(4, Part 1), 1041–1050 (2014)

8. Kiritchenko, S., Zhu, X., Cherry, C., Mohammad, S.: NRC-Canada-2014: detecting aspects and sentiment in customer reviews. In: Proceedings of SemEval, vol. 2014, pp. 437–442 (2014)

9. Liu, B.: Sentiment analysis and opinion mining. In: Synthesis Lectures on Human Language Technologies, vol. 5, no. 1, pp. 1–167 (2012)

10. Liu, Y., Bi, J.W., Fan, Z.P.: Ranking products through online reviews: a method based on sentiment analysis technique and intuitionistic fuzzy set theory. Inf. Fusion **36**, 149–161 (2017)

11. Peng, Y., Kou, G., Li, J.: A fuzzy PROMETHEE approach for mining customer reviews in Chinese. Arab. J. Sci. Eng. **39**(6), 5245–5252 (2014)

12. Pontiki, M., Papageorgiou, H., Galanis, D., Androutsopoulos, I., Pavlopoulos, J., Manandhar, S.: SemEval-2014 task 4: aspect based sentiment analysis. In: SemEval 2014, p. 27 (2014)

13. Shengli, L., Fan, L.: The interaction effects of online reviews and free samples on consumers' downloads: an empirical analysis. Inf. Process. Manage. **56**(6), 102071 (2019)

14. Simonson, I.: Imperfect progress: an objective quality assessment of the role of user reviews in consumer decision making, a commentary on de Langhe, Fernbach, and Lichtenstein. J. Consum. Res. **42**(6), 840–845 (2016)

15. Valls, A., Moreno, A., Pascual-Fontanilles, J.: Preference modeling tools for the decision support method ELECTRE-III-H. In: Proceedings of the 22nd CCIA, vol. 319, p. 295. IOS Press (2019)

16. Webb, E.C., Simonson, I.: Using reviews to determine preferences: how variance in customer-generated reviews affects choice. ACR North American Advances (2017)

17. Zha, Z.J., Yu, J., Tang, J., Wang, M., Chua, T.S.: Product aspect ranking and its applications. IEEE Trans. KDE **26**(5), 1211–1224 (2014)

18. Zhang, K., Narayanan, R., Choudhary, A.: Mining online customer reviews for ranking products (2009)

19. Zhang, M.-L., Li, Y.-K., Liu, X.-Y., Geng, X.: Binary relevance for multi-label learning: an overview. Front. Comput. Sci. **12**(2), 191–202 (2018). https://doi.org/10.1007/s11704-017-7031-7

20. Çali, S., Balaman, Ş.Y.: Improved decisions for marketing, supply and purchasing: mining big data through an integration of sentiment analysis and intuitionistic fuzzy multi criteria assessment. Comput. Industr. Eng. **129**, 315–332 (2019)

Efficient Detection of Byzantine Attacks in Federated Learning Using Last Layer Biases

Najeeb Jebreel, Alberto Blanco-Justicia$^{(\boxtimes)}$, David Sánchez, and Josep Domingo-Ferrer

Department of Computer Engineering and Mathematics,
CYBERCAT-Center for Cybersecurity Research of Catalonia,
UNESCO Chair in Data Privacy, Universitat Rovira i Virgili,
Av. Països Catalans 26, 43007 Tarragona, Catalonia
{najeebmoharramsalim.jebreel,alberto.blanco,
david.sanchez,josep.domingo}@urv.cat

Abstract. Federated learning (FL) is an alternative to centralized machine learning (ML) that builds a model across multiple decentralized edge devices (a.k.a. workers) that own the training data. This has two advantages: i) the data used for training are not uploaded to the server and ii) the server can distribute the training load across the workers instead of using its own resources. However, due to the distributed nature of FL, the server has no control over the behaviors of the workers. Malicious workers can, therefore, orchestrate different kinds of attacks against FL. Byzantine attacks are amongst the most common and straightforward attacks on FL. They try to prevent FL models from converging by uploading random updates. Several techniques have been proposed to detect such kind of attacks, but they usually entail a high cost for the server. This hampers one of the main benefits of FL, which is load reduction. In this work, we propose a highly efficient approach to detect workers that try to perform Byzantine FL attacks. In particular, we analyze the last layer biases of deep learning (DL) models on the server side to detect malicious workers. We evaluate our approach with two deep learning models on the MNIST and CIFAR-10 data sets. Experimental results show that our approach significantly outperforms current methods in runtime while providing similar attack detection accuracy.

Keywords: Federated learning · Security · Byzantine attacks

1 Introduction

Machine learning (ML) is extensively employed in areas such as computer vision, voice processing or natural language processing [14]. However, ML algorithms are highly *data-hungry*; that is, they require a large number of observations to reach high accuracy levels [13], and thus, substantial training costs. Intelligent

© Springer Nature Switzerland AG 2020
V. Torra et al. (Eds.): MDAI 2020, LNAI 12256, pp. 154–165, 2020.
https://doi.org/10.1007/978-3-030-57524-3_13

devices, such as smartphones, produce rich raw data in terms of quantity and quality and have ample unused storage and computing capabilities. Federated learning (FL) is an alternative to centralized ML that leverages the local data and the free computing resources of edge devices to train models in a distributed way [11]. Specifically, in FL a global shared model is learned cooperatively by edge devices (a.k.a. workers) without sharing their local private data. In this way, FL significantly decreases the training costs at the server side while providing privacy by design, because the clients' data do not need to be uploaded to the server.

An example application of FL is the generation of predictive text for smartphone keyboard apps. In this scenario, a service provider (the server) trains a predictive text model from a publicly available text corpus and distributes it among its users (the workers), along with the keyboard app. The keyboard app is capable of adapting (in other words, updating) its locally stored model with the user's inputs, so that it is able to predict the text that best resembles the way each user writes. From time to time, the service provider will randomly select a subset of its user base and collect their updated models. Then, the service provider will aggregate these updated models using some aggregation algorithm to obtain a new predictive text model which better matches the way its user base writes (the way groups of people write changes over time). The federated averaging (*FedAvg*) algorithm is widely used, which takes either the average of the difference between the global model and the locally updated models or the average of the local models themselves [7,11,16]. This training process is repeated over time, potentially indefinitely.

Although FL systems provide privacy and autonomy by design for workers, that same autonomy may be exploited by malicious adversaries. Since the server has no control over the behavior of the workers, any of them may deviate from the prescribed training protocol to conduct attacks such as Byzantine attacks [2]. In Byzantine attacks, attackers try to prevent the model from converging by sending maliciously crafted updates. In [3,5] the authors show that a single malicious worker can significantly degrade the accuracy of the model and even prevent the model from converging.

Several methods have been proposed to counter Byzantine attacks targeting FL. In [15] the authors propose to use the coordinate-wise median of the workers' updates to filter out malicious updates. Specifically, let w_t be the global model parameters at time t and $w_t^{(k)}$ be the local model parameters from worker k at time t. Additionally, let $w[i]$ refer to the value of w at coordinate i (a specific weight or bias). The server sets each global parameter as $w_t[i] = median(\{w_t^{(k)}[i]\}_{k=1}^m)$, where m is the number of participating workers at time t. Alternative methods, named *Krum* and *Multi-Krum*, were proposed in [5]. The Krum method obtains a new global model by selecting the local model that is most similar to the other local models received at a certain training round. For each $w^{(k)}$ (and assuming m' malicious workers in the federated learning system), Krum computes the sum of the distances between $w^{(k)}$ and its closest $m - m'$ local models. Then, Krum selects the local model with the smallest sum of distances as the new global model. In Multi-Krum, the $(m - m')$ local models with the smallest sum of distances are aggregated to yield the new global model. An improvement on Krum, named Bulyan, was proposed in [12] and proved to guarantee stronger Byzantine resilience than Krum.

All these defense methods are robust against Byzantine attacks up to a certain number of malicious workers. However, they incur high computational costs at the server side for large ML models due to the number of parameters to be analyzed (for example, the number of parameters in the ResNet deep learning model is about 25.6 million [8]). On top of that, the number of workers involved in federated learning systems may also reach tens of millions. Since these methods analyze *all* the model parameters for each worker update, the cost at the server side grows unaffordably with the size of the models and the number of involved workers. This goes against one of the fundamental motivations of FL, which is to alleviate the computational cost at the server side. Several recent works have proposed to reduce the computation cost at the server-side. For example, DRACO [6] uses coding theory to ensure robust aggregation of local gradients, and signSGD [4] with majority vote uses gradient compression to reduce communication overhead. However, DRACO requires heavy redundant computation from the workers while signSGD may diverge in practice because its convergence is based on unrealistic assumptions.

Contributions and Plan of This Paper

In this paper, we introduce a novel approach for detecting Byzantine attacks in FL with minimum computational cost at the server side. Our approach aims to detect and eliminate malicious workers' updates at every training round by using only the last layer biases of deep learning models. The intuitive idea is that, instead of analyzing the whole updates, it is possible to just analyze a small fraction of the updates that holds enough information to differentiate honest from malicious workers. In deep learning models (e.g. image classification models), the early layers extract features, while the final decision depends on the latter layers [8]. We exploit this to develop a robust aggregation method that achieves accurate detection of malicious workers with low computational cost. Specifically, for each worker, whatever the worker's update size is, we only compute the distance between the worker's biases and the geometric median of all workers' last layer biases. We use these distances to detect and filter out a malicious worker's update. As far as we know, this is the first study that applies a detection approach using the last layer biases in deep learning models.

We tested our method with two well-known data sets (MNIST and CIFAR-10) and by using two different architectures of deep learning models. We also compared our approach with three state-of-the-art methods: Median, Krum and Multi-Krum. Our results show that our method achieves similar attack detection performance as the best baseline (Multi-Krum) while significantly reducing the runtime required to verify and aggregate updates at the server side at each training round.

The remainder of the paper is organized as follows. Section 2 gives background on federated learning and deep learning. Section 3 details our approach. Section 4 describes and reports the results of our experiments. Finally, Section 5 gathers conclusions and proposes several lines of future research.

2 Background

2.1 Federated Averaging

In FL, the federated averaging algorithm (*FedAvg*) [11] iteratively computes a new shared global model, which is the result of averaging local models that have been trained on the workers' remote devices. The training process is coordinated by a central server that updates and distributes the shared global model w_t. More specifically, the server starts by randomly initializing the global model w_t. Then, at each training round of *FedAvg*, the server selects a subset of workers S of size $C \cdot K \geq 1$ where K is the total number of workers in the system and C is the fraction of workers that are selected in the training round. After that, the server distributes the current global model w_t to all workers in S. Besides the global model, the server sends a set of hyper-parameters to be used at the workers' side to train their model locally. The server sends the number of local epochs E, the local batch size BS and a learning rate η. After receiving the new shared model w_t, each worker divides its local data into batches of size BS and performs E local training epochs of stochastic gradient descent (SGD). Finally, workers upload their updated local models $w_{t+1}^{(k)}$ to the server, which then computes the new global model w_{t+1} by averaging all received local models. The aggregation framework in *FedAvg* is given by:

$$w_{t+1} = \sum_{k=1}^{K} \frac{n^{(k)}}{n} w_{t+1}^{(k)}. \tag{1}$$

Here, n denotes the total number of data points at K workers such that $\sum_{k=1}^{K} n^{(k)} = n$, and $n^{(k)}$ refers to the number of data points at the worker k.

2.2 Bias in Neural Networks

The biological neuron is known to fire only when its processed input exceeds a certain threshold value. The same mechanism is present in artificial neural networks [9], where the bias (a.k.a. threshold) is used to control the triggering of the activation function, i.e, it is used to tune the output along with the weighted sum of the inputs to the neuron as shown in Fig. 1. Thus, the bias helps the model in a way that it can best fit the given data.

Accordingly, biases for models sharing the same architecture and trained by several workers are assumed to be similar. This assumption is based on the similarity of the distribution of the local private data of each worker, and on the identical structure of the model and the hyper-parameters used during the

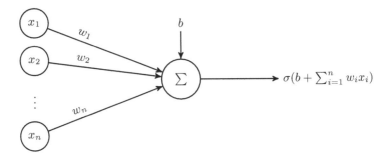

Fig. 1. Example of a neuron, showing inputs $x_1, ..., x_n$, weights $w_1, ..., w_n$, bias b and an activation function σ applied to the weighted sum to get the output.

training to optimize the model parameters. In our work, we take advantage of the similarity of the last layer biases of the honest workers' neural networks, which we use to differentiate malicious workers from honest ones. This is illustrated in Fig. 2, which shows how the biases for honest workers are close together, while the biases of malicious workers providing random updates follow a random and uncorrelated pattern.

We assume that, if there is a set $\{w^{(1)}, w^{(2)}, \ldots, w^{(K)}\}$ of workers' model updates and each $w^{(k)}$ has $L^{(k)}$ layers, there is a last layer $l_{-1}^{(k)} \in L^{(k)}$ that contains n neurons and a bias vector $b^{(k)} \in \mathbb{R}^n$. Then, let B be the set that contains all the last layer vectors $b^{(1)}, b^{(2)}, \ldots, b^{(K)}$ of the K workers' updates.

2.3   Geometric Median

For a given set of m points $x_1, x_2, ..., x_m$ with $x_i \in \mathbb{R}^n$, the geometric median is defined as:

$$\underset{y \in \mathbb{R}^n}{\operatorname{argmin}} \sum_{i=1}^{m} \|x_i - y\|_2 , \tag{2}$$

where the value of the argument y is the point that minimizes the sum of Euclidean distances from the x_i's to y. The geometric median does not change if up to half of the sample data are arbitrarily corrupted [10]. Therefore, it provides a robust way for identifying the biases of honest workers, because the geometric median will be closer to the majority who are honest.

3   Efficient Detection of Byzantine Attacks

We consider a typical FL scenario where there is one server who uses the *FedAvg* algorithm [11] to coordinate the training process of K workers. The workers cooperatively train a shared global model w by using their local data. A fraction of such workers may be malicious attackers. Honest workers compute their updates correctly and send the result to the server. However, malicious workers

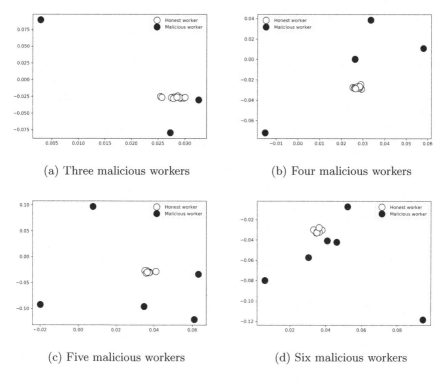

(a) Three malicious workers   (b) Four malicious workers

(c) Five malicious workers   (d) Six malicious workers

Fig. 2. Graphical representation of last layer biases

may deviate from the prescribed protocols and send random updates aiming at degrading the global model.

The key idea of our approach is that, at each global training round, the server assigns a score to each participating worker and uses that score to consider or not the worker's update at the aggregation phase. Unlike state-of-art methods, we do not analyze the entire workers' updates. Instead, we focus only on the last layer biases of workers' updates to filter out malicious updates.

Algorithm 1 formalizes the methodology we propose. The algorithm has a set of hyper-parameters K, C, BS, E, η and τ. K is the number of participating workers, C is the fraction of workers to be selected at each round, BS is the local batch size, E the number of local epochs, η is the learning rate and τ is the confidence value used to detect malicious workers.

At the first global training round, the server starts a federated learning task by initializing the global model w_0. Then it selects a random set S of m workers, where $m = max(C \cdot K, 1)$, transfers w_t to each worker in S, and asks them to train the model (locally) using the defined hyper-parameters. Each worker trains the model using her private local data and transfers the updates $w_{t+1}^{(k)}$ to the server, who aggregates updates to create a new global model for the next training round w_{t+1}.

The server then decides whether a worker is honest or malicious by using a scoring function. The scoring function **GetScores** receives the last layer biases B of all the updates of the workers, and computes the geometric median $GeoMed$ of the layer biases $b^{(k)}$ for each $k \in S$. After that, it finds the set of distances $Dist$ between the $GeoMed$ and each $b^{(k)}$. Then, the server sorts $Dist$ and computes $Q1, Q3, IQR = Q3 - Q1$. Next, by using the hyper-parameter τ, the scoring function assigns 1 to a worker k if $dist_k \leq Q3 + \tau \times IQR$; else 0.

Once the server has obtained the set $Scores$, it performs the aggregation phase of updates $w_{t+1}^{(k)}$ for each $k \in S$ such that $score^{(k)} = 1$. We propose replacing Equation (1) in $FedAvg$ by $w_{t+1} = \sum_{k \in S, score^{(k)}=1} \frac{n^{(k)}}{sn} w_{t+1}^{(k)}$, where $sn = \sum_{k \in S, score^{(k)}=1} n^{(k)}$. Note that, since $\sum_{k \in S, score^{(k)}=1} \frac{n^{(k)}}{sn} = 1$, the convergence of the proposed averaging procedure at the server side is guaranteed as long as $FedAvg$ converges.

4   Experimental Results

Experiments have been carried out on two public data sets: MNIST and CIFAR-10. MNIST consists of 70,000 images of handwritten digits (which correspond to classes), from 0 to 9. Each image is gray-scaled and 28×28 pixels in size. We used 60,000 of them for training and the remaining 10,000 for testing. The CIFAR-10 dataset contains of 60,000 32×32 colored images of 10 classes. Each class has 6,000 images. We used 50,000 images for training and the remaining 10,000 for testing. For both datasets, we divided the training data set into 100 equally sized shards, which were distributed among 100 simulated workers.

For MNIST, we used a small DL model with 21,840 trainable parameters. The earlier layers were convolutional layers that extracted the features and passed them to the last two fully connected layers: fully connected layer 1 (FC1), and fully connected layer 2 (FC2). FC1 contained 50 neurons with 1 bias for each neuron, while FC2 contained 10 neurons with 1 bias for each neuron. The output of FC2 was fed to a SoftMax layer that introduced the final probabilities of the 10 classes. For CIFAR-10, we used a large DL model that contained 14,728,266 trainable parameters in total. The earlier layers extracted the features and they had only one fully connected layer (FC1) that contained 10 neurons with 1 bias for each neuron. The output of the fully connected layer was fed to a SoftMax layer that introduced the final probabilities of the 10 classes.

We trained the small DL model for 30 global rounds. Each worker trained the model locally using the stochastic gradient descent (SGD) with 5 local epochs, local batch size = 20, learning rate = 0.001, and momentum = 0.9. We trained the large DL model for 60 global rounds. Each worker trained the model locally using the stochastic gradient descent (SGD) with 5 local epochs, local batch size = 50, learning rate = 0.01 and momentum = 0.9.

For both models, we used the precision as a performance metric. Precision is the amount of true positives TP divided by the sum of true positives and false positives FP, that is $TP/(TP + FP)$.

Algorithm 1: Byzantine-attack robust aggregation of updates using last layer biases scores.

Input: K, C, BS, E, η, τ

1   initialize w_0
2   **for** *each round $t = 0, 1, \ldots$* **do**
3     $m \leftarrow \max([C \cdot K], 1)$
4     $S =$ random set of m workers.
5   **for** *each worker $k \in S$* **in parallel** **do**
6       $w_{t+1}^{(k)} = \texttt{WorkerUpdate}(k, w_t)$
7   **end**
8     $Scores = \texttt{GetScores}(S)$
9     $w_{t+1} = \sum_{k \in S, score^{(k)} = 1} \frac{n^{(k)}}{sn} w_{t+1}^{(k)}$ where $sn = \sum_{k \in S, score^{(k)} = 1} n^{(k)}$
10   **end**
11   **Function** *WorkerUpdate(k, w_t)*
12     $w \leftarrow w_t$
13   **for** *each local epoch $i = 1, 2, \ldots E$* **do**
14   **for** *each batch β of size BS* **do**
15         $w \leftarrow w - \eta \nabla L(w, \beta)$
16   **end**
17   **end**
18   **return** w
19   **end**
20   **Function** *GetScores(S)*
21   Let B be the set of biases of the last layers for each worker $k \in S$
22     $GeoMed = GeometricMedian(B)$
23     $Dist =$ the distances between the $GeoMed$ and each $b^{(k)} \in B$
24   Compute $Q1, Q3$ of $Dist$
25     $IQR = Q3 - Q1$
26     $Scores = [\,]$
27   **for** *each $k \in S$* **do**
28   **if** $dist_k \leq Q3 + \tau \times IQR$ **then**
29   Add(1, Scores)
30   **else**
31   Add(0, Scores)
32   **end**
33   **end**
34   **return** Scores
35   **end**

In the experiments the server selected a fraction $C = 15\%$ of workers at each round. We limited the number of malicious updates per round between 1 to 6. This is consistent with the Byzantine resilience property, which states that a method is considered to be Byzantine resilient if it is robust against m' Byzantine workers where $(2m' + 2) < m$ [1].

Figure 3 shows the evolution of the classification precision achieved by the small DL model at each round when no detection is performed. The dashed line shows the precision when all the workers acted honestly. The global model achieves a precision around 96% after 30 rounds. In addition to the all-honest case, we considered different scenarios where a fraction of the workers were malicious. In our experiments, malicious workers trained their model honestly and, after that, they added some random noise to each parameter in their model from a Gaussian distribution with mean 0 and standard deviation 0.5 for the small DL model, and with mean of 0 and standard deviation 0.2 for the large DL model. It is noticeable that the precision of the model decreases as the number of malicious updates per round increases.

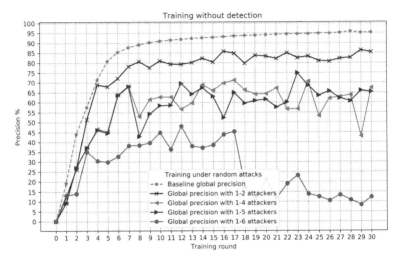

Fig. 3. Evolution of the small DL model precision with MNIST under random attacks when no detection is performed

We next considered training with detection of malicious workers. We chose τ to be -0.5 in this case, that is, we consider each worker has a distance greater than the median to be malicious. We compared the results of our method with three state-of-art methods: Median, Krum, and Multi-Krum. Figure 4 shows the precision of the small DL model when subjected to up to 6 random updates per round. We can see that both Median and Krum have similar performance, and that both Multi-Krum and our method outperform them and obtain virtually identical results. The reason behind this identical performance is that our method takes the most honest set of updates, like Multi-Krum.

In Fig. 5 we show the performance of our method with the large DL model and CIFAR-10 w.r.t. the values of τ. The dashed line shows the baseline precision when all workers acted honestly, which is around 84% after 60 rounds. With $\tau = 1$ our method filtered out most of the malicious workers when they were far enough from the majority. However, we can see that between training rounds 22 and 31 some malicious workers were able to escape detection, and that their

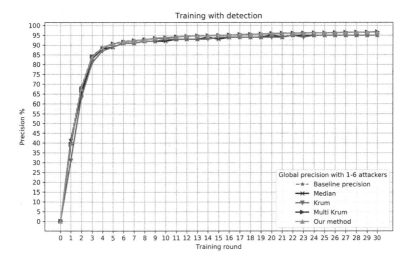

Fig. 4. Evolution of the small DL model precision with MNIST under random attacks with detection

effect in the model precision was devastating. With $\tau = 0.5$ most malicious workers were eliminated and with $\tau = -0.5$, all of them were eliminated.

Even more relevant, Fig. 6 shows that, in addition to our method offering detection as good as state-of-the-art methods, it is also significantly more *computationally efficient* than its counterparts. The reason is that for each worker our method only focuses on the last layer parameters.

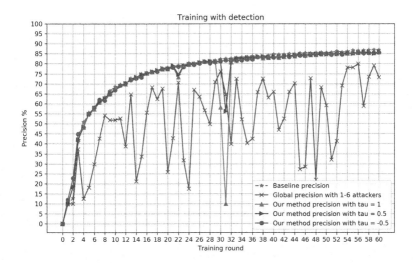

Fig. 5. Evolution of the large DL model precision with CIFAR-10 under random attacks with detection

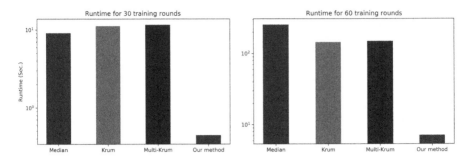

Fig. 6. Runtime (in seconds) for each method: small DL model on the left and large DL model on the right.

5   Conclusions and Future Work

We have shown that by focusing on a specific aspect of the learning model (biases) we can efficiently detect Byzantine attacks in FL. As shown in the experiments, our method achieves state-of-the-art accuracy while significantly decreasing the runtime of the attack detection at the server side. The scalability of our method is particularly useful when using large DL models with millions of parameters and workers. Although our results are promising, choosing the value of τ is challenging. The increase or the decrease of the value of τ must depend on the proportion of the malicious workers in the system. The higher the expected number of malicious workers the smaller the value of τ.

As future work, we plan to study the possibility of using our method to detect other kinds of attacks to FL, such as targeted and stealthy attacks.

Acknowledgements. Partial support is acknowledged from the European Commission (project H2020-871042 "SoBigData++"), the Government of Catalonia (ICREA Acadèmia Prize to J. Domingo-Ferrer and grant 2017 SGR 705), and the Spanish Government (projects RTI2018-095094-B-C21 "Consent" and TIN2016-80250-R "Sec-MCloud"). The authors are with the UNESCO Chair in Data Privacy, but the views in this paper are their own and are not necessarily shared by UNESCO.

References

1. Alistarh, D., Allen-Zhu, Z., Li, J.: Byzantine stochastic gradient descent. In: Advances in Neural Information Processing Systems, pp. 4613–4623 (2018)
2. Bagdasaryan, E., Veit, A., Hua, Y., Estrin, D., Shmatikov, V.: How to backdoor federated learning. arXiv preprint arXiv:1807.00459 (2018)
3. Bhagoji, A.N., Chakraborty, S., Mittal, P., Calo, S.: Analyzing federated learning through an adversarial lens. arXiv preprint arXiv:1811.12470 (2018)
4. Bernstein, J., Zhao, J., Azizzadenesheli, K., Anandkumar, A.: signSGD with majority vote is communication efficient and fault tolerant. arXiv preprint arXiv:1810.05291 (2018)

5. Blanchard, P., Guerraoui, R., Stainer, J., et al.: Machine learning with adversaries: byzantine tolerant gradient descent. In: Advances in Neural Information Processing Systems, pp. 119–129 (2017)
6. Chen, L., Wang, H., Papailiopoulos, D.: Draco: robust distributed training against adversaries. In: 2nd SysML Conference (2018)
7. Chen, X., Chen, T., Sun, H., Wu, Z. S., Hong, M.: Distributed training with heterogeneous data: bridging median and mean based algorithms. arXiv preprint arXiv:1906.01736 (2019)
8. Khan, A., Sohail, A., Zahoora, U., Qureshi, A.S.: A survey of the recent architectures of deep convolutional neural networks. arXiv preprint arXiv:1901.06032 (2019)
9. Leshno, M., Lin, V.Y., Pinkus, A., Schocken, S.: Multilayer feedforward networks with a nonpolynomial activation function can approximate any function. Neural Networks $\mathbf{6}(6)$, 861–867 (1993)
10. Lopuhaa, H.P.: On the relation between S-estimators and M-estimators of multivariate location and covariance. Ann. Stat. $\mathbf{17}(4)$, 1662–1683 (1989)
11. McMahan, H.B., Moore, E., Ramage, D., Hampson, S., Arcas, B.A.Y.: Communication-efficient learning of deep networks from decentralized data. arXiv preprint arXiv:1602.05629 (2016)
12. Mhamdi, E.M.E., Guerraoui, R., Rouault, S.: The hidden vulnerability of distributed learning in byzantium. arXiv preprint arXiv:1802.07927 (2018)
13. Obermeyer, Z., Emanuel, E.J.: Predicting the future-big data, machine learning, and clinical medicine. New Engl. J. Med. $\mathbf{375}(13)$, 1216 (2016)
14. Wang, X., Han, Y., Wang, C., Zhao, Q., Chen, X., Chen, M.: In-edge AI: intelligentizing mobile edge computing, caching and communication by federated learning. IEEE Network $\mathbf{33}(5)$, 156–165 (2019)
15. Yin, D., Chen, Y., Ramchandran, K., Bartlett, P.: Byzantine-robust distributed learning: towards optimal statistical rates. arXiv preprint arXiv:1803.01498 (2018)
16. Yu, H., Yang, S., Zhu, S.: Parallel restarted SGD with faster convergence and less communication: demystifying why model averaging works for deep learning. In: Proceedings of the AAAI Conference on Artificial Intelligence, vol. 33, pp. 5693–5700 (2019)

Multi-object Tracking Combines Motion and Visual Information

Fan Wang[1], En Zhu[1], Lei Luo[1(✉)], and Jun Long[2]

[1] School of Computer, National University of Defense Technology,
Changsha, China
{wangfan10,enzhu,l.luo}@nudt.edu.cn
[2] Jide Company, Guangzhou, China

Abstract. Real-time online multi-object tracking is a fundamental task in video analysis applications. A major challenge in the tracking by detection paradigm is how to deal with missing detections. Visual single object trackers (SOTs) have been introduced to make up for the poor detectors in the assumption of appearance continuity of tracking objects. However, visual SOTs may easily be confused by the invaded foreground when occlusion occurs. In this paper, we propose to combine object motion information and appearance feature to improve the performance of object tracker. We use a lightweight re-identification feature to monitor occlusion. A Kalman filter, as the motion predictor, and a visual SOT, as the appearance model are worked together to estimate the new position of the occluded object. Experimental evaluation on MOT17 dataset shows that our online tracker reduces the number of ID switches by 26.5% and improves MOTA by 1–2% compared to the base intersection-over-union (IOU) tracker. The effectiveness of our method is also verified on MOT16 datasets. At the same time, the tracking speed can reach 29.4 fps which can basically achieve real-time tracking requirement while ensuring accuracy.

Keywords: Multi-object tracking · Single object tracking · Kalman filter · Re-identification feature

1 Introduction

Object tracking is a fundamental task in computer vision with many applications such as video surveillance, automatic driving and robot navigation. The research on object tracking can be mainly divided into two categories: single object tracking (SOT) and multi-object tracking (MOT). MOT can be viewed as multiple parallel SOT tasks in many simple scenarios with few objects. But in crowded scenarios, MOT not only includes the difficulties of SOT, but also faces the challenges of frequent occlusion and interference from the surroundings.

Benefiting from advances in object detection in the past decade, *tracking-by-detection* is becoming the preferred paradigm of MOT. The typical process of this method is detecting objects on each frame and then applying the data

© Springer Nature Switzerland AG 2020
V. Torra et al. (Eds.): MDAI 2020, LNAI 12256, pp. 166–178, 2020.
https://doi.org/10.1007/978-3-030-57524-3_14

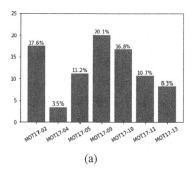
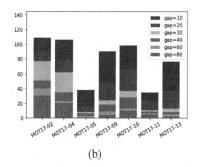

(a)   (b)

Fig. 1. (a) We view occlusion as the period when the object disappears and reappears, and calculate the ratio of object occlusion for each sequence on MOT17 training set. The higher the ratio, the more difficult it is to track. (b) We define "gap" as the interval between frames when the object disappears and reappears, and use it to measure the degree of occlusion. The figure shows the number of occlusion occur on MOT17 training sequences under different gap value.

association from trajectories. Detection missing caused by occlusion or false negative detections will lead to fragmentary trajectories. The statistical results on MOT17 dataset in Fig. 1(a) show that the occlusion problem is widespread and inevitable in multi-object tracking task. Multiple cues, including motion model, object appearance feature are proposed to mitigate this problem. [7,30] employ Kalman filter as motion model to predict possible positions of objects in future frames. [10,32] employ visual SOT method, such as KCF tracker [15], to locate the object position based on object's appearance feature. We found that the V-IOU tracker [10] is simple in these methods, but has advantages in speed and accuracy. We tried to further improve the processing of object occlusion in the V-IOU tracker and implement an online multi-object tracker.

In this work, we approach the occlusion problem by employing Kalman motion model into the V-IOU tracker to predict the position of object when occlusion occurs. The V-IOU tracker continues each track by a visual SOT if no new detection can be associated. The employed visual SOT fills the "gaps" between fragmented trajectories (named tracklets) which can reduce ID switches. However, object occlusion lowers visual SOT's confidence which will lead to confusion while stitching tracklets. The basic reason is that visual SOT is essentially a template matching process, which depends on the appearance feature of an object. When occlusion occurs, the appearance feature changes drastically, and the benefit of visual SOT deteriorates. We try to use a Kalman motion model to complement the visual SOT to predict the new position of the object when detection missing. The latter one performs as the default model in prediction, while it is substituted by Kalman motion model if occlusion occurs. We use the cosine distance of an object's re-identification (ReID) feature between consecutive two frames to detect occlusion. If the distance is greater than a certain threshold indicating the occurrence of occlusion, the Kalman filter replaces the visual SOT to estimate the new position of the object. The experimental results on MOT17 dataset show our online tracker outperforms some state-of-the-art MOT methods and maintains a real-time speed. The performance on MOT16 sequences filmed by static camera also presents some advantages.

The paper is further organized in this manner. We first introduce related work in recent years in Sect. 2. Section 3 describes our proposed multi-object tracking algorithm in detail. Section 4 shows the effectiveness of our method on MOT16 and MOT17 datasets and analyzes the experimental results. Finally, Sect.5 summarizes the main contributions of this paper and presents some final remarks.

2   Related Work

2.1   Single Object Tracking

Single object tracking (SOT) is to track an object in a video which is given in the first frame. In the past few years, the correlation filter and deep learning methods have been well developed in SOT. Although the deep learning methods have achieved remarkable results in real-time SOT with the emergence of siamese neural network [6], the correlation filter still has its unique advantages. The original correlation filter named MOSSE is proposed in [11], which is a high-speed SOT based on a linear classifier. As real-time trackers, CSK [14] and KCF [15] introduce a circular matrix to achieve dense sampling of training samples, which greatly improves the accuracy of tracking. HCF [23] combines features extracted from different layers in the convolutional neural network to better describe the appearance information of an object. [8,29] make further improvements and enhancements to maximize the tracking performance of correlation filters.

2.2   Multiple Object Tracking

Multiple object tracking (MOT) is to integrate object trajectory based on detections. It mainly focuses on data association problem. With this paradigm, [25,26] find object trajectories in a global optimization problem that processes entire video batches at once. However, these methods are not applicable in online scenarios. To achieve real-time requirements, [9] proposed an IOU tracker without using image information which is simple enough with a extremely high speed. [10] poses a V-IOU tracker based on IOU tracker by introduce a visual SOT into the tracking framework to reduce ID switch problem. Although the V-IOU tracker has no complicated structure, it could achieve state-of-art performance and is ranked high in MOT Challenge 2019[1].

---

[1] https://motchallenge.net.

2.3   Appearance Models

In the person re-identification task, it is important to learn distinguish appearance features to re-identify objects. Therefore, object appearance feature can also be used to improve multi-object tracking performance. [16] uses the color-based feature for pedestrian tracking. However, this may not always reliable because pedestrians often wear the same color clothes, background and illumination changes also influence a lot. [19,27] borrow ideas from person re-identification to apply ReID feature to multi-object tracking and achieve obvious advantage.

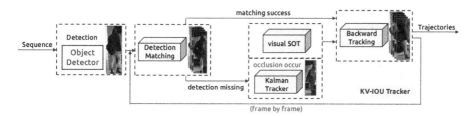

Fig. 2. The framework of KV-IOU tracker. It is mainly composed of three phases: *detection matching, tracker extension* and *backward tracking*. 1) *detection matching* phase is the main process of IOU tracker; 2) *tracker extension* phase includes a visual SOT and a Kalman tracker, which hands matching failures caused by detection missing; 3) *backward tracking* phase focuses on tracklets integration.

2.4   Motion Models

Motion model is also an important part of MOT task. The most common assumption is the objects with constant velocity [2,12]. However, these may fail in crowded scenarios since pedestrians' motions get more complex. Although the linear motion model is simple, it has great advantage in short-term prediction. Therefore, [7,30] use Kalman motion model to predict new position of an object. [21,31] turn to learn more expensive Social Force Model to describe the motion information of an object. Deep learning methods [1,28] are also extended to learn motion trend for pedestrian behavior understanding and prediction.

3   The Proposed KV-IOU Tracker

The Kalman filter based visual intersection-over-union (KV-IOU) tracker focuses on trajectory continuity when object-object occlusion occurs. The V-IOU tracker is designed to reduce the high amount of ID switches when detection missing. However, visual SOT in V-IOU can easily be confused in crowded scenarios. The proposed KV-IOU tracker employs Kalman motion model and deep appearance feature to overcome the occlusion problem. The framework of KV-IOU tracker is shown in Fig. 2, which is composed of three phases: detection matching, tracker extension and backward tracking. Detection matching phase is the main process of IOU tracker. The detections are connected by IOU to form trajectories, which is discussed in Sect. 3.1. Tracker extension phase is the core part of our tracker. It employes a visual SOT and a Kalman tracker to deal with detection missing, which is discussed in Sect. 3.2. Backward tracking phase is used to stitching tracklets based on probability calculated by Kalman filter, which is discussed in Sect. 3.3.

3.1 Detection Matching

Based on the assumption that objects in the video do not move at a very high speed, detection matching phase aims to connect detections to form trajectories based on IOU since the same object in consecutive frames is high overlapped. For each frame t, we continue a trajectory by associating the detection with the highest IOU to the last detection in the previous frame $t-1$ if a certain threshold σ_{IOU} is met. The tracker performance can be guaranteed when the frame rate is high enough and the accuracy of the detection participating in the matching is high. Thus, in order to improve the accuracy of detection matching phase, we filter out the detections whose scores low than threshold σ_l. However, in crowd and other complicated scenarios, object detections are inevitably lost, which will cause detection matching phase failure. The missed detections just like "gap" between trajectories which will lead to amount ID switches.

3.2 Tracker Extension

Detection missing caused by occlusion or false negative detections is inevitable and will lead to the failure of detection matching phase mentioned above. We introduce a Kalman motion model to complement the visual SOT to deal with this situation, especially for the case of object occlusion. We initialize a visual SOT and a Kalman tracker simultaneously when detection matching fails. The visual SOT performs as the default model in prediction to deal with the detection missing. The prediction result will replace the missing detection and matches the current trajectory. We use a lightweight re-identification feature to monitor occlusion. If the prediction result of visual SOT and the current trajectory are significantly different in ReID features, occlusion occurs at this time. The Kalman motion model will replace visual SOT as position predictor. With the short-term tracking capability, the prediction result of the Kalman filter has a higher confidence than visual SOT when occlusion occurs. Kalman tracker relies on the update of object motion information, and visual SOT relies on the changes of the object appearance model. If these two extended trackers cannot be updated through detections for a long time, the accuracy of the position prediction will become lower and lower. We define the age of a trajectory as the number of times it uses tracker extension. So, we stop these two trackers if the age of a trajectory larger than m_{age}, or they will be reinitialized if detection matching successed.

3.3   Backward Tracking

We define an active trajectory as a tracking sequence longer than t_{min} with at least one detection score larger than σ_h, and define a finished trajectory as a tracking sequence that is no longer updated. Finished trajectories does not mean that the object has permanently left the picture. It may reappear in the future. Backward tracking phase tries to connect new objects with previous finished trajectories. The V-IOU tracker performances visual SOT backward on the new emerging objects. If the overlap criteria is met for an existing finished trajectory, V-IOU tracker will merge them. Different from V-IOU tracker, we use the objects' position probability based on Kalman filter to narrow the matching scope instead of directly doing backward tracking with visual SOT. The main process is shown in Fig. 3. Finished trajectories $\mathcal{T}_f$ only try to connect to new detections in the area predicted by Kalman filter (read circle). The cosine distance between finished trajectories and new detections in ReID features low than threshold λ_{dist} means matching successful. The core of Kalman filter is described by five equations, which are mainly divided into two steps, i.e., time update and status update. Time update is to predict the new object status $\hat{x}_t$ and covariance P_t at frame t based on the previous frame $t-1$. We define object state as $(u, v, a, h, \dot{u}, \dot{v}, \dot{a}, \dot{h})$ which contains the bounding box center position (u, v), aspect ratio a, height h, and their respective velocities in image coordinates. The change of object status is described as:

$$\hat{x}_t = A\hat{x}_{t-1} , \quad P_t = AP_{t-1}A^T + Q, \tag{1}$$

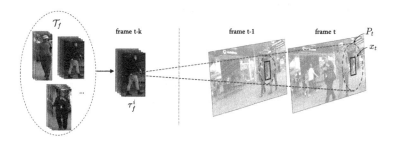

Fig. 3. Backward tracking phase attempts to match some new emerging objects to the temporary finished trajectories. Different from the method used in V-IOU tracker, we use Kalman filter to narrow the scope of backward tracking. Assuming that we have got some finished trajectories $\mathcal{T}_f$ at frame t. Considering one of the finished trajectories τ_f^i which finishes at frame $t-k$, we try to connect the new emerging objects at frame t to the finished trajectory τ_f^i. We initialize a Kalman filter at frame $t-k$ and will get the possible position of trajectory τ_f^i at frame t after k times prediction. The red circle represents the covariance P_t of Kalman filter and the blue rectangle represents the position x_t of the object in current frame t. We only need to consider the detections within the scope of the red circle in the phase of backward tracking.

Algorithm 1: KV-IOU tracker

Input: A video sequence v with F frames and its detections
$$\mathcal{D} = \{\mathcal{D}_k \mid \mathcal{D}_k = \{d_0, d_1, ...\}\}$$
Output: Trajectories T_f of the video

1 **Initialization**: $T_a = \emptyset$, $T_f = \emptyset$
2 **for** *frame f_k in v* **do**
3   **for** τ *in* T_a **do**
4   Select d_j from $\mathcal{D}_k$ where $\max(\text{IOU}(d_j, \tau))$
        // detection matching
5   **if** $\max(\text{IOU}(d_j, \tau)) > \sigma_{IOU}$ **then** $\tau \leftarrow \tau \cup d_j$, $\mathcal{D}_k \leftarrow \mathcal{D}_k - d_j$
        // tracker extension
6   **else if** $Age(\tau) < m_{age}$ **then**
7              $(p_v, p_k) \leftarrow \text{Predict}(T_v, T_k)$
8              $(\mathcal{F}_v, \mathcal{F}_\tau) \leftarrow \text{Extract}(T_v, \tau)$
9              $C \leftarrow \text{Filter}(p_v, p_k, \mathcal{F}_v, \mathcal{F}_\tau, \lambda_{dist})$
10             $\tau \leftarrow \tau \cup C$
11   **else if** $max\_score(\tau_i) \geq \sigma_h$ *and* $len(\tau_i) \geq t_{min}$ **then**
12             $T_f \leftarrow T_f \cup \tau$, $T_a \leftarrow T_a - \tau$
13   **end**
14   **end**
    // backward tracking
15     $\mathcal{D}_k \leftarrow \text{InArea}(\mathcal{D}_k)$
16   **for** $d_j \in \mathcal{D}_k$ *and* $\tau_j \in T_f$ **do**
17         $(\mathcal{F}_{det}, \mathcal{F}_\tau) \leftarrow \text{Extract}(d_j, \tau_j)$
18   **if** $distance(\mathcal{F}_{det}, \mathcal{F}_\tau) < \lambda_{dist}$ **then** $\tau_j \leftarrow \tau_j \cup d_j$, $T_a \leftarrow T_a \cup \tau_j$
19         $\mathcal{D}_k \leftarrow \mathcal{D}_k - d_j$, $T_f \leftarrow T_f - \tau_j$
20   **end**
21   Start new trajectories τ with remaining detections and add it to T_a
22 **end**
23 **return** T_f

---

where A is a constant matrix defined as (2), denoting state transition matrix from frame $t - 1$ to t, Q is used to describe the uncertainty outside the system, which is only related to height h.

$$A = \begin{pmatrix} I_{4 \times 4} & I_{4 \times 4} \\ O_{4 \times 4} & I_{4 \times 4} \end{pmatrix}, \quad H = (I_{4 \times 4} \ O_{4 \times 4}), \tag{2}$$

status update is to update parameters $\hat{x}_t$ and P_t based on detections which have been merged into the current trajectories. The update of object status is described as:

$$K_t = P_t H^T (H P_t H^T + R)^{-1}, \quad \hat{x}_t = \hat{x}_t + K_t (y_t - H \hat{x}_t), \quad P_t = (I - K_t H) P_t, \tag{3}$$

where R is the uncertainty of the measure sensor, K_t donates Kalman gain which used to balance predicted and measured values, H is the transfer matrix from state value to measured value, y_t represents the measured value at time t, here is the object detection results at frame t. If detecting matching is successful, Kalman filter will be updated by detections. Therefore, the covariance P_t

will reduce gradually according to formula (3). If detection is lost, Kalman filter cannot be adjusted based on measured values which means covariance P_t will increase according to formula (1). We accordingly use the area determined by Kalman filter to reduce the spatial scope of the tracklets matching. The detections in the matching area are the candidates which will be connected to existing finished trajectories.

A detailed description of the method is shown in Algorithm 1, where $\mathcal{T}_a$ and $\mathcal{T}_f$ donate active and finished trajectories respectively. Line 5 is the process of detection matching. The detection with the largest overlap with one of active trajectories will be connected to the corresponding trajectory. σ_{IOU} donates the threshold of IOU value. After detection matching, detections from frame 1 to frame $t-1$ will form some short tracklets. Line 6 to line 13 are the process of tracker extension designed to deal with detection matching failure, especially for failures caused by occlusion. If the age of current active trajectory $\mathtt{Age}(\tau)$ is less than the max age m_{age}, we start a visual SOT T_v and a Kalman tracker T_k. These two trackers predict positions p_v and p_k respectively. The function $\mathtt{Extract}(d_j, \tau_i)$ is to extract the ReID feature from detection d_j and trajectory τ_i. The function $\mathtt{Filter}(p_v, p_k, \mathcal{F}_v, \mathcal{F}_t, \lambda_{dist})$ is used to judge whether occlusion occurs according to the parameter λ_{dist}, and chooses p_v or p_k to put into candidate C. The position C will finally connect to current trajectory. Line 15 to line 19 are the process of backward tracking which is used for tracklets stitching. Only the detections within the area determined by $\mathtt{InArea}(\mathcal{D}_k)$ will participate in this process. Finally, the remaining detections will be started as new trajectories. It should be noted that the extraction of the object's appearance feature is only performed when occlusion and tracklets stitching occurs. Therefore, our method can maintain high accuracy while reach the real-time requirements.

4   Experiments

We evaluate our KV-IOU tracker on the MOT16 and MOT17 tracking benchmarks. The common metric is multi-object tracking accuracy: $MOTA = 1 - \frac{\sum_t (FP_t + FN_t + IDS_t)}{\sum_t GT_t}$ [20], where GT_t, FP_t, FN_t, and IDS_t are the number of ground-truth bounding boxes, false positives, false negatives, and identity switches in frame t, respectively. In the experiments, we also consider the effect of static and moving cameras on the tracking performance.

Table 1. Best parameters for all detectors for the MOT16 and MOT17 datasets.

Dateset	Detector	σ_l	σ_h	σ_{IOU}	t_{min}	m_{age}	ttl	λ_{dist}
MOT16	DPM	0.3	0.5	0.3	5	15	12	0.2
MOT17	DPM	−0.5	0.5	0.4	4	15	12	0.2
	Faster R-CNN	0.0	0.98	0.6	3	8	6	0.2
	SDP	0.4	0.5	0.2	2	6	5	0.2

4.1 Implementation Details

The parameters in each phase of the tracking framework shown in Table 1 are determined by grid search. Detection matching is the main process of IOU tracker, we choose $\sigma_l, \sigma_h, \sigma_{IOU}$ and t_{min} according to [9] to adapt to different detectors. When detection matching fails, i.e., the detection is lost, the visual SOT tracker and the Kalman tracker are simultaneously launched to predict the possible position of an object. In this phase, it is determined whether the occlusion has occurred by comparing ReID feature of an object between two consecutive frames. We use the same network structure as in [30] to extract the ReID feature of an object and choose features' cosine distance value threshold $\lambda_{dist} = 0.2$ to judge if the appearance has changed a lot. According to the analysis of the occurrence of occlusion in the tracking sequences in Fig. 1(b), it is most common that the object is occluded within 0–20 frames. We use this to set the length of backward tracking ttl ranged from 0 to 20 and it is smaller for better detectors. The life cycle m_{age} of extended trackers (a visual SOT and a Kalman tracker) is set from 0 to 20 according to difference detectors. We choose the CSRT tracker comes with OpenCV as visual SOT.

4.2 Benchmark Evaluations

We evaluate our proposed online tracker quantitatively and compare it with other state-of-the-art trackers in Table 2 on MOT17 benchmark dataset. The great difference among the KV-IOU, IOU, V-IOU trackers is that the proposed KV-IOU tracker is an online tracker while the other two are offline trackers. The online tracker only uses current and previous frames while offline tracker can use future frames. Despite it, our tracker performs competitive compared to the offline trackers from Table 2. Compared to the base IOU tracker, the MOTA improves by an additional 1–2% whereas the number of ID switches is reduced by up to 26.5%, the IDF1, MT, ML and FN metrics have also been improved. Although low-quality detection results combined with a motion camera limit our tracker's performance, experiments show that our tracker is still effective on sequences filmed by stationary cameras. We performed experiments on three stationary camera sequences on the MOT16 data set and the results are shown in Fig. 4. The MOTA, IDF1, FP and IDS metrics have been improved to some extent. The experimental results show that our fusion of motion and visual information is simple and effective, especially on sequences with simple scenes filmed by stationary camera.

Our gains come from two sources. Firstly, in tracker extension phase, the object occlusion was further processed. By effectively combining the Kalman filter and the visual SOT, the short-term and long-term tracking capabilities of the KV-IOU tracker are enhanced which results in a much lower rate of false negatives. And second, backward tracking phase based on Kalman motion prediction reduces the amount of calculation and accelerates the speed of object tracking. Since our tracker does not specifically deal with camera motion, the performance increase on the sequences filmed by moving camera is not significant.

Table 2. Tracking performance of representative trackers developed using both online and offline methods. All trackers are evaluated on the test data set of the **MOT17** benchmark using public detections. The results of * are taken from https://motchallenge.net and results of # are from [3]. Top values are highlighted by bold font for each metric. Metrics with (↑) show that higher is better, and with (↓) denote lower is better. N/A means not available.

Tracker	Mode	MOTA↑	MOTP↑	IDF1↑	MT(%)↑	ML(%)↓	FP↓	FN↓	IDS↓	Frag↓	FPS↑
MHT-DAM* [16]	Offline	**50.7**	**77.5**	47.2	20.8	**36.9**	22875	**252889**	2314	2865	0.9
MHT-bLSTM* [17]	Offline	47.5	77.5	51.9	18.2	41.7	25981	268042	2069	3124	1.9
IOU17* [9]	Offline	45.5	76.9	39.4	15.7	40.5	19993	281643	5988	7404	**1522.9**
SAS-MOT17* [24]	Offline	44.2	76.4	**57.2**	16.1	44.3	29473	283611	**1529**	**2644**	4.8
DP-NMS# [26]	Offline	43.7	76.9	N/A	12.6	46.5	**10048**	302728	4942	5342	N/A
V-IOU [10]	Offline	46.2	76.1	44.0	**21.0**	35.6	44854	254438	4258	6307	32
GM-PHD* [13]	Online	36.2	76.1	33.9	4.2	56.6	23682	328526	8025	11972	38.4
GMPHD-KCF* [18]	Online	40.3	75.4	36.6	8.6	43.1	47056	283923	5734	7576	3.3
GMPHD-N1Tr* [4]	Online	42.1	77.7	33.9	11.9	42.7	18214	297646	10694	10864	9.9
EAMTT* [19]	Online	42.6	76.0	41.8	12.7	42.7	30711	288474	4488	5720	12
SORT17* [7]	Online	43.1	**77.8**	39.8	12.5	42.3	28398	287582	4852	7147	**143.3**
HISP-T* [5]	Online	44.6	77.2	38.8	15.1	38.8	25478	276395	10617	7487	4.7
OTCD-1* [22]	Online	44.9	77.4	42.3	14	44.2	**16280**	291136	**3573**	**5444**	46.5
HISP-DAL# [3]	Online	45.4	77.3	39.9	14.8	39.2	21820	277473	8727	7147	3.2
KV-IOU (ours)	Online	**46.6**	76.3	**44.0**	**17.3**	**38.1**	34838	**262008**	4379	7884	29.4

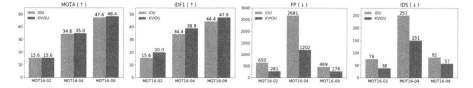

Fig. 4. Performance of IOU and KV-IOU on MOT16 training sequences with static camera. It is shown that the KV-IOU tracker has significant performance improvement on MOTA, IDF1, FP and IDS metrics.

5   Conclusions

By analyzing the video sequences in MOT17 dataset, we find that occlusion problem is inevitable in multi-object tracking. The better the occlusion problem is solved, the better the effect of multi-object tracking. In this paper, we proposed an online tracker combines motion and appearance information to deal with the situation of detection missing, especially for which caused by occlusion. We use a lightweight re-identification feature to monitor occlusion. The Kalman tracker shows the superiority of its short-term tracking ability when the object is occluded. The experimental results on MOT17 test dataset show that the proposed tracker has improved performance on most metrics, such as MOTA, IDF1 and IDS. Experimental results on the MOT16 dataset show that our tracker is able to deliver performance improvements on the sequences filmed by the static camera even tracks by poor detections. At the same time, the tracking speed can reach 29.4 fps on the Nvidia GTX1060, which can basically achieve real-time tracking requirement while ensuring accuracy.

References

1. Alahi, A., Goel, K., Ramanathan, V., Robicquet, A., Fei-Fei, L., Savarese, S.: Social LSTM: Human trajectory prediction in crowded spaces. In: Proceedings of the IEEE conference on computer vision and pattern recognition. pp. 961–971 (2016)
2. Andriyenko, A., Schindler, K.: Multi-target tracking by continuous energy minimization. In: CVPR 2011, pp. 1265–1272. IEEE (2011)
3. Baisa, N.L.: Robust online multi-target visual tracking using a HISP filter with discriminative deep appearance learning. arXiv preprint arXiv:1908.03945 (2019)
4. Baisa, N.L., Wallace, A.: Development of a N-type GM-PHD filter for multiple target, multiple type visual tracking. J. Vis. Commun. Image Represent. **59**, 257–271 (2019)
5. Baisa, N.L., et al.: Online multi-target visual tracking using a HISP filter. In: VISIGRAPP (5: VISAPP), pp. 429–438 (2018)
6. Bertinetto, L., Valmadre, J., Henriques, J.F., Vedaldi, A., Torr, P.H.S.: Fully-convolutional siamese networks for object tracking. In: Hua, G., Jégou, H. (eds.) ECCV 2016. LNCS, vol. 9914, pp. 850–865. Springer, Cham (2016). https://doi.org/10.1007/978-3-319-48881-3_56
7. Bewley, A., Ge, Z., Ott, L., Ramos, F., Upcroft, B.: Simple online and realtime tracking. In: 2016 IEEE International Conference on Image Processing (ICIP), pp. 3464–3468. IEEE (2016)
8. Bhat, G., Johnander, J., Danelljan, M., Shahbaz Khan, F., Felsberg, M.: Unveiling the power of deep tracking. In: Proceedings of the European Conference on Computer Vision (ECCV), pp. 483–498 (2018)
9. Bochinski, E., Eiselein, V., Sikora, T.: High-speed tracking-by-detection without using image information. In: 2017 14th IEEE International Conference on Advanced Video and Signal Based Surveillance (AVSS), pp. 1–6. IEEE (2017)
10. Bochinski, E., Senst, T., Sikora, T.: Extending IOU based multi-object tracking by visual information. In: 2018 15th IEEE International Conference on Advanced Video and Signal Based Surveillance (AVSS), pp. 1–6. IEEE (2018)
11. Bolme, D.S., Beveridge, J.R., Draper, B.A., Lui, Y.M.: Visual object tracking using adaptive correlation filters. In: 2010 IEEE Computer Society Conference on Computer Vision and Pattern Recognition, pp. 2544–2550. IEEE (2010)
12. Choi, W., Savarese, S.: Multiple target tracking in world coordinate with single, minimally calibrated camera. In: Daniilidis, K., Maragos, P., Paragios, N. (eds.) ECCV 2010. LNCS, vol. 6314, pp. 553–567. Springer, Heidelberg (2010). https://doi.org/10.1007/978-3-642-15561-1_40
13. Eiselein, V., Arp, D., Pätzold, M., Sikora, T.: Real-time multi-human tracking using a probability hypothesis density filter and multiple detectors. In: 2012 IEEE Ninth International Conference on Advanced Video and Signal-Based Surveillance, pp. 325–330. IEEE (2012)
14. Henriques, J.F., Caseiro, R., Martins, P., Batista, J.: Exploiting the circulant structure of tracking-by-detection with kernels. In: Fitzgibbon, A., Lazebnik, S., Perona, P., Sato, Y., Schmid, C. (eds.) ECCV 2012. LNCS, vol. 7575, pp. 702–715. Springer, Heidelberg (2012). https://doi.org/10.1007/978-3-642-33765-9_50

15. Henriques, J.F., Caseiro, R., Martins, P., Batista, J.: High-speed tracking with kernelized correlation filters. IEEE Trans. Pattern Anal. Mach. Intell. **37**(3), 583–596 (2014)
16. Kim, C., Li, F., Ciptadi, A., Rehg, J.M.: Multiple hypothesis tracking revisited. In: Proceedings of the IEEE International Conference on Computer Vision, pp. 4696–4704 (2015)
17. Kim, C., Li, F., Rehg, J.M.: Multi-object tracking with neural gating using bilinear LSTM. In: Proceedings of the European Conference on Computer Vision (ECCV), pp. 200–215 (2018)
18. Kutschbach, T., Bochinski, E., Eiselein, V., Sikora, T.: Sequential sensor fusion combining probability hypothesis density and kernelized correlation filters for multi-object tracking in video data. In: 2017 14th IEEE International Conference on Advanced Video and Signal Based Surveillance (AVSS), pp. 1–5. IEEE (2017)
19. Leal-Taixé, L., Canton-Ferrer, C., Schindler, K.: Learning by tracking: siamese CNN for robust target association. In: Proceedings of the IEEE Conference on Computer Vision and Pattern Recognition Workshops, pp. 33–40 (2016)
20. Leal-Taixé, L., Milan, A., Schindler, K., Cremers, D., Reid, I., Roth, S.: Tracking the trackers: an analysis of the state of the art in multiple object tracking. arXiv preprint arXiv:1704.02781 (2017)
21. Leal-Taixé, L., Pons-Moll, G., Rosenhahn, B.: Everybody needs somebody: Modeling social and grouping behavior on a linear programming multiple people tracker. In: 2011 IEEE International Conference on Computer Vision Workshops (ICCV Workshops), pp. 120–127. IEEE (2011)
22. Liu, Q., Liu, B., Wu, Y., Li, W., Yu, N.: Real-time online multi-object tracking in compressed domain. IEEE Access **7**, 76489–76499 (2019)
23. Ma, C., Huang, J.B., Yang, X., Yang, M.H.: Hierarchical convolutional features for visual tracking. In: Proceedings of the IEEE International Conference on Computer Vision, pp. 3074–3082 (2015)
24. Maksai, A., Fua, P.: Eliminating exposure bias and metric mismatch in multiple object tracking. In: Proceedings of the IEEE Conference on Computer Vision and Pattern Recognition, pp. 4639–4648 (2019)
25. Milan, A., Schindler, K., Roth, S.: Detection-and trajectory-level exclusion in multiple object tracking. In: Proceedings of the IEEE Conference on Computer Vision and Pattern Recognition, pp. 3682–3689 (2013)
26. Pirsiavash, H., Ramanan, D., Fowlkes, C.C.: Globally-optimal greedy algorithms for tracking a variable number of objects. In: CVPR 2011, pp. 1201–1208. IEEE (2011)
27. Ristani, E., Tomasi, C.: Features for multi-target multi-camera tracking and re-identification. In: Proceedings of the IEEE Conference on Computer Vision and Pattern Recognition, pp. 6036–6046 (2018)
28. Sadeghian, A., Alahi, A., Savarese, S.: Tracking the untrackable: learning to track multiple cues with long-term dependencies. In: Proceedings of the IEEE International Conference on Computer Vision, pp. 300–311 (2017)
29. Wang, M., Liu, Y., Huang, Z.: Large margin object tracking with circulant feature maps. In: Proceedings of the IEEE Conference on Computer Vision and Pattern Recognition, pp. 4021–4029 (2017)
30. Wojke, N., Bewley, A., Paulus, D.: Simple online and realtime tracking with a deep association metric. In: 2017 IEEE International Conference on Image Processing (ICIP), pp. 3645–3649. IEEE (2017)

31. Yamaguchi, K., Berg, A.C., Ortiz, L.E., Berg, T.L.: Who are you with and where are you going? In: CVPR 2011, pp. 1345–1352. IEEE (2011)
32. Yang, Y., Bilodeau, G.A.: Multiple object tracking with kernelized correlation filters in urban mixed traffic. In: 2017 14th Conference on Computer and Robot Vision (CRV), pp. 209–216. IEEE (2017)

Classifying Candidate Axioms via Dimensionality Reduction Techniques

Dario Malchiodi[1]([⊠]) [iD], Célia da Costa Pereira[2] [iD],
and Andrea G. B. Tettamanzi[2] [iD]

[1] DSRC and Dipartimento di Informatica, Università degli Studi di Milano,
Milan, Italy
dario.malchiodi@unimi.it
[2] Université Côte d'Azur, CNRS, I3S, Sophia-Antipolis, France
{celia.pereira,andrea.tettamanzi}@unice.fr

Abstract. We assess the role of similarity measures and learning methods in classifying candidate axioms for automated schema induction through kernel-based learning algorithms. The evaluation is based on (i) three different similarity measures between axioms, and (ii) two alternative dimensionality reduction techniques to check the extent to which the considered similarities allow to separate true axioms from false axioms. The result of the dimensionality reduction process is subsequently fed to several learning algorithms, comparing the accuracy of all combinations of similarity, dimensionality reduction technique, and classification method. As a result, it is observed that it is not necessary to use sophisticated semantics-based similarity measures to obtain accurate predictions, and furthermore that classification performance only marginally depends on the choice of the learning method. Our results open the way to implementing efficient surrogate models for axiom scoring to speed up ontology learning and schema induction methods.

Keywords: Possibilistic axiom scoring · Dimensionality reduction

1 Introduction

Among the various tasks relevant to ontology learning [10], schema enrichment is a hot field of research, due to the ever-increasing amount of Linked Data published on the semantic Web. Its goal is to extract schema axioms from existing ontologies (typically expressed in OWL) and instance data (typically represented in RDF) [7]. To this aim, induction-based methods akin to inductive logic programming and data mining are exploited. They range from using statistical schema induction to enrich the schema of an RDF dataset with property axioms [5] or of the DBpedia ontology [24] to learning datatypes within ontologies [6] or even developing light-weight methods to create a schema of any knowledge base accessible via SPARQL endpoints with almost all types of OWL axioms [3].

All these approaches critically rely on (candidate) axiom scoring. In practice, testing an axiom boils down to computing an acceptability score, measuring the

V. Torra et al. (Eds.): MDAI 2020, LNAI 12256, pp. 179–191, 2020.
https://doi.org/10.1007/978-3-030-57524-3_15

extent to which the axiom is compatible with the recorded facts. Methods to approximate the semantics of given types of axioms have been thoroughly investigated in the last decade (e.g., approximate subsumption [20]) and some related heuristics have been proposed to score concept definitions in concept learning algorithms [19]. The most popular candidate axiom scoring heuristics proposed in the literature are based on statistical inference (see, e.g., [3]), but alternative heuristics based on possibility theory have also been proposed [22]. While it appears that these latter may lead to more accurate results, their heavy computational cost makes it hard to apply them in practice. However, a promising alternative to their direct computation is to train a surrogate model on a sample of candidate axioms for which the score is already available, to learn to *predict* the score of a novel, unseen candidate axiom.

In [11], two of us proposed a semantics-based similarity measure to train surrogate models based on kernel methods. However, a doubt remained whether the successful training of the surrogate model really depended on the choice of such a measure (and not, for example, on the choice of the learning method). Furthermore, it was not clear if a similarity measure has to capture the semantics of axioms to give satisfactory results or if any similarity measure satisfying some minimal requirements would work equally well. The goal of this paper is, therefore, to shed some light on these issues.

To this aim, we (i) introduce three different syntax-based measures, having an increasing degree of problem-awareness, computing the similarity/distance between axioms; (ii) starting from these measures, we use two alternative kernel-based dimensionality reduction techniques to check how well the images of true and false axioms are separated; and (iii) we apply a number of supervised machine learning algorithms to these images in order to learn how to classify axioms. This allows us to determine the combinations of similarity measure, dimensionality reduction technique, and classification method giving the most accurate results. It should be stressed that we do not address here the broader topic of how axiom scoring should be used to perform knowledge base enrichment, axiom discovery, or ontology learning. Here, we focus specifically on the problem of learning good surrogate models of an axiom scoring heuristics whose exact computation has proven to be extremely expensive [21].

2   The Hypothesis Language

The hypotheses we are interested in classifying as true or false are OWL 2 axioms, expressed in functional-style syntax [14], and their negations. In particular, we focus on subsumption axioms, whose syntax is described by the production: `Axiom := 'SubClassOf' '(' ClassExpression ' ' ClassExpression ')'`. Although a `ClassExpression` can be quite complicated, according to the full OWL syntax, here, for the sake of simplicity, we will restrict ourselves to class expressions consisting of just one atomic class, represented by its internationalized resource identifier (IRI), possibly abbreviated as in SPARQL, i.e., as a prefix, followed by the class name, like `dbo:Country`. The formulas we consider are thus described by the production `Formula := Axiom | '-' Axiom`. If a formula consists of an axiom (the first alternative in the above production) we will say it is *positive*; otherwise (the second alternative), we will say it is *negative*. If ϕ is a formula, we define $\text{sgn}(\phi) = 1$ if ϕ is positive, -1 otherwise. Alongside the "sign" of a formula, it is also convenient to define a notation for the OWL axiom from which the formula is built, i.e., what remains if one removes a minus sign prepended to it: $\text{abs}(\phi) = \phi$ if ϕ is positive, ψ if $\phi = -\psi$, with ψ an axiom.

Because the OWL syntax tends to be quite verbose, when possible we will write axioms in description logic (DL) notation. In such notation, a subsumption axiom has the form $A \sqsubseteq B$, where A and B are two class expressions. For its negation, we will write $\neg(A \sqsubseteq B)$, although this would not be legal DL syntax.[1] With this DL notation, the definition of $\text{abs}(\phi)$ can be rewritten as follows: $\text{abs}(\phi) = \phi$ if $\text{sgn}(\phi) = 1$, $\neg\phi$ if $\text{sgn}(\phi) = -1$.

We will denote with $[C] = \{a : C(a)\}$ the extension of an OWL class C in an RDF dataset, and with $\|E\|$ the cardinality of set E.

In particular, the dataset we processed has been built considering 722 formulas, together with their logical negations, thus gathering a total of 1444 elements. Each item is associated to a score inspired by possibility theory [4], called *acceptance-rejection index* (ARI for short), introduced in [11] and numerically summarizing the suitability of that formula as an axiom within a knowledge base expressed through RDF. More precisely, the ARI of a formula ϕ has been defined as the combination of the possibility and necessity measures of ϕ as follows: $\text{ARI}(\phi) = \Pi(\phi) - \Pi(\neg\phi) \in [-1, 1]$, so that a negative $\text{ARI}(\phi)$ suggests rejection of ϕ ($\Pi(\phi) < 1$), whilst a positive $\text{ARI}(\phi)$ suggests its acceptance ($N(\phi) > 0$),[2] with a strength proportional to its absolute value. A value close to zero reflects ignorance about the status of ϕ. For all ϕ, $\text{ARI}(\neg\phi) = -\text{ARI}(\phi)$.

The 722 formulas of our dataset were exactly scored against DBpedia, which required a little less than 290 days of CPU time on quite a powerful machine [23].

3   Formula Translation

Among the landscape of dimensionality reduction techniques (see for instance [13,17]), formulas have been processed using (i) kernel-based Principal Component Analysis (PCA) [18] and (ii) t-distributed Stochasting Neighbor Embedding (t-SNE) [9]. The first technique applies standard PCA [15] to data nonlinarly mapped onto a higher-dimensional space. On the other hand, t-SNE describes data using a probability distribution linked to a similarity measure, minimizing its Kullback-Leibler divergence with an analogous distribution in map space. We did not use such techniques to reduce data dimensionality [25], but as a way to map formulas into $\mathbb{R}^d$, for arbitrary choices of d, based on the similarity measures detailed below. The result of this mapping is considered as an intermediate representation to be further processed as explained in the rest of the paper.

---

[1] It should always be borne in mind that this is just a shorthand notation for the underlying OWL 2 functional-style syntax extended with the "minus" operator as explained above.

[2] We recall that $N(\phi) = 1 - \Pi(\neg\phi)$ and $\Pi(\phi) = 1 - N(\neg\phi)$.

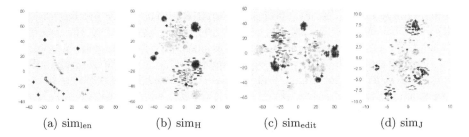

Fig. 1. t-SNE-based scatter plots for the considered similarity measures. Positive and negative formulas are marked using + and −, respectively, using a gray shade reflecting ARI (dark shades for low ARI and vice versa.)

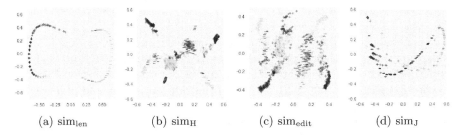

Fig. 2. PCA-based scatter plots. Same notations as in Fig. 1.

Length-Based Similarity. This similarity is obtained by comparing the textual representation length of two formulas ϕ_1 and ϕ_2 as follows:

$$s_{\text{len}}(\phi_1, \phi_2) = 1 - \frac{|\#\phi_1 - \#\phi_2|}{\max\{\#\phi_1, \#\phi_2\}}, \tag{1}$$

where $\#\phi$ denotes the length of the textual representation of ϕ. Normalization is required in order to transform the distance $|\#\phi_1 - \#\phi_2|$ into a similarity.[3] Such measure, however, cannot be reasonably expected to be able to capture meaningful information, as it merely relies on string length. For instance, two formulas whose formal description have equal length will always exhibit maximal similarity, regardless of the concepts they express. Nonetheless, such similarity has been included as a sort of *litmus test*. Anyhow, (1) is excessively naive, for it would assign quasi-maximal similarity to a formula and its logical negation: indeed, their descriptions would differ only for a trailing minus character. This inconvenience can be easily fixed by complementing to 1 the above definition when the signs of the two formulas are opposed and limiting the comparison to the base axiom only:

$$\text{sim}_{\text{len}}(\phi_1, \phi_2) = \begin{cases} 1 - s_{\text{len}}(\text{abs}(\phi_1), \text{abs}(\phi_2)) & \text{if } \text{sgn}(\phi_1) \neq \text{sgn}(\phi_2), \\ s_{\text{len}}(\phi_1, \phi_2) & \text{otherwise.} \end{cases} \tag{2}$$

---

[3] The used implementations of t-SNE and PCA in scikit-learn [16] accept, respectively, a distance and a similarity matrix. Thus we normalized all similarities.

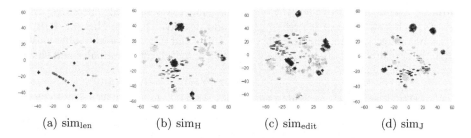

(a) sim$_\text{len}$ (b) sim$_\text{H}$ (c) sim$_\text{edit}$ (d) sim$_\text{J}$

Fig. 3. Scatter plots of the output of PCA, followed by t-SNE. Same notations as in Fig. 1.

This basically ensures, as one would intuitively expect, that the similarity between a formula and its negation be zero.

Hamming Similarity. Despite its simplicity, length-based similarity catches very simple instances of *syntactical* similarities (e.g. when identical—or almost identical—formulas only differ from synonymous class names); in this respect, it is reasonable to obtain better results if the distance is defined in a smarter way. This can be achieved for instance considering $\text{sim}_\text{H}(\phi_1, \phi_2) = \text{H}(\text{abs}(\phi_1), \text{abs}(\phi_2))$, if $\text{sgn}(\phi_1) \neq \text{sgn}(\phi_2)$, $1 - \text{H}(\phi_1, \phi_2)$ otherwise, where H is the normalized Hamming distance between the textual representations of two formulas, that is the fraction of positions where such strings contain a different character, aligning them on the left and ignoring the extra characters of the longest string.

Levenshtein Similarity. We can aim at grasping more sophisticated forms of syntactical similarities, and even simple *semantic* similarities. To such extent, we considered $\text{sim}_\text{edit}(\phi_1, \phi_2) = \text{Lev}(\text{abs}(\phi_1), \text{abs}(\phi_2))$ if $\text{sgn}(\phi_1) \neq \text{sgn}(\phi_2)$, $1 - \text{Lev}(\phi_1, \phi_2)$ otherwise, where Lev is the normalized Levenshtein distance [8] between strings representing two formulas, intended as the smallest number of atomic operations allowing to transform one string into the other one.

Jaccard Similarity. Semantics heavily depends on context. This is why we also consider the similarity measure introduced in [11] exploiting notation of Sect. 1:

$$\text{sim}_\text{J}(\phi_1, \phi_2) = \frac{\|[A] \cap [B] \cup [C] \cap [D]\|}{\|[A] \cup [C]\|}, \tag{3}$$

where ϕ_1 is the subsumption $A \sqsubseteq B$, and ϕ_2 is $C \sqsubseteq D$. Note that among all the considered similarities, this is the only one taking into account both the specific form of the formulas (subsumptions) and their meaning within a dataset. It has been termed *Jaccard similarity* because (3) racalls the Jaccard similarity index.

Figures 1 and 2 show the obtained set of points in $\mathbb{R}^2$ when applying t-SNE and kernel PCA, coupled with the above similarities, to the considered subsumption formulas[4]. In each plot, + and − denote the images of positive and negative

---

[4] Code and data to replicate all experiments described in the paper is available at https://github.com/dariomalchiodi/MDAI2020.

Table 1. Radius statistics for PCA-based clusters. Rows: similarity measures; Dist: average Euclidean distance between positive and negative clusters; Max, m, IQR, μ, and σ: maximum, median, interquartile range, mean, and standard deviation for cluster radius. Indices are computed separately for positive and negative formulas and shown in columns marked with $+$ and $-$.

	Dist	Max$^+$	m^+	IQR$^+$	μ^+	σ^+	Max$^-$	m^-	IQR$^-$	μ^-	σ^-
sim$_{\mathrm{len}}$	0.05	5.0e−1	1.7e−1	0.26	2.0e−1	0.16	5.0e−1	1.7e−1	0.25	2.0e−1	0.16
sim$_{\mathrm{H}}$	0.45	8.2e−2	3.3e−2	0.03	3.3e−2	0.02	8.2e−2	3.2e−2	0.03	3.3e−2	0.02
sim$_{\mathrm{edit}}$	0.41	3.1e−1	1.1e−1	0.14	1.2e−1	0.09	3.0e−1	1.0e−1	0.14	1.1e−1	0.09
sim$_{\mathrm{J}}$	0.30	5.2e−1	2.3e−2	0.39	1.8e−1	0.23	5.2e−1	2.3e−2	0.39	1.8e−1	0.23

formulas in $\mathbb{R}^2$ through the reduction technique; the color of each bullet reflects the ARI associated to a formula (cf. Sect. 2), with minimal and maximal values being mapped onto dark and light gray. Figure 3 shows the same scatter plots when PCA and t-SNE are applied in chain, as commonly advised [9]. In particular, not having here an explicit dimension for the original space of formulas, we extracted 300 principal components via PCA and considered the cumulative fraction of explained variance (EV) for each of the extracted components, after the latter were sorted by nondecreasing EV value. In such cases, the number of components to be considered is the one explaining a fixed fraction of the total variance, and we set such fraction to 75%, which led us to roughly 150 components, to be further reduced onto $\mathbb{R}^2$ via t-SNE.

Recalling that our final aim is to tell "good" formulas from "bad" ones, where "good" means a high ARI, i.e., light gray symbols in the scatter plots, the obtained results highlight the following facts.

- In all sim$_{\mathrm{len}}$ plots, light and dark bullets tend to heavily overlap, thus confirming that sim$_{\mathrm{len}}$ is not a suitable choice for our classification purposes.
- A similar behavior is generally exhibited by PCA; however, shapes in Fig. 2(a) and 2(b) strongly recall the projection onto $\mathbb{R}^2$ of a saddle-shaped manifold, thus suggesting a low-dimensional intrinsic dimensionality.

Leaving apart the length-based similarity and the sole kernel PCA technique, we are left with six possibilities, which are hard to rank via qualitative judgment only. We can only remark that sim$_{\mathrm{H}}$ tends to generate plots where the two classes do not appear as sharply distinguished as in the remaining cases. However, this criterion is too weak and, therefore, a quantitative approach is needed. This is why we considered the *radius statistics* of the clusters of points in $\mathbb{R}^2$ referring to a same *concept*. More precisely:

Table 2. Radius statistics for clusters of points obtained through t-SNE. Same notations as in Table 1.

	Dist	Max$^+$	m^+	IQR$^+$	μ^+	σ^+	Max$^-$	m^-	IQR$^-$	μ^-	σ^-
sim$_{len}$	15.80	7.8e+1	3.4e+1	36.62	3.4e+1	23.12	8.0e+1	3.7e+1	37.35	3.6e+1	24.01
sim$_H$	60.29	1.2e+1	3.0e+0	7.10	4.4e+0	3.98	1.0e+1	2.1e+0	5.26	3.4e+0	3.23
sim$_{edit}$	57.66	2.7e+1	8.3e+0	11.42	9.6e+0	7.62	2.6e+1	7.6e+0	9.31	8.9e+0	6.96
sim$_J$	2.33	3.8e+0	9.1e−1	2.75	1.5e+0	1.44	3.3e+0	6.3e−1	2.54	1.3e+0	1.33

Table 3. Radius statistics for clusters of points obtained chaining kernel PCA and t-SNE. Same notations as in Table 1.

	Dist	Max$^+$	m^+	IQR$^+$	μ^+	σ^+	Max$^-$	m^-	IQR$^-$	μ^-	σ^-
sim$_{len}$	17.76	7.9e+1	3.5e+1	36.68	3.5e+1	3.65	8.0e+1	3.7e+1	37.87	3.7e+1	24.45
sim$_H$	56.19	3.3e+0	1.1e+0	1.51	1.3e+0	0.91	3.1e+0	1.1e+0	1.41	1.2e+0	0.87
sim$_{edit}$	60.56	3.1e+1	1.0e+1	14.28	1.2e+1	9.22	3.2e+1	1.1e+1	13.97	1.2e+1	9.45
sim$_J$	13.02	2.3e+1	1.3e+0	16.47	7.8e+0	9.72	2.6e+1	1.4e+0	16.46	7.9e+0	10.05

- each concept is identified by an OWL class (such as for instance Coach, TennisLeague, Eukaryote, and so on);
- given a concept, we identify all positive formulas having it as antecedent,[5] and consider the *positive cluster* of corresponding points in $\mathbb{R}^2$; we proceed analogously for the *negative cluster*;
- we subsequently obtain the centroids of the two clusters and compute their Euclidean distance; moreover, we consider the population of distances between each point in a cluster and the corresponding centroid. Such population is described in terms of basic descriptive indices (namely, maximum, mean and standard deviation, median and interquartile range (IQR)).

Thus, for each combination of reduction technique, similarity measure, and concept, we obtain a set of measurements. In order to reduce this information to a manageable size, we average all indices across concepts: Tables 1, 2, 3 summarize the results, for which we propose the following interpretation:

- the use of kernel PCA results in loosely decoupled clusters: indeed, independently of the considered similarity measure, the distance between positive and negative formulas is generally smaller than the sum of the radii of the corresponding clusters (identified with the max index); moreover, all measured indices tend to take up similar values;
- as already found out in the qualitative analysis of Figs. 1, 2, 3, a weakness analogous to that illustrated in the previous point tends to be associated to sim$_{len}$, regardless of the chosen reduction technique;
- quite surprisingly, sim$_H$ is the only similarity measure which results in sharply distinguished clusters when coupled with t-SNE.

---

[5] We also repeated all experiments considering both antecedents and consequents, obtaining comparable results.

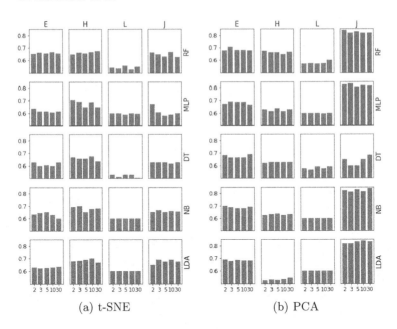

Fig. 4. Comparison of the more significative results in learning ARI when using t-SNE (a) and PCA (b). Rows are associated to learning algorithms, using the abbreviations introduced in Sect. 4, while columns refer to similarity measures (E: sim_{edit}, H: sim_H, L: sim_{len}, J: sim_J). Each graph plots test set accuracy vs. number of dimensions.

4   Learning ARI

As all considered formulas have been mapped to points in a Euclidean space, it is now easy to use the images of the considered mappings as patterns to be fed to a supervised learning algorithm, whose labels are the ARI of the relevant formulas. More precisely, we considered the following models (and, specifically, their implementation in scikit-learn [16]), each described together with the hyperparameters involved in the model selection phase:

- Decision trees (DT), parameterized on purity index, maximal number of leaves, maximal number of features, and maximal tree depth;
- Random forests (RF), parameterized on the number of trees, as well as on the same quantities as DT;
- Naive Bayes with Gaussian priors (NB), without hyperparameters;
- Linear Discriminant Analysis (LDA), without hyperparameters;
- Three-layered feed-forward neural networks (NN), parameterized on the number of hidden neurons;
- Support vector classifiers (SVC), parameterized on the tradeoff constant and on the used kernel.

When applicable, model selection was carried out via a repeated holdout consisting of five iterations, each processing a grid of the above-mentioned values for hyperparameters, shuffling available data, and considering a split assigning 80%, 10%, and 10% of points, respectively, to train, validation, and test. When model selection was not involved, respectively 80% and 20% of the available examples where assigned to training and test.

In order to build examples to be fed to learning algorithms, each formula needed to be associated with a binary label. That was done through binarization of the ARI values, using a threshold $\alpha \in \{i/10$ for $i = 1, \ldots, 9\}$. More precisely, an experiment was carried out for each of the possible thresholding levels. Finally, the transformation of formulas into vectors was done by considering different values for the dimension d of the resulting Euclidean space, namely the values in $\{2, 3, 5, 10, 30\}$ were tested. Summing up, for each reduction technique R (that is, either PCA or t-SNE coupled with any similarity measure) and considered dimension d, the following holdout protocol was iterated ten times:

1. R was applied to the available data in order to obtain a set of vectors in $\mathbb{R}^d$;
2. for each model M such vectors were randomly shuffled and subsequently divided into training, validation (if needed), and test set;
3. for each threshold value α, a model selection for M using the data in training and validation set was carried out, testing the selected model against generalization using the test set;
4. the model whose value of α gave the best generalization was retained.

Table 4. Test set accuracy for the considered learning algorithms, when fed with points in $\mathbb{R}^2$ representing the original formulas transformed through kernel PCA using sim$_J$.

Model	Test accuracy				Model	Test accuracy			
	α	mean	median	σ		α	mean	median	σ
DT	0.8	0.71	0.64	0.12	LDA	0.9	0.82	0.82	0.01
RF	0.8	0.84	0.83	0.01	MLP	0.2	0.83	0.82	0.02
NB	0.9	0.82	0.82	0.00	SVC	0.9	0.60	0.60	0.00

Table 5. Comparison between test set median accuracy of algorithms using PCA. Rows: models; columns: similarity; d: space dimension.

	$d = 2$			$d = 3$			$d = 5$			$d = 10$		
	H	J	Edit	H	J	Edit	H	J	Edit	H	J	Edit
DT	0.62	0.64	**0.68**	0.62	0.60	0.66	0.62	0.60	0.66	0.62	0.64	0.66
RF	0.67	**0.83**	0.67	0.66	0.81	0.70	0.66	0.82	0.68	0.64	0.81	0.68
NB	0.62	0.82	0.69	0.63	0.80	0.68	0.63	**0.83**	0.68	0.62	0.81	0.68
LDA	0.52	0.82	0.68	0.52	0.82	0.67	0.52	**0.83**	0.68	0.53	**0.83**	0.68
MLP	0.62	0.82	0.66	0.61	**0.83**	0.68	0.63	0.80	0.68	0.61	0.82	0.68
SVC	0.60	0.60	0.60	0.60	0.60	0.60	0.60	0.60	0.60	0.60	0.60	0.60

Table 6. Comparison between algorithms using t-SNE. Same notations as in Table 5.

	$d = 2$			$d = 3$			$d = 5$			$d = 10$		
	H	J	Edit	H	J	Edit	H	J	Edit	H	J	Edit
DT	0.66	0.62	0.62	0.65	0.62	0.60	0.65	0.62	0.60	**0.67**	0.61	0.60
RF	0.64	**0.66**	0.65	**0.66**	0.64	**0.66**	0.65	0.63	0.65	**0.66**	**0.66**	**0.66**
NB	**0.69**	0.65	0.63	0.69	0.66	0.64	0.65	0.65	0.65	0.67	0.65	0.62
LDA	0.67	0.65	0.62	0.68	0.68	0.62	0.68	0.67	0.62	**0.70**	0.68	0.62
MLP	**0.70**	0.66	0.63	0.68	0.60	0.61	0.64	0.57	0.61	0.68	0.58	0.60
SVC	0.60	0.60	0.60	0.60	0.60	0.60	0.60	0.60	0.60	0.60	0.60	0.60

Table 4 shows the results obtained when using kernel PCA with Jaccard similarity. For each model, the optimal value of α is shown, as well as the estimated generalization capability, in terms of mean and median accuracy on the test set.

Tables 5 and 6 show a more compact representation of the corresponding results when considering the possible combinations of learning algorithm, similarity measure, reduction technique, and dimension of the space onto which formulas are mapped. Because, as expected, the length-based distance always gave the worst results, it has been omitted from the tables to save space. The same applies to the results obtained when setting $d = 30$, which in this case did not give any significant improvement w.r.t. the figures for $d = 10$.

Looking at these results, it is immediately clear that SVC is systematically outperformed by the remaining models, and the results it obtains are independent of the combination of reduction technique, similarity measure, and space dimension. Moreover, the best results (highlighted using boldface in the tables) are almost always higher when using kernel PCA, thus somehow in contrast with the preliminary results obtained in Sect. 3. In order to further analyze this trend, Fig. 4 graphically shows a subset of the obtained results, where SVC is left out. The graphs highlight that the space dimension d has a weak influence on the results, and that sim_{len} always attains a low score, thus it has been excluded from the rest of the analysis. Experiments can be divided into three groups:

- all combinations of LDA with sim_H, getting a bottom-line performance,
- a majority of cases where median accuracy lays roughly between 60 and 70%, notably containing all remaining results based on t-SNE, and
- a top set of experiments with median accuracy > 80%, including all those involving kernel PCA with sim_J and any learning algorithm but DT.

In order to verify that there is a significant difference between the experiments in these groups, we executed the Cramér-Von Mises (CVM) test [2]. For the sake of conciseness, we will call a *case* any combination of learning algorithm, dimensionality reduction technique, dimension of the resulting space, and similarity measure (excluding, as previously stated, SVC and sim_{len}). For each case, we repeated the experiments ten times, thus getting a sample of ten values for the median test set error. We then gathered cases in the following categories:

$G_{\text{PCA}}^{\text{top}}$, $G_{\text{PCA}}^{\text{mid}}$, $G_{\text{PCA}}^{\text{btm}}$: respectively the best, middle, and low performing cases using PCA (corresponding to the three items in the previous list); G_{tSNE}: all cases using tSNE, i.e. those shown in Fig. 4(a). Now, given a generic pair of cases taken from $G_{\text{PCA}}^{\text{top}}$ and $G_{\text{PCA}}^{\text{mid}}$, the hypothesis that the corresponding test error samples are drawn from the same distribution is strongly rejected by the CVM test, with $p < 0.01$. The same happens when considering any other pair of groups related to PCA, and these results do not change if $G_{\text{PCA}}^{\text{mid}}$ is swapped with G_{tSNE}. On the other hand, the same test run on two cases from $G_{\text{PCA}}^{\text{top}}$ didn't reject the same hypothesis in 81% of the times.

5 Conclusions

Our results show that it is possible to learn reasonably accurate predictors of the score of candidate axioms without having to resort to sophisticated (and expensive-to-compute) semantics-based similarity measures. Indeed, we showed that using a dimensionality-reduction technique to map candidate axioms to a low-dimension space allows supervised learning algorithms to effectively train predictive models of the axiom score. This agrees with the observation that there are no obvious indicators to inform the decision to choose between a cheap or expensive similarity measure based on the properties of an ontology [1].

Our findings can contribute to dramatically speed up ontology-learning approaches based on exhaustive or stochastic generate-and-test strategies, like [3,12]. As a further development, we plan to apply to the same dataset more sophisticated techniques able to learn the ARI without the need of binarizing labels.

Acknowledgments. Part of this work was done while D. Malchiodi was visiting scientist at Inria Sophia-Antipolis/I3S CNRS Université Côte d'Azur. This work has been supported by the French government, through the 3IA Côte d'Azur "Investments in the Future" project of the Nat'l Research Agency, ref. no. ANR-19-P3IA-0002.

References

1. Alsubait, T., Parsia, B., Sattler, U.: Measuring conceptual similarity in ontologies: how bad is a cheap measure? In: Description Logics, pp. 365–377 (2014)
2. Anderson, T.W.: On the distribution of the two-sample Cramer-von Mises criterion. Ann. Math. Stat. **33**(3), 1148–1159 (1962)
3. Bühmann, L., Lehmann, J.: Universal OWL axiom enrichment for large knowledge bases. In: ten Teije, A., et al. (eds.) EKAW 2012. LNCS (LNAI), vol. 7603, pp. 57–71. Springer, Heidelberg (2012). https://doi.org/10.1007/978-3-642-33876-2_8
4. Dubois, D., Prade, H.: Possibility Theory–An Approach to Computerized Processing of Uncertainty. Plenum Press, New York (1988)
5. Fleischhacker, D., Völker, J., Stuckenschmidt, H.: Mining RDF data for property axioms. In: Meersman, R., et al. (eds.) OTM 2012. LNCS, vol. 7566, pp. 718–735. Springer, Heidelberg (2012). https://doi.org/10.1007/978-3-642-33615-7_18

6. Huitzil, I., Straccia, U., Díaz-Rodríguez, N., Bobillo, F.: Datil: learning fuzzy ontology datatypes. In: Medina, J., et al. (eds.) IPMU 2018. CCIS, vol. 854, pp. 100–112. Springer, Cham (2018). https://doi.org/10.1007/978-3-319-91476-3_9

7. Lehmann, J., Völker, J. (eds.): Perspectives on Ontology Learning, Studies on the Semantic Web, vol. 18. IOS Press, Amsterdam (2014)

8. Levenshtein, V.I.: Binary codes capable of correcting deletions, insertions, and reversals. Sov. Phys. Dokl. **10**(8), 707–710 (1966)

9. Maaten, L.V.D., Hinton, G.: Visualizing data using t-SNE. J. Mach. Learn. Res. **9**(Nov), 2579–2605 (2008)

10. Maedche, A., Staab, S.: Ontology learning for the semantic web. IEEE Intell. Syst. **16**(2), 72–79 (2001)

11. Malchiodi, D., Tettamanzi, A.G.B.: Predicting the possibilistic score of OWL axioms through modified support vector clustering. In: SAC 2018, pp. 1984–1991 (2018)

12. Nguyen, T.H., Tettamanzi, A.G.B.: Learning class disjointness axioms using grammatical evolution. In: Sekanina, L., Hu, T., Lourenço, N., Richter, H., García-Sánchez, P. (eds.) EuroGP 2019. LNCS, vol. 11451, pp. 278–294. Springer, Cham (2019). https://doi.org/10.1007/978-3-030-16670-0_18

13. Nonato, L.G., Aupetit, M.: Multidimensional projection for visual analytics: linking techniques with distortions, tasks, and layout enrichment. IEEE Trans. Visual. Comput. Graph. **25**(8), 2650–2673 (2018)

14. Parsia, B., Motik, B., Patel-Schneider, P.: OWL 2 web ontology language structural specification and functional-style syntax, 2nd edn. W3C recommendation, W3C, December 2012. http://www.w3.org/TR/2012/REC-owl2-syntax-20121211/

15. Pearson, K.: LIII. on lines and planes of closest fit to systems of points in spaceLIII on lines and planes of closest fit to systems of points in space. Lond. Edinb. Dublin Philos. Mag. J. Sci. **2**(11), 559–572 (1901)

16. Pedregosa, F., et al.: Scikit-learn: machine learning in Python. J. Mach. Learn. Res. **12**, 2825–2830 (2011)

17. Sacha, D., et al.: Visual interaction with dimensionality reduction: a structured literature analysis. IEEE Trans. Vis. Comput. Graph. **23**(1), 241–250 (2016)

18. Schölkopf, B., Smola, A., Müller, K.-R.: Kernel principal component analysis. In: Gerstner, W., Germond, A., Hasler, M., Nicoud, J.-D. (eds.) ICANN 1997. LNCS, vol. 1327, pp. 583–588. Springer, Heidelberg (1997). https://doi.org/10.1007/BFb0020217

19. Straccia, U., Mucci, M.: pFOIL-DL: learning (fuzzy) EL concept descriptions from crisp OWL data using a probabilistic ensemble estimation. In: SAC 2015, pp. 345–352 (2015)

20. Stuckenschmidt, H.: Partial matchmaking using approximate subsumption. In: Proceedings of the Twenty-Second AAAI Conference on Artificial Intelligence, 22–26 July 2007, Vancouver, British Columbia, Canada, pp. 1459–1464 (2007)

21. Tettamanzi, A.G.B., Faron-Zucker, C., Gandon, F.: Dynamically time-capped possibilistic testing of subclass of axioms against RDF data to enrich schemas. In: K-CAP 2015. Article No. 7 (2015)

22. Tettamanzi, A.G.B., Faron-Zucker, C., Gandon, F.: Possibilistic testing of OWL axioms against RDF data. Int. J. Approximate Reasoning **91**, 114–130 (2017)

23. Tettamanzi, A.G.B., Faron-Zucker, C., Gandon, F.: Testing OWL axioms against RDF facts: a possibilistic approach. In: Janowicz, K., Schlobach, S., Lambrix, P., Hyvönen, E. (eds.) EKAW 2014. LNCS (LNAI), vol. 8876, pp. 519–530. Springer, Cham (2014). https://doi.org/10.1007/978-3-319-13704-9_39

24. Töpper, G., Knuth, M., Sack, H.: DBpedia ontology enrichment for inconsistency detection. In: I-SEMANTICS, pp. 33–40 (2012)

25. Yin, H.: Nonlinear dimensionality reduction and data visualization: a review. Int. J. Autom. Comput. **4**(3), 294–303 (2007). https://doi.org/10.1007/s11633-007-0294-y

Sampling Unknown Decision Functions to Build Classifier Copies

Irene Unceta[1,2] , Diego Palacios[1], Jordi Nin[3(✉)] , and Oriol Pujol[2]

[1] BBVA Data & Analytics, Barcelona, Spain
{irene.unceta,diego.palacios.contractor}@bbvadata.com
[2] ESADE, Universitat Ramon Llull, Barcelona, Spain
oriol_pujol@ub.edu
[3] Universitat de Barcelona, Barcelona, Spain
jordi.nin@esade.edu

Abstract. Copies have been proposed as a viable alternative to endow machine learning models with properties and features that adapt them to changing needs. A fundamental step of the copying process is generating an unlabelled set of points to explore the decision behavior of the targeted classifier throughout the input space. In this article we propose two sampling strategies to produce such sets. We validate them in six well-known problems and compare them with two standard methods in terms of both their accuracy performance and their computational cost.

Keywords: Bayesian sampling · Classification · Copies · Synthetic data

1 Introduction

As the use of machine learning models continues to grow and evolve, so does the need to adapt them to a changing context. In many everyday situations, constraints related to production software requirements [17], external threats [4, 19] or specific regulations [7] limit the performance of decision systems. In those cases, an originally functional solution may no longer meet the requirements of its operational environment.

Under such circumstances, model re-training is often neither advisable nor possible. Take for example company production environments, where model performance needs to be maintained in time. Or scenario where the training data is no longer available and access to the model is limited or its internals unknown. In such cases, copies [21] have been proposed as a suitable alternative to correct performance or behavioral inefficiencies by substituting existing models with new ones that operate roughly in the same way and are yet enhanced with new features and characteristics.

Copying corresponds to the problem of building a machine learning classifier that replicates the decision behavior of another. While the literature has traditionally focused on scenarios where the training data distribution is directly

© Springer Nature Switzerland AG 2020
V. Torra et al. (Eds.): MDAI 2020, LNAI 12256, pp. 192–204, 2020.
https://doi.org/10.1007/978-3-030-57524-3_16

[8] or indirectly [1] known and where rich information outputs are used as soft targets [12,24], copying envisages a more general setting. The overall decision behavior needs to be preserved and access to the model is limited to a membership query interface. Moreover, all information regarding the training data distribution is assumed to be lost or unknown. In its place, copies are built on synthetic data points sampled from the original problem domain and annotated using the existing model as an oracle.

In this article, we explore different methods to generate such sets of synthetic data. We propose two sampling techniques: an exploration-exploitation policy using a Boundary sampling model and a modified fast Bayesian sampling algorithm that uses Gaussian processes to reduce the uncertainty of the decision function. For comparative purposes, we also use a modified Jacobian sampling strategy based on [16] and uniform sampling. We conduct experiments on well-established datasets and classifiers. Copies built using the artificially generated data converge to an optimal solution for a sufficiently large number of samples, independently of the sampling technique. While Boundary sampling is better suited to learn linear problems, it is Fast Bayesian sampling which yields the best performance for less points.

The rest of this paper is organized as follows. We discuss related work in Sect. 2. We provide an overview of the copying problem in Sect. 3. In Sect. 4 we present the proposed methods and we validate them through a series of experiments in Sect. 5. Section 6 discusses our main findings. The paper concludes with a summary of our main results and an outline of future research.

2 Related Work

Drawing samples from arbitrary probability distributions is a well known problem in mathematics and statistics. Research has been traditionally focused on extracting representative data samples from a given population. In machine learning, sampling methods are widely used when training probabilistic models, for tasks such as data augmentation. Previous research includes Markov Chain Monte Carlo methods [6,20], sampling of specific probability distributions [15] and methods for rejection [5,14] and importance sampling to reduce the variance in Monte Carlo estimation [3]. Contrary to traditional sampling methods, however, when copying sampling of an original non-probabilistic hard decision function cannot be performed under an implicit probability density function.

Several works have demonstrated the utility of active learning strategies in this context [2,23]. Here, a query selection protocol minimizes the amount of samples that are passed along to a human annotator [18]. Particularly relevant to this paper are works on uncertainty sampling [10,11] and Bayesian active learning [9]. When selecting the samples to be annotated, the informativeness of unlabeled instances is evaluated in terms of class probability outputs. When copying, however, only hard predictions are available. Moreover, the cost of annotation in this context is zero.

Finally, a seemingly related problem is that of *transferability*-based adversarial learning [13,16], where a malicious adversary exploits samples crafted from a local substitute of a model to compromise it. Adversarial attacks focus on a local approximation of the target decision boundary to foil the system. Our aim, however, is to obtain an unlabeled set of samples to build a copy.

3   Preliminaries

Let us take a classifier $f_O : \mathcal{X} \to \mathcal{T}$, with $\mathcal{X}$ and $\mathcal{T}$ the sample and label spaces, respectively; $\mathcal{D} = \{(\boldsymbol{x}_i, t_i)\}_{i=1}^M$ the training data and M the number of samples. We restrict to case where $\mathcal{X} = \mathbb{R}^d$ and $\mathcal{T} = \mathbb{Z}_k$ for k the number of classes, i.e. classification of real-valued attributes. Assume that f_O is only accessible through a membership query interface and that its internals and training data remain unknown. Copying [21] refers to the process of globally replicating f_O without loss of accuracy and with the added benefit of endowing it with new properties.

A copy is a new classifier, $f_C(\theta)$, parameterized by θ, such that its decision function mimics f_O all over the space. The optimal parameter values θ^* are those such that $\theta^* = \arg\max_\theta P(\theta|f_O)$. However, because the training data $\mathcal{D}$ is unknown we can not resort to it, nor can we estimate its distribution. Hence, finding θ^* requires that we sample the input space to gain information about the specific form of f_O. We define synthetic data points $\boldsymbol{z}_j \in \mathcal{X}$ so that

$$\theta^* = \arg\max_\theta \int_{z \sim P_Z} P(\theta|f_O(\boldsymbol{z}))dP_Z \tag{1}$$

for an arbitrary generating probability distribution P_Z, from which the synthetic samples are independently drawn. In using P_Z to model the spatial support of the copy, we ensure the copying process is agnostic to both the specific form of f_O and the training data distribution. If we assume an exponential family form for all probability distributions, we can rewrite (1) as

$$\theta^* = \arg\min_\theta \left[\int_{z \sim P_Z} \gamma_1 \ell_1(f_C(\boldsymbol{z}, \theta), f_O(\boldsymbol{z}))dP_Z + \gamma_2 \ell_2(\theta, \theta^+) \right], \tag{2}$$

for ℓ_1 and ℓ_2 a measure of the disagreement between the two models, and θ^+ our prior knowledge of θ. Since we cannot draw infinite samples to explore the space in full, direct computation of (2) is not possible. Instead, we recall regularized empirical risk minimization [22] and approximate the expression above as

$$(\theta^*, \boldsymbol{Z}^*) = \arg\min_{\theta, z_j \in Z} \left[\frac{1}{N} \sum_{j=1}^N \gamma_1 \ell_1(f_C(\boldsymbol{z}_j, \theta), f_O(\boldsymbol{z}_j)) + \gamma_2 \ell_2(\theta, \theta^+) \right] \tag{3}$$

for the optimal set $\boldsymbol{Z}^* = \{\boldsymbol{z}_j\}_{j=1}^N$. We label this set according to the class prediction outputs of f_O and define the synthetic dataset $\mathcal{Z} = \{(\boldsymbol{z}_j, f_O(\boldsymbol{z}_j))\}_{j=1}^N$. According to (3), copying is a dual optimization problem, where we simultaneously optimize the parameters θ and the set $\boldsymbol{Z}$.

In the simplest approach, we cast this problem into one where we use a single iteration of an alternating projection optimization scheme: the *single-pass copy*

[21]. We split the dual optimization in two independent sub-problems. We first find an optimal set of unlabelled data points $\boldsymbol{Z}^*$ and then optimize for the copy parameters θ^{*1}. In this article we address the first part of this optimization to answer the question: *which set of synthetic points minimizes the error when substituting the integral in (2) with a finite sum over N elements?*

4   Methods

With the aim of answering this question, we propose two methods to generate unlabeled data: Boundary sampling and Fast Bayesian sampling.

4.1   Boundary Sampling

This method combines uniform exploration with a certain amount of exploitation. The main idea is to conduct a targeted exploration of the space until the decision boundary is found. The area around the boundary is then exploited by alternatively sampling points at both sides. Because different decision regions are to be expected, this process is repeated several times to ensure a proper coverage of the whole decision space.

We generate samples uniformly at random until we find a sample whose predicted class label differs from the others. We then proceed to do a binary search in the line that connects the last two samples. This binary search is stopped when a pair of points $(\boldsymbol{z}_a, \boldsymbol{z}_b)$ is found such that $\|\boldsymbol{z}_a - \boldsymbol{z}_b\|_2 < \varepsilon$ with $f_{\mathcal{O}}(\boldsymbol{z}_a) \neq f_{\mathcal{O}}(\boldsymbol{z}_b)$ for a given tolerance ε, *i.e.* points located at a distance from the boundary no larger than ε. We draw samples at a constant step distance λ in the direction of the unitary random vector defined taking one of these two points $\boldsymbol{z}$ as a starting point. We stop when we find a point $\boldsymbol{z}'$ such that $f_{\mathcal{O}}(\boldsymbol{z}) \neq f_{\mathcal{O}}(\boldsymbol{z}')$, and repeat the process.

The number of samples in the binary search increases with the logarithm of $1/\varepsilon$. The value of ε must be small compared to the boundary exploration step λ, which determines the Euclidean distance between two consecutive samples. The higher the value of λ, the faster the boundary will be covered with less resolution. If λ is small, a large proportion of the boundary may remain unexplored.

This process results in 1-dimensional thread that alternates the two sides of the decision boundary, with distance to the boundary bounded by λ. A thread contains a predefined number of steps $\mathcal{N}$, which at any given time depends on the number of generated samples. Threads are stopped when out of range or when no other samples are found in the given direction. To ensure a good coverage of the space, we allow $\mathcal{T}$ threads to be created from each point $\boldsymbol{z}$. When this number is reached, a new binary search starts. For high values of $\mathcal{T}$, the boundary is well

---

[1] The resulting optimization problem is always separable and, given enough capacity, zero training error is always achievable without hindering the generalization performance of the copy. In this context, we argue that ad hoc techniques can be used to better exploit these properties when building a copy [21].

sampled in certain areas, but many regions are left uncovered. For lower values, more regions of the boundary are explored with less intensity.

Each thread generates other threads with a frequency modelled by a Poisson distribution parameterized by λ'. We perform $\mathcal{I}$ independent runs, increasing the maximum number of threads from run to run. Given a desired number of samples N, we generate half of them following the Boundary sampling algorithm and the other half using random sampling. The theoretical computational cost of this method is $\mathcal{O}(Nd)$.

4.2   Fast Bayesian Sampling

In this method, the function to optimize is assumed to be a random process and samples are generated maximizing an acquisition function. We start assigning a large uncertainty to the whole input space and reduce it everytime a new sample is generated. We reduce the global uncertainty by guiding future sampling towards the most uncertain areas.

Let us define a Gaussian Process $g \sim \mathcal{GP}(0, k_{SE})$ with mean 0 and a squared exponential kernel k_{SE} with length scale l and variance σ^2. Every realization g_i of the stochastic process g is such that $g_i : \mathcal{X} \rightarrow \mathbb{R}$. We treat $f_{\mathcal{O}}$ as one of such realizations[2]. Our objective is to find a set of points $\mathbf{Z}$ such that the function $\lfloor \mathbb{E}[g^{\mathbf{Z}}] \rceil$[3], where $g^{\mathbf{Z}} = (g \mid g(\mathbf{z}) = f_{\mathcal{O}}(\mathbf{z}), \forall \mathbf{z} \in \mathbf{Z})$, is similar enough to $f_{\mathcal{O}}$. We propose an acquisition function $f : \mathcal{GP} \times \mathcal{X} \rightarrow \mathbb{R}$ of the form

$$f(g, \mathbf{z}) = \mathbb{V}_{ar}[g(\mathbf{z})] \cdot [1 + \tau \cdot frac(\mathbb{E}[g(\mathbf{z})])^2 (1 - frac(\mathbb{E}[g(\mathbf{z})]))^2],$$

where $frac(\mathbf{z})$ stands for the decimal part of $\mathbf{z}$. The first term involves the variance of the process. When choosing the next sample, it gives a higher priority to points with a bigger variance, *i.e.* located in a region where we have little information about $f_{\mathcal{O}}$. The second term focuses on exploring the boundary. Its maximum is located at 0.5. This encourages refining areas close to a transition between classes. Parameter τ governs the trade-off between the two terms.

As it is, finding the *a posteriori* distribution of the Gaussian Process has a high computational cost due to the estimation of the mean and the covariance matrix. Moreover, this cost is increased when optimizing for the maximum of the acquisition function. The total cost is roughly $\mathcal{O}(dN^3)$. To overcome this limitation, we propose a faster version, where we find the *a posteriori* Gaussian distribution in a single optimization, by limiting the number of samples used to compute the *a posteriori* process to b. The lower the value of b, the faster the algorithm converges, but also the less accurate it is. While this approach is notably faster, it is not warranted to find an optimal solution.

When computing the maximum of the acquisition function, the starting point is set to $z_0 \sim Uniform(\mathcal{X})$ and the total number of iterations to $\mathcal{N}$, the number of random samples used to compute the first a posteriori distribution. Given the no

---

[2] There exist realizations of g as close to $f_{\mathcal{O}}$ as desired.
[3] $\lfloor x \rceil$ rounds x to the nearest integer.

Algorithm 1. Fast Bayesian Sampling(**int** N, **Classifier** $f_\mathcal{O}$)

1: $Z \leftarrow \{(z, f_\mathcal{O}(z)) \mid 10 \times z \sim Uniform(\mathcal{D})\}$
2: **while** $|Z| < N$ **do**
3:     **if** $|Z| \leq b$ **then**
4:         $Z_r \leftarrow Z$
5:     **else** ▷ Limit to b the number of samples to calculate the posterior distribution
6:         $Z_r \subseteq Z$ s.t. $|Z_r| = b$
7:     **end if**
8:     $g^{Z_r} \leftarrow g \mid g(z) = y \, \forall (z,y) \in Z_r$ ▷ A posteriori Gaussian process
9:     $T \leftarrow \emptyset$
10:     **repeat**
11:         $z_0 \sim Uniform(\mathcal{D})$
12:         $z \leftarrow \text{argmax}_{z \in \mathcal{V}(z_0) \subseteq \mathcal{D}} \, f(g^{Z_r}, z)$
13:         $y \leftarrow f_\mathcal{O}(z)$
14:         $T \leftarrow T \cup (z, y)$
15:     **until** $|T| = \lfloor |Z_r|/\text{sf} \rfloor$ ▷ sf: slowness factor
16:     $Z \leftarrow Z \cup T$
17: **end while**
18: **return** S

---

convexity (in general) of the acquisition function, we can identify the next point to sample to be $z = \text{argmax}_{z \in \mathcal{V}(z_0) \subseteq \mathcal{D}} \, f(g^Z, z)$ for $\mathcal{V}(z_0)$ a neighbourhood of z_0 not necessarily open. The number of points generated by this Gaussian process is limited by slowness factor sf, the inverse of the fraction of samples generated from the gaussian process with respect to those used to compute the *a posteriori*, without recalculation. This factor exploits the fact that the optimization of the acquisition function only finds local minima, i.e. it does not generate the same samples. A low value makes the algorithm faster, but less precise.

These modifications allow us to sample the acquisition function without having to constantly recalculate the posterior. In all cases, the number of points calculated without reoptimizing the Gaussian process is proportional to the number of samples used to optimize the previous Gaussian process. The full algorithm for Fast Bayesian sampling is depicted in Algorithm 1. This method has linear complexity with respect to the number of samples, roughly $\mathcal{O}(Ndb^2)$. Unless otherwise specified, we use the term Bayesian sampling to refer to this faster version.

5 Experiments

We use the datasets described in Table 1, a hetereogeneous sample of binary and multiclass problems for a real feature space. We assume data are normally distributed and apply a linear transformation such that, when variables are not correlated, 0.99^d samples lay inside the $[0, 1]^d$ hypercube. We split data into stratified 80/20 training and test sets and train artificial neural networks with a single hidden layer of 5 neurons.

Table 1. Description of the 6 selected datasets from the UCI machine learning repository.

Dataset	Features	Classes	Samples
bank	16	2	3616
ilpd-indian-liver	9	2	466
magic	10	2	15216
miniboone	50	2	104051
seeds	7	3	168
synthetic-control	60	6	480

Table 2. Parameters settings.

Algorithm	Parameters
Boundary	$\varepsilon = 0.01$, $\lambda = 0.05$, $\lambda' = 5$, $\mathcal{I} = \text{round}(2 + \log(N))$
	$\mathcal{T} = \text{round}(8 + 4\log(N))$, $\mathcal{N} = 5 + 2.6\log(N)$
Bayesian	$l = 0.5\sqrt{d}$, $\sigma^2 = 0.25k^2$, $\tau = 10$, $sf = 20$ $b = 1000$, $\mathcal{N} = 10$
Jacobian	$\mathcal{I} = \min(100, \text{round}(5 + N/4)$, $\mathcal{T} = 50$, $\lambda = 0.05$, $\mathcal{P} = 5$

We generate synthetic sets of size 10^6 in the restricted input space $[0, 1]^d$. For comparative purposes, we also generate samples using random sampling and an adapted version of the Jacobian-based Dataset Augmentation algorithm, proposed by [16]. The choice of parameters for each method is specified in Table 2. Because all algorithms produce new samples in an accumulative way, we generate smaller sets by selecting the first j points. We also generate balanced reference sample sets $\mathcal{W} = \{\boldsymbol{w}_i, f_{\mathcal{O}}(\boldsymbol{w}_i)\}_{i=1}^{L}$, comprised of $L = 10^7$ data points sampled uniformly at random in the $[0, 1]^d$ hypercube.

5.1   Evaluation Metrics

We evaluate sampling strategies in terms of the performance of copies built on the resulting synthetic data. We build copies using an ANN with the same architecture as above (ANN), a logistic regression (LR), a decision tree classifier (DT) and a deeper ANN with 3 hidden layers with 50 neurons each (ANN2).

We define the empirical fidelity error $R_{\mathcal{F}}$ as the disagreement between $f_{\mathcal{O}}$ and $f_{\mathcal{C}}$ over a given set of data points [21]. To compensate for the potential under-representation of one or more classes in the synthetic dataset, we also compute the *balanced empirical fidelity error*, $R_{\mathcal{F}_b}$. We report metrics averages over 10 repetitions, except for Bayesian sampling, for which we use 5 repetitions. We also provide the execution times of the different methods and sample sizes. All experiments are carried out in a single m4.16xlarge Amazon EC2 instance with 64 cores, 256 GB of RAM and 40 GB of SSD storage.

5.2   Intuition

Before discussing our results we provide an intuition of how the different methods perform on a toy dataset. Examples of synthetic sets for each algorithm are displayed in Fig. 1 for different number of samples.

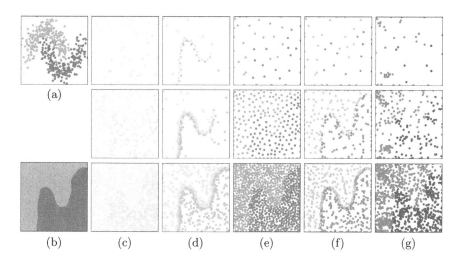

(a)

(b)   (c)   (d)   (e)   (f)   (g)

Fig. 1. (a) Training dataset and (b) decision boundary learned by an SVM with a radial basis function kernel. Synthetic datasets of sizes 50, 250 and 1000 generated using (c) Random sampling, (d) Boundary sampling, (e) Fast Bayesian sampling, (f) reoptimized Bayesian sampling and (g) adapted Jacobian sampling.

When comparing results for Boundary sampling in Fig. 1(d) to the case of exclusive uniform random sampling in Fig. 1(c), the boundary is sampled with more emphasis, although a large number of samples is required to properly cover this region. Indeed, the main advantage of Random sampling is that it scatters the samples across the space with equal probability. However, this method is oblivious to the structures of interest, *i.e.* it retains no knowledge about the form of the decision boundary for future sampling steps.

Samples generated from Bayesian sampling, in Fig. 1(e), are well distributed, without forming clusters. The importance given to the boundary on the acquisition function is not manifested in a big scale due to the simplifications. If we let the Gaussian process re-optimize several times, it correctly samples the boundary as well as the rest of the space, as shown in Fig. 1(f). Finally, synthetic data points generated with the adapted Jacobian sampling, in Fig. 1(g), form diagonal lines due to the use of the sign of the gradient.

6   Discussion of Results

In what follows we discuss our main experimental results. We first validate the generated reference sample set and then discuss the performance of the different methods, as well as their associated computational cost.

6.1   Reference Set Evaluation

We propose two checks to validate the generated reference sample sets $\mathcal{W}$. First, we fit the original architecture to the reference data and compute the corresponding balanced empirical fidelity error, $\mathcal{R}_{\mathcal{F}_b}^{\mathcal{W}}$. As a complementary check, we also evaluate the empirical fidelity error over the original set, $\mathcal{R}_{\mathcal{F}_b}^{\mathcal{D}}$. Results are shown in Table 3 with the accuracies of the original models denoted as $\mathcal{A}(f_{\mathcal{O}})$. Most values are close to 0, which we take as an indication that the reference sample sets are a suitable baseline with which to compare our proposed sampling strategies. We note the exception of the *ilpd-indian-liver* dataset, for which we are not confident enough of our evaluation.

Table 3. Quality checks for the reference sample sets.

	bank	*ilpd*	*magic*	*miniboone*	*seeds*	*synthetic*
$\mathcal{R}_{\mathcal{F}_b}^{\mathcal{W}}$	0.023	0.080	0.001	0.009	0.020	0.010
$\mathcal{R}_{\mathcal{F}_b}^{\mathcal{D}}$	0.021	0.385	0.001	0.168	0.000	0.000
$\mathcal{A}(f_{\mathcal{O}})$	0.8829	0.6410	0.8562	0.9119	0.8095	0.6750

6.2   Algorithm Evaluation

In Fig. 2 we report the balanced empirical fidelity error for the different copy architectures, sampling strategies and datasets, measured on the reference sample sets. Plots show the 20, 50 and 80 percentiles of the multiple realizations.

Boundary sampling performs well for copies based on LR, since there is a lot of information to find the optimal decision hyperplane. However, Bayesian sampling performs comparably better with fewer samples *i.e.* displays the fastest growth. This is because it focuses on globally reducing the uncertainty during the first steps. In the case of LR, it learns fast until it reaches its capacity limit. For DT models, Random sampling displays the best behavior. This may be because DTs work well when there is a sample in each region of the space in order to create the leaves. In high dimensionality, the coverage of $\mathcal{X}$ with DTs is costly, which seems to be in accordance with their slowly increasing score.

Copies based on ANN and specially on ANN2 achieve the best scores in general. We highlight the cases of *bank* and *ilpd-indian-liver* datasets for which the simpler ANN performs significantly worst, indicating that the use of matching architectures does not guarantee a good performance. This may happen

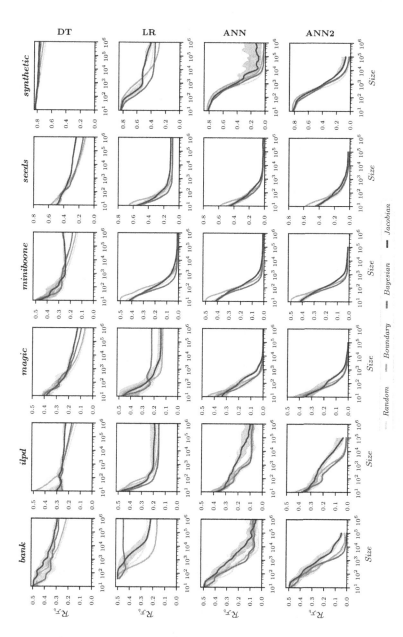

Fig. 2. Median and 20–80 percentil band. The similarity axis starts from $1/k$ for k the number of classes: the expected score for a classifier that has not learned anything. For ANN2 models, we only show results for 10^5 samples, due to the high training times.

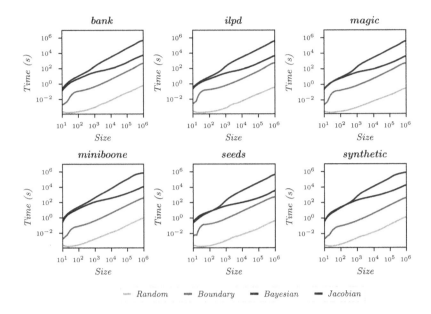

Fig. 3. Execution time of the different sampling strategies as a function of dataset size.

because the characteristics of the problem change when using the synthetic dataset instead of the training data. This, together with the fact that the original architecture has just enough degrees of freedom to replicate the decision boundary, deters the copy from converging to the same solution.

In terms of the aggregated comparison among techniques, Jacobian sampling seems to perform the worst. This method generates linear structures which contain a large number of samples. As a result, it has a wide uncertainty band. For a large number of samples Random sampling gathers the greatest number of victories. Closely behind, Boundary and Bayesian sampling are both reasonably similar in terms of their averaged performance.

6.3   Computational Cost

Figure 3 shows the computational cost of the different algorithms for copies based on ANN2, i.e. the worst-case scenario. The execution time is asymptotically linear. Bayesian sampling and Random sampling are the slowest and the fastest, respectively. A great advantage of Random sampling is its simplicity, and consequently its low cost. Its main drawback is, however, that it samples points with no regards to the form of the decision function or the resulting class distribution. In high dimensional problems, Boundary sampling may be a good compromise between time and accuracy. In the absence of any time constrain, however, Bayesian sampling ensures a more reliable exploration.

7   Conclusions and Future Work

In this paper we evaluate different algorithms to sample unknown decision functions. We conclude that Bayesian sampling is the most promising, despite its high computational cost in high dimensional datasets. Strategies that focus on boundaries tend to outperform the rest when building copies with low capacity models, such as LR. Random sampling exhibits an adequate behavior in general settings. However, its use is not encouraged when the volume of one or more classes is substantially smaller than that of the rest.

References

1. Bucila, C., Caruana, R., Niculescu-Mizil, A.: Model Compression. In: SIGKDD, pp. 535–541 (2006)
2. Dagan, I., Engelson, S.P.: Committee-based sampling for training probabilistic classifiers. In: ICML, pp. 150–157 (1995)
3. Doucet, A., de Freitas, N., Gordon, N.: An introduction to sequential Monte Carlo methods. In: Doucet, A., de Freitas, N., Gordon, N. (eds.) Sequential Monte Carlo Methods in Practice. Statistics for Engineering and Information Science, pp. 3–14. Springer, New York (2001). https://doi.org/10.1007/978-1-4757-3437-9_1
4. Fredrikson, M., Jha, S., Ristenpart, T.: Model inversion attacks that exploit confidence information and basic countermeasures. In: SIGSAC (2015)
5. Gilks, W.R., Wild, P.: Adaptive rejection sampling for Gibbs sampling. J. Roy. Stat. Soc.: Ser. C (Appl. Stat.) 41(2), 337–348 (1992)
6. Gilks, W., Richardson, S., Spiegelhalter, D.: Markov Chain Monte Carlo in Practice. Chapman and Hall/CRC, London (1996)
7. Goodman, B., Flaxman, S.: European Union regulations on algorithmic decision-making and a right to explanation. AI Mag. 38(3), 50–57 (2017)
8. Hinton, G., Vinyals, O., Dean, J.: Distilling the knowledge in a neural network. In: NIPS Deep Learning and Representation Learning Workshop (2015)
9. Houlsby, N., Huszar, F., Ghahramani, Z., Lengyel, M.: Bayesian active learning for classification and preference learning. eprint arXiv:1112.5745 (2011)
10. Lewis, D., Catlett, J.: Heterogeneous uncertainty sampling for supervised learning. In: Proceedings of International Conference on Machine Learning (ICML), pp. 148–156 (1994)
11. Lewis, D., Gale, W.: A sequential algorithm for training text classifiers. In: Croft, B.W., van Rijsbergen, C.J. (eds.) SIGIR 1994, pp. 3–12. Springer, London (1994). https://doi.org/10.1007/978-1-4471-2099-5_1
12. Liu, R., Fusi, N., Mackey, L.: Teacher-student compression with generative adversarial networks. eprint arXiv:1812.02271 (2018)
13. Liu, Y., Chen, X., Liu, C., Song, D.: Delving into transferable adversarial examples and black-box attacks. In: ICLR (2017)
14. Mansinghka, V., Roy, D., Jonas, E., Tenenbaum, J.: Exact and approximate sampling by systematic stochastic search. In: AISTATS, pp. 400–407 (2009)
15. Papandreou, G., Yuille, A.L.: Gaussian sampling by local perturbations. In: NIPS, pp. 1858–1866 (2010)
16. Papernot, N., McDaniel, P., Goodfellow, I., Jha, S., Berkay Celik, Z., Swami, A.: Practical black-box attacks against ML. In: ASIACCS, pp. 506–519 (2017)

17. Sculley, D., et al.: Hidden technical debt in machine learning systems. In: NIPS, pp. 2503–2511 (2015)
18. Settles, B.: Active learning literature survey. Computer Sciences Technical report 1648, University of Wisconsin-Madison (2009)
19. Shokri, R., Tech, C., Stronati, M., Shmatikov, V.: Membership inference attacks against machine learning models. In: S&P, pp. 3–18 (2017)
20. Tierney, L.: Markov chains for exploring posterior distributions. Ann. Stat. **22**, 1701–1762 (1996)
21. Unceta, I., Nin, J., Pujol, O.: Copying machine learning classifiers. eprint arXiv:1903.01879 (2019)
22. Vapnik, V.N.: The Nature of Statistical Learning Theory. Springer, New York (2000). https://doi.org/10.1007/978-1-4757-3264-1
23. Willett, R., Nowak, R., Castro, R.M.: Faster rates in regression via active learning. In: NIPS, pp. 179–186 (2006)
24. Yang, C., Xie, L., Qiao, S., Yuille, A.L.: Knowledge distillation in generations: more tolerant teachers educate better students. eprint arXiv:1711.09784 (2018)

Towards Analogy-Based Explanations in Machine Learning

Eyke Hüllermeier[(✉)]

Heinz Nixdorf Institute, Department of Computer Science, Intelligent Systems and
Machine Learning Group, Paderborn University, Paderborn, Germany
eyke@upb.de

Abstract. Principles of analogical reasoning have recently been applied
in the context of machine learning, for example to develop new methods
for classification and preference learning. In this paper, we argue that,
while analogical reasoning is certainly useful for constructing new learning algorithms with high predictive accuracy, is arguably not less interesting from an interpretability and explainability point of view. More
specifically, we take the view that an analogy-based approach is a viable
alternative to existing approaches in the realm of explainable AI and
interpretable machine learning, and that analogy-based explanations of
the predictions produced by a machine learning algorithm can complement similarity-based explanations in a meaningful way. To corroborate
these claims, we outline the basic idea of an analogy-based explanation
and illustrate its potential usefulness by means of some examples.

Keywords: Machine learning · Analogy · Similarity · Classification ·
Ranking · Interpretability · Explainability

1 Introduction

Over the past couple of years, the idea of explainability and related notions such
as transparency and interpretability have received increasing attention in artificial intelligence (AI) in general and machine learning (ML) in particular. This is
mainly due to the ever growing number of real-world applications of AI technology and the increasing level of autonomy of algorithms taking decisions on behalf
of people, and hence of the social responsibility of computer scientists developing
these algorithms. Meanwhile, algorithmic decision making has a strong societal
impact, which has led to the quest for understanding such decisions, or even
to claiming a "right to explanation" [12]. Explainability is closely connected to
other properties characterizing a "responsible" or "trustworthy" AI/ML, such
as fairness, safety, robustness, responsibility, and accountability, among others.

Machine learning models, or, more specifically, the predictors induced by a
machine learning algorithm on the basis of suitable training data, are not immediately understandable most of the time. This is especially true for the most
"fashionable" class of ML algorithms these days, namely deep neural networks.

V. Torra et al. (Eds.): MDAI 2020, LNAI 12256, pp. 205–217, 2020.
https://doi.org/10.1007/978-3-030-57524-3_17

On the contrary, a neural network is a typical example of what is called a "black-box" model in the literature: It takes inputs and produces associated outputs, often with high predictive accuracy, but the way in which the inputs and outputs are related to each other, and the latter are produced on the basis of the former, is very intransparent, as it involves possibly millions of mathematical operations and nonlinear transformations conducted by the network (in an attempt to simulate the neural activity of a human brain). A lack of transparency and interpretability is arguably less problematic for other ML methodology with a stronger "white-box" character, most notably symbol-oriented approaches such as rules and decision trees. Yet, even for such methods, interpretability is far from being guaranteed, especially because accurate models often require a certain size and complexity. For example, even if a decision tree might be interpretable in principle, a tree with hundreds of nodes will hardly be understandable by anyone.

The lack of transparency of contemporary ML methodology has triggered research that is aimed at improving the interpretability of ML algorithms, models, and predictions. In this regard, various approaches have been put forward, ranging from "interpretability by design", i.e., learning models with in-built interpretability, to model-agnostic explanations—a brief overview will be given in the next section. In this paper, we propose to add principles of *analogical reasoning* [11] as another alternative to this repertoire. Such an approach is especially motivated by so-called example-based explanations, which refer to the notion of *similarity*. Molnar [16] describes the blueprint of such explanations as follows: "Thing B is similar to thing A and A caused Y, so I predict that B will cause Y as well". In a machine learning context, the "things" are data entities (instances), and the causes are predictions. In (binary) classification, for example, the above pattern might be used to explain the learner's prediction for a query instance: A belongs to the positive class, and B is similar to A, hence B is likely to be positive, too. Obviously, this type of explanation is intimately connected to the nearest neighbor estimation principle [8].

Now, while similarity establishes a relationship between pairs of objects (i.e. tuples), an analogy involves four such objects (i.e. quadruples). The basic regularity assumption underlying analogical reasoning is as follows: Given objects A, B, C, D, if A relates to B as C relates to D, then this "relatedness" also applies to the properties caused by these objects (for example, the predictions produced by an ML model). Several authors have recently elaborated on the idea of using analogical reasoning for the purpose of (supervised) machine learning [2,5,6], though without raising the issue of interpretability. Here, we will argue that analogy-based explanations can complement similarity-based explanations in a meaningful way.

The remainder of the paper is organized as follows. In the next section, we give a brief overview of different approaches to interpretable machine learning. In Sect. 3, we provide some background on analogy-based learning—to this end, we recall the basics of a concrete method that was recently introduced in [2]. In Sect. 4, we elaborate on the idea on analogy-based explanations in machine learning, specifically focusing on classification and preference learning.

2   Interpretable Machine Learning

In the realm of interpretable machine learning, two broad approaches are often distinguished. The first is to learn models that are inherently interpretable, i.e., models with in-built transparency that are interpretable by design. Several classical ML methods are put into this category, most notably symbol-oriented approaches like decision trees, but also methods that induce "simple" (typically linear) mathematical models, such as logistic regression. The second approach is to extract interpretable information from presumably intransparent "black-box" models. Within this category, two subcategories can be further distinguished.

In the first subcategory, *global* approximations of the entire black-box model are produced by training more transparent "white-box" models as a surrogate. This can be done, for example, by using the black-box model as a teacher, i.e., to produce training data for the white-box model [3]. In the second subcategory, which is specifically relevant for this paper, the idea is to extract interpretable *local* information, which only pertains to a restricted region in the instance space, or perhaps only to a single instance. In other words, the idea is to approximate a black-box model only locally instead of globally, which, of course, can be accomplished more easily, especially by simple models. Prominent examples of this approach are LIME [17] and SHAP [14]. These approaches are qualified as *model agnostic*, because they use the underlying model only as a black-box that is queried for the purpose of data generation.

In addition to generic, universally applicable methods of this kind, there are various methods for extracting useful information that are specifically tailored to certain model classes, most notably deep neural networks [18]. Such methods seek to provide some basic understanding of how such a network connects inputs with outputs. To this end, various techniques for making a network more transparent have been proposed, many of them based on the visualization of neural activities.

Interestingly, the focus in interpretable machine learning has been very much on classification so far, while other ML problems have been considered much less. In particular, there is very little work on interpretable preference learning and ranking [10]. As will be argued later on, the idea of analogy-based explanation appears to be especially appealing from this point of view.

3   Analogy-Based Learning

In this section, we briefly recall the basic ideas of an analogy-based learning algorithm that was recently introduced in [2]. This will set the stage for our discussion of analogy-based explanation in the next section, and provide a basis for understanding the main arguments put forward there.

The mentioned approach proceeds from the standard setting of supervised learning, in which data objects (instances) are described in terms of feature vectors $x = (x_1, \ldots, x_d) \in \mathcal{X} \subseteq \mathbb{R}^d$. The authors are mainly interested in the problem of *ranking*, i.e., in learning a ranking function ρ that accepts any (query) subset $Q = \{x_1, \ldots, x_n\} \subseteq \mathcal{X}$ of instances as input. As output, the function produces a ranking in the form of a total order of the instances, which can be represented by a permutation π, with $\pi(i)$ the rank of instance x_i. Yet, the algorithmic principles underlying this approach can also be used for the purpose of classification. In the following, to ease explanation, we shall nevertheless stick to the case of ranking.

3.1 Analogical Proportions

The approach essentially builds on the following inference pattern: If object a relates to object b as c relates to d, and knowing that a is preferred to b, we (hypothetically) infer that c is preferred to d. This principle is formalized using the concept of analogical proportion [15]. For every quadruple of objects a, b, c, d, the latter provides a numerical degree to which these objects are in analogical relation to each other. To this end, such a degree is first determined for each attribute value (feature) separately, and these degrees are then combined into an overall degree of analogy.

More specifically, consider four values a, b, c, d from an attribute domain $\mathbb{X}$. The quadruple (a, b, c, d) is said to be in analogical proportion, denoted by $a : b :: c : d$, if "a relates to b as c relates to d", or formally:

$$E\big(\mathcal{R}(a, b), \mathcal{R}(c, d)\big), \tag{1}$$

where the relation E denotes the "as" part of the informal description. $\mathcal{R}$ can be instantiated in different ways, depending on the underlying domain $\mathbb{X}$:

- In the case of Boolean variables, where $\mathbb{X} = \{0, 1\}$, there are $2^4 = 16$ instantiations of the pattern $a : b :: c : d$, of which only the following 6 satisfy a set of axioms required to hold for analogical proportions:

a	b	c	d
0	0	0	0
0	0	1	1
0	1	0	1
1	0	1	0
1	1	0	0
1	1	1	1

This formalization captures the idea that a differs from b (in the sense of being "equally true", "more true", or "less true", if the values 0 and 1 are interpreted as truth degrees) exactly as c differs from d, and vice versa.
- In the numerical case, assuming all attributes to be normalized to the unit interval $[0, 1]$, the concept of analogical proportion can be extended on the basis of generalized logical operators [6,9]. In this case, the analogical proportion will become a matter of degree, i.e., a quadruple (a, b, c, d) can be in analogical proportion *to some degree* between 0 and 1. An example of such a proportion, with $\mathcal{R}$ being the arithmetic difference, i.e., $\mathcal{R}(a, b) = a - b$, is the following:

$$v(a, b, c, d) = \begin{cases} 1 - |(a - b) - (c - d)|, & \text{if } \text{sign}(a - b) = \text{sign}(c - d) \\ 0, & \text{otherwise.} \end{cases} \quad (2)$$

Note that this formalization indeed generalizes the Boolean case (where $a, b, c, d \in \{0, 1\}$). Another example is geometric proportions $\mathcal{R}(a, b) = a/b$.

To extend analogical proportions from individual values to complete feature vectors, the individual degrees of proportion can be combined using any suitable aggregation function, for example the arithmetic mean:

$$v(\boldsymbol{a}, \boldsymbol{b}, \boldsymbol{c}, \boldsymbol{d}) = \frac{1}{d} \sum_{i=1}^{d} v(a_i, b_i, c_i, d_i).$$

3.2   Analogical Prediction

The basic idea of analogy-based learning is to leverage analogical proportions for the purpose of *analogical transfer*, that is, to transfer information about the target of prediction. In the case of preference learning, the target could be the preference relation between two objects $\boldsymbol{c}$ and $\boldsymbol{d}$, i.e., whether $\boldsymbol{c} \succ \boldsymbol{d}$ or $\boldsymbol{d} \succ \boldsymbol{c}$. Likewise, in the case of classification, the target could be the class label of a query object $\boldsymbol{d}$ (cf. Fig. 1 for an illustration).

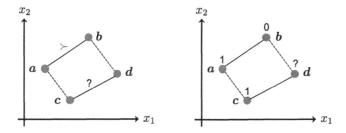

Fig. 1. Illustration of analogy-based prediction: The four objects $\boldsymbol{a}$, $\boldsymbol{b}$, $\boldsymbol{c}$, $\boldsymbol{d}$ are in analogical proportion to each other. In the case of preference learning, the known preference $\boldsymbol{a} \succ \boldsymbol{b}$ would hence be taken as an indication that $\boldsymbol{c} \succ \boldsymbol{d}$ (left). Likewise, in the case of binary classification, knowing that $\boldsymbol{a}$ and $\boldsymbol{c}$ are positive while $\boldsymbol{b}$ is negative, analogical inference suggests that $\boldsymbol{d}$ is negative, too (right).

In the context of preference learning, the authors in [2] realize analogical transfer in the style of the k-nearest neighbor approach: Given a query pair $(\boldsymbol{c}, \boldsymbol{d})$, they search for the tuples $(\boldsymbol{a}_i, \boldsymbol{b}_i)$ in the training data producing the k highest analogies $\{\boldsymbol{a}_i : \boldsymbol{b}_i :: \boldsymbol{c} : \boldsymbol{d}\}_{i=1}^{k}$. Since the preferences between $\boldsymbol{a}_i$ and $\boldsymbol{b}_i$ are given as part of the training data, each of these analogies suggests either $\boldsymbol{c} \succ \boldsymbol{d}$ or $\boldsymbol{d} \succ \boldsymbol{c}$ by virtue of analogical transfer, i.e., each of them provides a vote in favor of the first or the second case. Eventually, the preference with the higher number of votes is adopted, or the distribution of votes is turned into an estimate of the probability of the two cases.

Obviously, a very similar principle could be invoked in the case of (binary) classification. Here, given a query instance d, one would search for triplets (a_i, b_i, c_i) in the training data forming strong analogies $a_i : b_i :: c_i : d$, and again invoke the principle of analogical transfer to conjecture about the class label of d. Each analogy will suggest the positive or the negative class, and the corresponding votes could then be aggregated in one way or the other.

3.3 Feature Selection

Obviously, the feature representation of objects will have a strong effect on whether, or to what extent, the analogy assumption applies to a specific problem, and hence influence the success of the analogical inference principle. Therefore, prior to applying analogical reasoning methods, it could make sense to find an embedding of objects in a suitable space, so that the assumption of the above inference pattern holds true in that space. This is comparable, for example, to embedding objects in $\mathbb{R}^d$ in such a way that the nearest neighbor rule with Euclidean distance yields good predictions in a classification task.

In [1], the authors address the problem of *feature selection* [13] in analogical inference, which can be seen as a specific type of embedding, namely a projection of the data from the original feature space to a subspace. By ignoring irrelevant or noisy features and restricting to the most relevant dimensions, feature selection can often improve the performance of learning methods. Moreover, feature selection is also important from an explainability point of view, because the representation of data objects in terms of meaningful features is a basic prerequisite for the interpretability of a machine learning model operating on this representation. In this regard, feature selection is also more appropriate than general feature embedding techniques. The latter typically produce new features in the form of (nonlinear) combinations of the original features, which lose semantic meaning and are therefore difficult to interpret.

4   Analogy-Based Explanation

To motivate an analogy-based explanation of predictions produced by an ML algorithm, let us again consider the idea of similarity-based explanation as a starting point. As we shall argue, the former can complement the latter in a meaningful way, especially because it refers to a different type of "knowledge transfer". As before, we distinguish between two exemplary prediction tasks, namely classification and ranking. This distinction is arguably important, mainly for the following reason: In classification, a property (class membership) is assigned to a *single* object x, whereas in ranking, a property (preference) is ascribed to a *pair* of objects (c, d). Moreover, in the case of ranking, the property is in fact a relation, namely a binary preference relation. Thus, since

analogy-based inference essentially deals with "relatedness", ranking and preference learning lends itself to analogy-based explanation quite naturally, perhaps
even more so than classification.

For the purpose of illustration, we make use of a data set that classifies 172
scientific journals in the field of pure mathematics into quality categories A^*, A,
B, C [4]. Each journal is moreover scored in terms of 5 criteria, namely

- cites: the total number of citations per year;
- IF: the well-known impact factor (average number of citations per article
  within two years after publication);
- II: the immediacy index measures how topical the articles published in a
  journal are (cites to articles in current calendar year divided by the number
  of articles published in that year);
- articles: the total number of articles published;
- half-line: cited half-life (median age of articles cited).

In a machine learning context, a classification task may consist of predicting the
category of a journal, using the scores on the criteria as features. Likewise, a
ranking task may consist of predicting preferences between journals, or predicting an entire ranking of several journals.

4.1   Explaining Class Predictions

Similarity-based explanations typically "justify" a prediction by referring to local
(nearest neighbor) information in the vicinity of a query instance x. In the
simplest case, the nearest neighbor of x is retrieved from the training data, and
its class label is provided as a justification (cf. Fig. 2): "There is a case x' that
is similar to x, and which belongs to class y, so x is likely to belong to y as
well". For example, there is another journal with similar scores on the different
criteria, and which is ranked in category A, which explains why this journal is
also put into this category.

A slightly more general approach is to retrieve, not only the single nearest
neighbor but the k nearest neighbors, and to provide information about the
distribution of classes among this set of examples. Information of that kind is
obviously useful, as it conveys an idea of the confidence and reliability of a
prediction. If many neighbors are all from the same class, this will of course
increase the trust in a prediction. If, on the other side, the distribution of classes
in the neighborhood is mixed, the prediction might be considered as uncertain,
and hence the explanation as less convincing.

Similarity- or example-based explanations of this kind are very natural and
suggest themselves if a nearest neighbor approach is used by the learner to make
predictions. It should be mentioned, however, that similarity-based explanations
can also be given if predictions are produced by another type of model (like the
discriminative model indicated by the dashed decision boundary in Fig. 2). In
this case, the nearest neighbor approach serves as a kind of surrogate model. This
could be justified by the fact that most machine learning methods do indeed obey

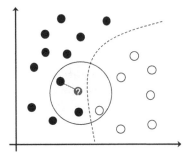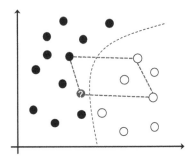

Fig. 2. Left: Illustration of similarity-based explanation in a binary classification setting (black points are positive examples, white ones negative). The shaded circle indicated the neighborhood (of size 3) of the query instance (red point). The dashed line is a discriminant function that could have been induced by another (global) learning method. Right: Illustration of analogy-based explanation. Again, the query point is shown in red. Together with the three other points, it forms an analogical relationship. (Color figure online)

the regularity assumption underlying similarity-based inference. For example, a discriminant function is more likely to assign two similar objects to the same class than to separate them, although such cases do necessarily exist as well.

Obviously, a key prerequisite of the meaningfulness of similarity-based explanations is a meaningful notion of similarity, formalized in terms of an underlying similarity or distance function. This assumption is far from trivial, and typically not satisfied by "default" metrics like the Euclidean distance. Instead, a meaningful measure of distance needs to properly modulate the influence of individual features, because not all features might be of equal importance. Besides, such a measure should be able to capture interactions between features, which might not be considered independently of each other. For example, depending on the value of one feature, another feature might be considered more or less important (or perhaps completely ignored, as it does not apply any more). In other words, a meaningful measure of similarity may require a complex aggregation of the similarities of individual features [7]. Needless to say, this may hamper the usefulness of similarity-based explanations: If a user cannot understand why two cases are deemed similar, she will hardly accept a similar case as an explanation.

Another issue of the similarity-based approach, which brings us to the analogy-based alternative, is related to the difficulty of interpreting a *degree* of similarity or distance. Often, these degrees are not normalized (especially in the case of distance), and therefore difficult to judge: How similar is similar? What minimal degree of similarity should be expected in an explanation? For example, when explaining the categorization of a journal as B by pointing to a journal with similar properties, which is also rated as B, one may wonder whether the difference between them is really so small, or not perhaps big enough to justify a categorization as A.

An analogy-based approach might be especially interesting in this regard, as it explicitly refers to this distance, or, more generally, the relatedness between data entities. In the above example, an analogy-based explanation given to the manager of a journal rated as B instead of A might be of the following kind: "There are three other journals, two rated A (a and c) and one rated B (b). The relationship between a and b is very much the same as the relationship between c and your journal, and b was also rated B". For example, a and b may have the same characteristics, except that b has 100 more articles published per year. The manager might now be more content and accept the decision more easily, because she understands that 100 articles more or less can make the difference.

A concrete example for the journal data is shown in Fig. 3. Here, an analogy is found for a journal with (normalized) scores of 0.03, 0.06, 0.08, 0.04, 1 on the five criteria. To explain why this journal is only put in category C but not in B, three other journals a, b, c are found, a from category A, and b and c from category B, so that $a : b :: c : d$ (the degree of analogical proportion is ≈ 0.98). Note that the score profiles of a and c resp. b and d are not very similar themselves, which is not surprising, because a and c are from different categories. Still, the four journals form an analogy, in the sense that an A-journal relates to a B-journal as a B-journal relates to a C-journal.

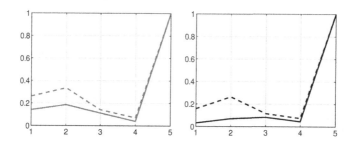

Fig. 3. Example of an analogy $a : b :: c : d$ in the journal data set, with a (dashed line) and b (solid) on the left panel, c (dashed line) and d (solid) on the right.

Note that this type of explanation is somewhat related to the idea of explaining with *counterfactuals* [19], although the cases forming an analogy are of course factual and not counterfactual. Nevertheless, an analogy-based explanation may give an idea of what changes of features might be required to achieve a change in the classification. On the other side, an analogy can of course also explain why a certain difference is *not* enough. For example, explaining the rating as B to a journal d by pointing to another journal c might not convince the journal manager, because she feels that, despite the similarity, her journal is still a bit better in most of the criteria. Finding an analogy $a : b :: c : d$, with journals a and b also categorized as B, may then be helpful and convince her that the difference is not significant enough.

Of course, just like the nearest neighbor approach, analogy-based explanations are not restricted to a single analogy. Instead, several analogies, perhaps

sorted by their strength (degree of analogical proportion), could be extracted
from the training data. In this regard, another potential advantage of an analogy-
based compared to a similarity-based approach should be mentioned: While both
approaches are local in the sense of giving an explanation for a specific query
instance, i.e., local with respect to the *explanandum*, the similarity-based approach is also local with respect to the *explanans*, as it can only refer to cases
in the vicinity of the query—which may or may not exist. The analogy-based
approach, on the other side, is not restricted in this sense, as the explanans is
not necessarily local. Instead, a triplet a, b, c forming an analogy with a query d
can be distributed in any way, and hence offers many more possibilities for constructing an explanation (see also Fig. 4). Besides, one should note that there are
much more triplets than potential nearest neighbors (the former scales cubicly
with the size of the training data, the latter only linearly).

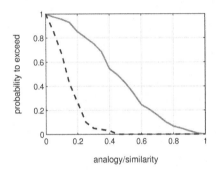

Fig. 4. Decumulative analogy/similarity distribution: The solid line shows the average
percentage of triplets (y-axis) a, b, c in the journal data set exceeding a degree of
analogical proportion (x-axis) with a specific query d. Likewise, the dashed line shows
the average percentage of examples in the training data exceeding a degree of similarity
$(1 - L_1\text{-distance})$ with a query instance. As can be seen, the latter drops much faster
than the former.

Last but not least, let us mention that analogy-based explanations, just like
similarity-based explanations, are in principle not limited to analogical learning,
but could also be used in a model-agnostic way, and as a surrogate for other
models.

4.2   Explaining Preference Predictions

In the case of classification, training data is given in the form of examples of
the form (x, y), where x is an element from the instance space $\mathcal{X}$ and y a class
label. In the case of preference learning, we assume training data in the form of
pairwise preferences $a \succ b$, and the task is to infer the preferential relationship
for a new object pair $(c, d) \in \mathcal{X} \times \mathcal{X}$ given as a query (or a ranking of more

than two objects, in which case the prediction of pairwise preferences could be an intermediate step). How could one explain a prediction $c \succ d$?

The similarity-based approach amounts to finding a "similar" preference $a \succ b$ in the training data. It is not immediately clear, however, what similarity of preferences is actually supposed to mean, even if a similarity measure on $\mathcal{X}$ is given. A natural solution would be a conjunctive combination: a preference $a \succ b$ is similar to $c \succ d$ if both a is similar to c and b is similar to d. This requirement might be quite strong, so that finding similar preferences gets difficult.

The analogy-based approach can be seen as a relaxation. Instead of requiring the objects to more or less coincide, they are only supposed to stand in a similar relationship to each other, i.e., $\mathcal{R}(a, b) \sim \mathcal{R}(c, d)$. The relationship $\mathcal{R}$ *can* mean similarity, but does not necessarily need to do so (as shown by the definition of $\mathcal{R}$ in terms of arithmetic or geometric proportions).

The explanation of preferences in terms of analogy appears to be quite natural. For the purpose of illustration, consider again our example: Why did the learner predict a preference $c \succ d$, i.e., that journal c is ranked higher (evaluated better than) journal d? To give an explanation, one could find a preference $a \succ b$ between journals in the training data, so that (a, b) is in analogical proportion to (c, d). In the case of arithmetic proportions, this means that the feature values (ratings of criteria) of a deviate from the feature values of d in much the same way as those of c deviate from those of d, and this deviation will then serve as an explanation of the preference.

5   Conclusion and Future Work

In this paper, we presented some preliminary ideas on leveraging the principle of analogy for the purpose of explanation in machine learning. This is essentially motivated by the recent interest in analogy-based approaches to ML problems, such as classification and preference learning, though hitherto without explicitly addressing the notion of interpretability. In particular, we tried to highlight the potential of an analogy-based approach to complement similarity-based (example-based) explanation in a reasonable way.

Needless to say, our discussion is just a first step, and the evidence we presented in favor of analogy-based explanations is more of an anecdotal nature. In future work, the basic ideas put forward need to be worked out in detail, and indeed, there are many open questions to be addressed. For example, which analogy in a data set is best suited for explaining a specific prediction? In the case of analogy, the answer appears to be less straightforward than in the case of similarity, where more similarity is simply better than less. Also, going beyond the retrieval of a single analogy for explanation, how should one assemble a good composition of analogies? Last but not least, it would of course also be important to evaluate the usefulness of analogy-based explanation in a more systematic way, ideally in a user study involving human domain experts.

References

1. Ahmadi Fahandar, M., Hüllermeier, E.: Feature selection for analogy-based learning to rank. In: Kralj Novak, P., Šmuc, T., Džeroski, S. (eds.) DS 2019. LNCS (LNAI), vol. 11828, pp. 279–289. Springer, Cham (2019). https://doi.org/10.1007/978-3-030-33778-0_22

2. Ahmadi Fahandar, M., Hüllermeier, E.: Learning to rank based on analogical reasoning. In: Proceedings AAAI-2018, 32th AAAI Conference on Artificial Intelligence, Louisiana, USA, New Orleans, pp. 2951–2958 (2018)

3. Andrews, R., Diederich, J., Tickle, A.: Survey and critique of techniques for extracting rules from trained artificial neural networks. Knowl. Based Syst. **8**(6), 373–389 (1995)

4. Beliakov, G., James, S.: Citation-based journal ranks: the use of fuzzy measures. Fuzzy Sets Syst. (2010)

5. Bounhas, M., Pirlot, M., Prade, H.: Predicting preferences by means of analogical proportions. In: Cox, M.T., Funk, P., Begum, S. (eds.) ICCBR 2018. LNCS (LNAI), vol. 11156, pp. 515–531. Springer, Cham (2018). https://doi.org/10.1007/978-3-030-01081-2_34

6. Bounhas, M., Prade, H., Richard, G.: Analogy-based classifiers for nominal or numerical data. Int. J. Approx. Reason. **91**, 36–55 (2017)

7. Cheng, W., Hüllermeier, E.: Learning similarity functions from qualitative feedback. In: Althoff, K.-D., Bergmann, R., Minor, M., Hanft, A. (eds.) ECCBR 2008. LNCS (LNAI), vol. 5239, pp. 120–134. Springer, Heidelberg (2008). https://doi.org/10.1007/978-3-540-85502-6_8

8. Cover, T., Hart, P.: Nearest neighbor pattern classification. IEEE Trans. Inf. Theory IT **13**, 21–27 (1967)

9. Dubois, D., Prade, H., Richard, G.: Multiple-valued extensions of analogical proportions. Fuzzy Sets Syst. **292**, 193–202 (2016)

10. Fürnkranz, J., Hüllermeier, E.: Preference Learning. Springer, Heidelberg (2011). https://doi.org/10.1007/978-3-642-14125-6

11. Gentner, D.: The mechanisms of analogical reasoning. In: Vosniadou, S., Ortony, A. (eds.) Similarity and Analogical Reasoning, pp. 197–241. Cambridge University Press, Cambridge (1989)

12. Goodman, R., Flaxman, S.: European Union regulations on algorithmic decision-making and a "right to explanation". AI Mag. **38**(3), 1–9 (2017)

13. Guyon, I., Elisseeff, A.: An introduction to variable and feature selection. J. Mach. Learn. Res. **3**, 1157–1182 (2003)

14. Lundberg, S.M., Lee, S.I.: A unified approach to interpreting model predictions. In: Processing of NeurIPS, Advances in Neural Information Processing Systems, pp. 4765–4774 (2017)

15. Miclet, L., Prade, H.: Handling analogical proportions in classical logic and fuzzy logics settings. In: Sossai, C., Chemello, G. (eds.) ECSQARU 2009. LNCS (LNAI), vol. 5590, pp. 638–650. Springer, Heidelberg (2009). https://doi.org/10.1007/978-3-642-02906-6_55

16. Molnar, C.: Interpretable Machine Learning: A Guide for Making Black Box Models Explainable (2018). http://leanpub.com/interpretable-machine-learning

17. Ribeiro, M.T., Singh, S., Guestrin, C.: "Why should I trust you?" Explaining the predictions of any classifier. In: Proceedings of the 22nd ACM SIGKDD International Conference on Knowledge Discovery and Data Mining, pp. 1135–1144 (2016)

18. Samek, W., Montavon, G., Vedaldi, A., Hansen, L.K., Müller, K.-R. (eds.): Explainable AI: Interpreting, Explaining and Visualizing Deep Learning. LNCS (LNAI), vol. 11700. Springer, Cham (2019). https://doi.org/10.1007/978-3-030-28954-6
19. Van Looveren, A., Klaise, J.: Interpretable counterfactual explanations guided by prototypes. CoRR abs/1907.02584 (2019). http://arxiv.org/abs/1907.02584

An Improved Bi-level Multi-objective Evolutionary Algorithm for the Production-Distribution Planning System

Malek Abbassi[✉], Abir Chaabani, and Lamjed Ben Said

SMART Lab, ISG, University of Tunis (Université de Tunis), Tunis, Tunisia
malekkabbassi@gmail.com

Abstract. Bi-level Optimization Problem (BOP) presents a special class of challenging problems that contains two optimization tasks. This nested structure has been adopted extensively during recent years to solve many real-world applications. Besides, a number of solution methodologies are proposed in the literature to handle both single and multi-objective BOPs. Among the well-cited algorithms solving the multi-objective case, we find the Bi-Level Evolutionary Multi-objective Optimization algorithm (BLEMO). This method uses the elitist Non-dominated Sorting Genetic Algorithm (NSGA-II) with the bi-level framework to solve Multi-objective Bi-level Optimization Problems (MBOPs). BLEMO has proved its efficiency and effectiveness in solving such kind of NP-hard problem over the last decade. To this end, we aim in this paper to investigate the performance of this method on a new proposed multi-objective variant of the Bi-level Multi Depot Vehicle Routing Problem (Bi-MDVRP) which is a well-known problem in combinatorial optimization. The proposed BLEMO adaptation is further improved combining jointly three techniques in order to accelerate the convergence rate of the whole algorithm. Experimental results on well-established benchmarks reveal a good performance of the proposed algorithm against the baseline version.

Keywords: Bi-level combinatorial optimization · MOEA · BLEMO · Bi-level production-distribution planning problem

1 Introduction

Bi-level modelling emerges commonly in many practical optimization problems involving two nested structured optimization tasks, where the resolution of the first level, hereafter the upper level, should be performed under the given response of the second level, hereafter the lower level. The decision process incorporates two types of Decision Makers (DMs): the leader who optimizes his upper objective function(s) using the upper variables, and the follower returns optimal feasible solutions. Moreover, the realized outcome of any solution or decision taken by the upper level authority to optimize his objective(s), is affected by the response of lower level entities, who try to optimize their own outcomes.

© Springer Nature Switzerland AG 2020
V. Torra et al. (Eds.): MDAI 2020, LNAI 12256, pp. 218–229, 2020.
https://doi.org/10.1007/978-3-030-57524-3_18

According to this description, we conclude that BOPs face enormous difficulties due primarily to the nested structure that characterizes this type of problem. A challenging variant of BOP is the multi-objective BOP (MBOP) which brings more complexities associated with the multiple conflicting entities.

Resolution methods dedicated to solve BOPs are classified into two categories: (1) the classical methods that apply mathematical derivations based on an exact resolution process, and (2) the evolutionary ones which apply approximate solving processes. The first family guarantees optimal solutions, but this involves enormous processing. Typical classical methods include: penalty functions [3], Karush–Kuhn–Tucker approach [5], branch-and-bound techniques [4], etc.

Evolutionary methods evolve a set of solutions in order to find the approximate best ones. The strength of evolutionary algorithms lies in their ability to provide an approximation of optimal solutions with less computational time cost. A number of research works are proposed in this category, for instance, we cite: nested evolutionary algorithms [1,10], Bi-level Particle Swarm Optimization (Bi-level PSO) for both single and multi-objective BOPs [7,16]. BLEMO (Bi-Level Evolutionary Multi-objective Optimization algorithm) is proposed as an evolutionary algorithm able to solve efficiently MBOP in the continuous area [10].

The bi-level framework has been successfully adopted to model many real-life applications such as network topology [16], active distribution networks [15], power systems [11], banking operations [14], etc. In this paper, we aim to solve the production-distribution planning system a well-known NP-hard problem in supply chain management. The main interest of this paper is to exploit the main features of the BLEMO method in order to solve efficiently an NP-hard combinatorial MBOP. To this end, the principal contributions of this paper are the following:

1. Proposing a new multi-objective bi-level formulation of the Bi-MDVRP [6].
2. Proposing an improved BLEMO algorithm, called A-BLEMO, by incorporating two main mechanisms: (1) a convergence-selection technique and (2) a threshold control technique.
3. Reporting a comparative comparison between the new variant and the baseline one on the multi-objective Bi-MDVRP using a statistical validation test.

2   Bi-level Programming: Basic Definitions

BOP presents two levels of optimization tasks in which the set of feasible solutions of the upper level is determined only by the optimal feasible candidates of the lower level. In this way, the lower level problem is parameterized by decision variables of the upper level one. Besides, the leader objective function value depends on the solution of the follower optimization problem. Based on such structure, the leader starts by making his decision and fixes his variables' values without considering the lower level part of the problem. The follower now perceives the leader's action and then tries to optimize his own objective function, subject to the decision variables of the leader and the lower constraints. Finally, the leader should take the follower's reaction into account. This scenario can be described by the following mathematical formulation.

$$Min\ F(x_u, x_l)$$

$$\text{s.t } G(x_u, x_l) \le 0 \begin{cases} Min\ f(x_u, x_l) \\ \text{s.t } g(x_u, x_l) \le 0 \end{cases} \tag{1}$$

Where F, f represent the upper and the lower level objective functions, respectively. (x_u, x_l) are the upper level and the lower level variables. $G(x_u, x_l)$ and $g(x_u, x_l)$ denote the upper and the lower constraints, respectively. MBOP is a special class of BOP where at least one task is presented as a multi-objective optimization problem. This type of optimization problem includes three variants: (1) multiple conflicting objectives are only handled by the upper level, (2) multiple conflicting objectives are only handled by the lower level, and (3) multiple conflicting objectives are handled by both levels. We mention that we treat through this work MBOPs considering multi-objective problems at the upper and the lower levels. Figure 1 illustrates the interaction between the two nested levels in a MBOP, where each vector-valued solution at the upper level $x_u = (x_{ui}, x_{uj}, x_{uk})$ needs the execution of the lower task to obtain the follower reaction aiming at finding the lower response $x_l = (x_{li}, x_{lj}, x_{lk})$. To this end, a new general formulation of the problem can be presented as follows:

$$Min\ F(x_u, x_l) = (F_1(x_u, x_l), ..., F_n(x_u, x_l))$$

$$\text{s.t } G(x_u, x_l) \le 0 \begin{cases} Min f(x_u, x_l) = (f_1(x_u, x_l), ..., f_n(x_u, x_l)) \\ \text{s.t } g\ (x_u, x_l) \le 0 \end{cases} \tag{2}$$

Where the optimization of the upper objective functions $F_1(x_u, x_l), ..., F_n(x_u, x_l)$ depends on the lower optimal solution for $f_1(x_u, x_l), ..., f_n(x_u, x_l)$ (which are the lower objective functions constrained by $g\ (x_u, x_l)$).

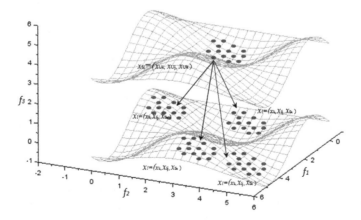

Fig. 1. Multi-objective bi-level programming: multiple conflicting objectives are handled by upper and lower levels. Thus, a lower Pareto front is expected for each upper level decision variable x_u.

3   Case Study on Production-Distribution Systems in Supply Chain Management

Production-distribution planning is a complex system requiring various integrated entities dispersed on the complete manufacturing process that can handle different DMs, related hierarchically to each other. This system may describe the control of the production and distribution processes at two firms: a leader company ensures transportation from depots to retailers. The follower company includes producing and manufacturing goods for depots. Moreover, retailers are served by a set of homogeneous capacitated vehicles belonging to several depots. The aim is to minimize the total cost of serving goods for costumers. To this end, the principal firm tries to reduce the total cost including the manufacturing (at the follower company) and the transportation costs. Once the leader makes the decision and obtains the total amount of retailer demands, the follower starts then to minimize the production cost. This model is described in the literature by [12] as a bi-level multi depot vehicle routing problem. In this work, we propose a multi-objective variant of the considered problem in which the first company tries to minimize both: the distance of transportation, and the number of vehicles used by each depot. The inner company optimizes the production process regarding retailer demands, and the direct transportation cost. To achieve these targets, DMs have to implement wise strategies integrating different levels of the previously described operating systems in the supply chain. We give in Table 1 the parameters and decision variables used in our mathematical model. Equations (3)–(11) present our proposed Multi-objective Bi-MDVRP (M-Bi-MDVRP) formulation:

Table 1. Notations of the mathematical model.

Notations	Parameters
$Dist_{p,d}$	Distance between depot d and plant p
C_v	Capacity of vehicle v
d_r	Demand of retailer r
PC_p	Production capacity of plant p
$c_{i,j}^a$	Upper level transportation cost between nodes i and j
$c_{p,d}^b$	Production cost of plant p to depot d
$c_{p,d}^c$	Lower level direct transportation cost between p and d
Notations	Variables
q_p	Quantity of goods manufactured from plant p to d
$P_{v,r}$	Passage order of vehicle v to retailer r
x_{ij}^v	Upper level decision variable
$y_{p,d}$	Lower level decision variable

$$Min \sum_{v \in V} \sum_{(i,j) \in E} c_{i,j}^a x_{i,j}^v + \sum_{v \in V} \sum_{(i,j) \in E} v\, x_{i,j}^v \qquad (3)$$

$$\sum_{i \in R} \sum_{v \in V} x_{ij}^v = 1, \quad \forall\, 1 \le j \le R \tag{4}$$

$$\sum_{i \in E \setminus \{j\}} x_{ij}^v = \sum_{i \in E \setminus \{j\}} x_{ji}^v, \quad \forall\, j \le E,\ \forall\, v \in V \tag{5}$$

$$\sum_{i \in E} \sum_{j \in E} d_r x_{ij}^v \le C_v, \quad \forall\, v \in V \tag{6}$$

$$\sum_{i,j \in E} x_{ij}^v \le |S| - 1,\ (i,j) \in S^2,\ S \subseteq R \tag{7}$$

$$where,\ for\ given\{x_{ij}^v\},\ \{y_{p,d}\}\ solves : min \sum_{p \in P} c_{p,d}^b\, y_{p,d} + \sum_{d \in D} c_{p,d}^c\, y_{p,d} \tag{8}$$

$$s.t \quad \sum_{d \in D, p \in P} y_{p,d} = 1 \tag{9}$$

$$\sum_{d \in D} q_p\, y_{p,d} \le PC_p, \quad \forall\, 1 \le p \le P \tag{10}$$

$$\sum_{p \in P} q_p\, y_{p,d} \ge \sum_{v \in V} \sum_{r \in R} d_r,\ 1 \le d \le D \tag{11}$$

Where P, D, R, and V design the set of plants, depots, retailers, and vehicles, respectively. Constraint (4) imposes that every retailer should be served exactly once by a vehicle. Constraint (5) guarantees the flow conservation. Constraint (6) defines that the total demand of retailers on one particular route must not exceed the capacity of the vehicle assigned to this route. Constraint (7) presents sub-tour elimination. Constraint (9) is a lower level binary requirement. Constraint (10) implies that the total demands of assigned depots should not exceed production availability of plant p. Constraint (11) ensures that the demands of retailers should be satisfied by associating a part of plant production to depots.

4   A-BLEMO for the Bi-level Multi-objective Production Distribution Planning System

As we mentioned previously, BLEMO is an evolutionary algorithm proposed to solve bi-level multi-objective optimization problems with continuous decision variables. This algorithm applies the NSGA-II scheme to optimize multiple conflicting objectives in both decision levels. It uses the principle of genetic algorithm, which is extensively used for solving real-life problems [13]. Motivated by this observation, we aim to adopt the BLEMO scheme to solve the previously discussed multi-objective variant of the Bi-MDVRP. In this paper, we further improve upon that method (cf. Fig. 2) by incorporating: (1) a denying archive, (2) a threshold control technique, and (3) a convergence-selection technique, the mechanisms 2 and 3 are devoted to improve the selection procedure. The first technique (1) uses an archive to prevent duplicate solutions from participating in the next iteration of the algorithm. This fact gives rise to a well-trained population during different iterations of the algorithm. The second technique (2) is proposed to promote convergence by choosing only promising solutions that should participate in the current genetic operations. A promising solution is detected through a deterioration control index NB_s detailed in the following and expressed in formula 12.

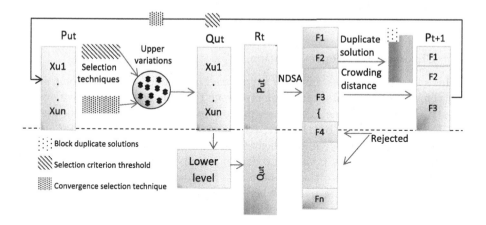

Fig. 2. Upper level scheme of the improved BLEMO.

This technique is proposed to prevent the genetic algorithm from falling into a local optimum. Since the later presents a major obstruction that drops and deteriorates the algorithm performance. In this regard, the suggested threshold control technique tries to verify if a candidate solution aiming to perform the mating process, may present a promising candidate for variations. For this reason, we mark each individual among the total N_u solutions with an index NB_s presenting the number of non-better-situations performed by the solution during the search. To this end, a candidate is allowed to be selected if:

$$NB_s \leq \varepsilon \tag{12}$$

Where ε represents a threshold parameter to accept a non-promising solution. ε is determined based on the problem dimension. We mention that a non-better-situation should be detected through the following comparison in the case of multi-objective search procedure:

$$S_O \prec S_P \tag{13}$$

Where S_O designs the offspring solution, S_P denotes the parent solution and '$\prec$' is the Pareto-dominance comparison operator indicating if S_O is dominated by S_P or not. The third incorporated improvement, which is the convergence selection technique, also adopts the NB_s indicator through the following formula:

$$\alpha - NB_s \geq \frac{\alpha}{2} \tag{14}$$

Where α, is the number of movements performed by a solution s. In fact, this formula indicates that accepted variations are considered if the performed best movements of s are bigger than half of its total movements. To conclude the ultimate goal of this modified BLEMO version is to improve the convergence rate of the algorithm by controlling the search directions basing on solution behaviour (deteriorating and promising one).

4.1 Chromosome Representation

In this subsection, we describe how we proceed to decode the chromosome solution. This step represents a fundamental issue to implement an effective evolutionary algorithm scheme. Therefore, we conceive a generic vector-valued solution for the M-Bi-MDVRP. In fact, an upper level solution is designed using a matrix (M_1) where $M_1[v,r]$ indicates the passage order of a vehicle v to a retailer r. $M_1[v,r]$ takes: 0 value if v does not support r. $M_1[v,r] = P_{v,r}$ indicates the passage order to the retailer r. To consider all the objective functions for a solution we design a matrix where the cells define both the traveling distance $Dist_{i,j}$ and the quantity q_p of goods manufactured by the plant p to the depot d. However, to facilitate the readability of the solution representation we divide the solution into two parts presenting each objective details. We give in Table 2 an example of the used encoding solution with nine retailers, four depots and four plants.

4.2 Upper Level Related Components

- **Step 1 (Initialization):** In this step, we randomly generate the initial population in the search space as shown in Table 2. Retailers are randomly assigned to vehicles (and therefore to depots). Then, the lower level task is executed for each upper level individual to obtain the lower Pareto front.
- **Step 2 (Selection):** First, the threshold control parameter and the convergence-selection technique are used to allow only promising solutions to perform this step. Similarly to the BLEMO, we use the binary tournament selection operator to choose the best solutions to participate in the mating pool.

Table 2. Example of solution representation.

		R_1	R_2	R_3	R_4	R_5	R_6	R_7	R_8	R_9
	V_1	0	$P_{v,r}$	$P_{v,r}$	$P_{v,r}$	0	0	0	0	$P_{v,r}$
(M_1)	V_2	$P_{v,r}$	0	0	0	0	$P_{v,r}$	0	0	0
	V_3	0	0	0	0	0	0	0	0	0
	V_4	0	0	0	0	$P_{v,r}$	0	$P_{v,r}$	$P_{v,r}$	0

	P_1	P_2	P_3	P_4
D_1	$Dist_{p,d}$	$Dist_{p,d}$	$Dist_{p,d}$	$Dist_{p,d}$
D_2	$Dist_{p,d}$	$Dist_{p,d}$	$Dist_{p,d}$	$Dist_{p,d}$
D_3	$Dist_{p,d}$	$Dist_{p,d}$	$Dist_{p,d}$	$Dist_{p,d}$
D_4	$Dist_{p,d}$	$Dist_{p,d}$	$Dist_{p,d}$	$Dist_{p,d}$

(M_3) is to the left of the first table above; (M_2) refers to the table below.

	P_1	P_2	P_3	P_4	Total
D_1	q_p	q_p	0	q_p	Q_{total}
D_2	0	0	0	q_p	Q_{total}
D_3	0	0	q_p	q_p	Q_{total}
D_4	q_p	0	q_p	0	Q_{total}

- **Step 3 (Variation at the upper level):** We choose to use in this work the Sequence-Based Crossover (SBX) a well-used diversification operator in the literature for routing problems [2]. First, a sub-route is selected randomly and then removed from the selected parent solutions. Next, an insertion is performed as linking the initial retailers of parent 1 to the remaining retailers of parent 2. The obtained route replaces the old one in the first parent. Similarly, the second offspring is obtained using the same procedure but inverting the order of parents. After the SBX procedure, the obtained offspring will be mutated with the polynomial mutation operator. We notice that any other genetic operators can be used instead.
- **Step 4 (Offspring upgrading and evaluation):** Execute the lower NSGA-II procedure for each upper offspring during T_l lower generations. After that, the lower level vectors are combined with their corresponding upper level ones to form the complete offspring solutions. The algorithm is able now to perform the evaluation step on the entire solutions. Next, we increment the NB_s index for each child solution that is dominated by its parent one.
- **Step 5 (Elitism):** The parent and offspring populations are combined to form a larger population.
- **Step 6 (Environmental selection):** Fill the new population based on the Non-Dominated Sorting Algorithm (NDSA) and the Crowding Distance (CD) of [10]. The duplicate solutions are forbidden from participating in the next iteration. If the stopping criterion is met the algorithm returns the obtained Pareto front. Otherwise, a new iteration is repeated from Step 2.

4.3   Lower Level Related Components

- **Step 1 (Initialization):** The initial population generation is triggered once a fixed upper decision variable x_u needs to be evaluated. Thus, all the vectors valued solution members are parameterized by this fixed x_u. To this end, N_l lower level vector-valued solutions are created based on the passed x_u.
- **Step 2 (Selection):** We use our proposed selection improvements to identify promising individuals. Then, the binary tournament selection operator is performed to choose the parent solutions that fit the lower mating pool.
- **Step 3 (Variation at the lower level):** The two-point crossover is applied here to produce the lower offspring individuals. Moreover, the selected blocks of genes should be exchanged between parent solutions when using this operator. Regarding the mutation operator, we choose to implement the polynomial mutation. We note that we can choose any other variation operators.
- **Step 4 (Lower evaluation):** Evaluate the obtained lower offspring individuals according to their fitness values. Then, we mark child solutions with their NB_s if a non-better fitness is obtained regarding their parent solution.
- **Step 5 (Elitism):** Combine the parent and offspring populations to ensure elitism.
- **Step 6 (Environmental selection):** The best N_l solutions are stored in the new lower level population based on the NDSA strategy replacement and CD measure. The duplicate solutions are rejected. The overall process is repeated from Step 2 until reaching the stopping criteria.

5   Experimental Study

In this section, we assess the relevance of A-BLEMO to solve the multi-objective bi-level optimization problems. We conduct a set of experiments to evaluate the performance of the algorithm with regards to the baseline version. A set of instances is generated from the 23 published Bi-MDVRP dataset (i.e., M-bip) provided by [9]. Considering these benchmarks, a set of factories, with the same number of depots, are added for each instance. As well, to make the analysis more realistic, the production capacity is fixed for each of the added plants and their geographic coordinates are randomly attributed.

5.1   Parameter Setting

The M-Bi-MDVRP was solved by the proposed algorithm and the baseline version with fixing the parameter values through a series of preliminary experiments. Since the lower level optimization procedure is executed interacting with the upper level one, we choose a fewer number of Function Evaluations (FEs) as stopping criteria and smaller population size at the lower level. To ensure a fair comparison, we keep the same design of experimentation for the algorithms under comparison as follows: population size = 50, crossover probability = 0.8, mutation probability = 0.2, and the number of FEs = 15000 for the upper level. Regarding the lower level we choose: population size = 30, crossover probability = 0.8, mutation probability = 0.1, and the number of FEs = 2500.

5.2   Statistical Analysis

To assess the statistical significance, we employed the Wilcoxon's rank-sum test at 5% level of confidence. We performed 31 runs for each couple (algorithm, problem). To describe statistically the difference between algorithm's results, we mark the results with "-: the difference is not significant" or "+: the difference is significant". The used Wilcoxon test allows only to verify if the generated results are statistically different or not. There is any indication of the magnitude of this difference. For this reason, it is relevant to indicate how accurate are the results. In this context, we find the Effect Size (ES) metric which is a measure quantifying the difference between algorithm results. In our study, we use the cohen's d, a commonly used measure that computes the effect size as follows [8]:

$$d = \frac{M_1 - M_2}{\sigma} \tag{15}$$

Where M_1, M_2 denote the population means and σ is the total standard deviation. The results can be interpreted by cohen's rules of thumbs: $d = 0.2$ represents a 'small' ES, $d = 0.5$ indicates a 'medium' ES and $d = 0.8$ is a 'large' ES. Based on this level of magnitude statistically significant difference can be trivial with 0.2 standard deviation and strong with 0.8 standard deviation.

5.3   Experimental Results

In this section, we evaluate the performance of our proposed algorithm regarding the baseline version. We conduct a set of experiments with respect to the following criteria: (1) the solution-quality at the upper level and (2) the Hypervolume (HV) to precise the convergence and diversity abilities of each algorithm. A good performance of A-BLEMO can be observed from Table 3 for average upper level results on the majority of used test problems. This can be explained by the use of the control technique that guides the search direction process to larger promising zones. Besides, the used selection criterion allows the algorithm to ignore non-promising solutions at an early stage. The discussed results are also valid using the ES measure (cf. Fig. 3). The obtained ES results are intensively distributed for the medium ES value (i.e., between the green lines [0.5,0.8]) for the majority of the M-bip used set of benchmarks. The remaining results are detected almost through a large ES (> 0.8).

The HV results are plotted in Fig. 4 and indicate the outperformance of A-BLEMO over the BLEMO algorithm on the majority of used test problems. Thus, we note a better convergence and diversity rates by A-BLEMO. This fact proves the efficiency of the proposed improved variant of the algorithm.

Table 3. Upper level average results on the M-Bi-MDVRP instances.

	A-BLEMO Fitness		BLEMO Fitness			A-BLEMO Fitness		BLEMO Fitness	
Instances	Distance	Vehicles	Distance	Vehicles	Instances	Distance	Vehicles	Distance	Vehicles
M-bip01	2937.02 +	14	2821.73	14	M-bip13	5389.88 +	10	5921.35	10
M-bip02	2654.55 +	14	2816.64	15	M-bip14	5453.37 +	10	5925.25	10
M-bip03	3807.17 +	14	3799.61	15	M-bip15	13966.23 +	20	14394.09	20
M-bip04	4931.19 +	16	4939.41	16	M-bip16	13871.49 +	20	14355.24	20
M-bip05	3916.89 +	10	4167.28	10	M-bip17	13475.24 +	19	14246.33	20
M-bip06	4647.34 +	18	4886.05	16	M-bip18	24790.61 +	29	25091.58	30
M-bip07	4355.96 +	16	4899.71	18	M-bip19	25088.72 −	30	25499.20	30
M-bip08	27471.71 +	28	27930.19	16	M-bip20	24908.07 +	30	25214.23	30
M-bip09	27190.92 +	35	27479.70	31	M-bip21	45239.5 +	45	45916.15	45
M-bip10	26861.08 −	31	27109.78	35	M-bip22	45680.44 −	45	46116.15	45
M-bip11	26991.01 +	30	27375.85	29	M-bip23	44702.05 +	45	45351.06	45
M-bip12	5739.89 +	31	5358.68	10					

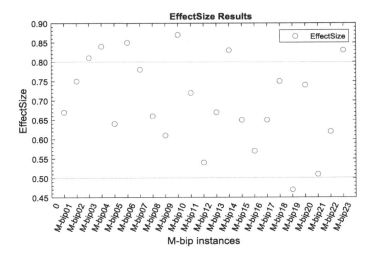

Fig. 3. ES results for M-bip instances.

6 Conclusion

In this work, we investigate an evolutionary approach to solve an NP-hard combinatorial problem. New techniques are proposed to enhance the BLEMO performance, particularly for the combinatorial MBOPs. Experiments reveal that the made improvements infer a better significant performance of the algorithm. These merits point to some future works: the model can be extended to include other socio-economical objectives. Also, a realization of a real-world case study could be an important subject for further study.

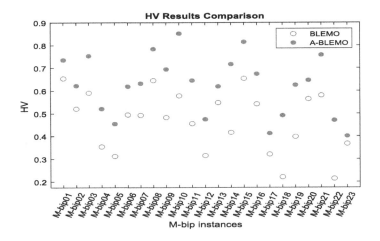

Fig. 4. HV results for M-bip instances.

References

1. Abbassi, M., Chaabani, A., Said, L.B.: An investigation of a bi-level non-dominated sorting algorithm for production-distribution planning system. In: Wotawa, F., Friedrich, G., Pill, I., Koitz-Hristov, R., Ali, M. (eds.) IEA/AIE 2019. LNCS (LNAI), vol. 11606, pp. 819–826. Springer, Cham (2019). https://doi.org/10.1007/978-3-030-22999-3_69

2. Agrawal, R.B., Deb, K., Agrawal, R.B.: Simulated binary crossover for continuous search space. Complex Syst. **9**(2), 115–148 (1995)

3. Aiyoshi, E., Shimizu, K.: Hierarchical decentralized systems and its new solution by a barrier method. IEEE Trans. Syst. Man Cybern. **6**, 444–449 (1981)

4. Bard, J.F., Falk, J.E.: An explicit solution to the multi-level programming problem. Comput. Oper. Res. **9**(1), 77–100 (1982)

5. Bianco, L., Caramia, M., Giordani, S.: A bilevel flow model for hazmat transportation network design. Transp. Res. Part C Emerg. Technol. **17**(2), 175–196 (2009)

6. Calvete, H.I., Galé, C.: A multiobjective bilevel program for production-distribution planning in a supply chain. In: Ehrgott, M., Naujoks, B., Stewart, T., Wallenius, J. (eds.) Multiple Criteria Decision Making for Sustainable Energy and Transportation Systems, pp. 155–165. Springer, Heidelberg (2010). https://doi.org/10.1007/978-3-642-04045-0_13

7. Carrasqueira, P., Alves, M.J., Antunes, C.H.: A bi-level multiobjective PSO algorithm. In: Gaspar-Cunha, A., Henggeler Antunes, C., Coello, C.C. (eds.) EMO 2015. LNCS, vol. 9018, pp. 263–276. Springer, Cham (2015). https://doi.org/10.1007/978-3-319-15934-8_18

8. Cohen, J.: The effect size index: d. Statistical power analysis for the behavioral sciences **2**, 284–288 (1988)

9. Cordeau, J.F., Gendreau, M., Laporte, G.: A tabu search heuristic for periodic and multi-depot vehicle routing problems. Netw. Int. J. **30**(2), 105–119 (1997)

10. Deb, K., Sinha, A.: Solving bilevel multi-objective optimization problems using evolutionary algorithms. In: Ehrgott, M., Fonseca, C.M., Gandibleux, X., Hao, J.-K., Sevaux, M. (eds.) EMO 2009. LNCS, vol. 5467, pp. 110–124. Springer, Heidelberg (2009). https://doi.org/10.1007/978-3-642-01020-0_13

11. Lv, T., Ai, Q., Zhao, Y.: A bi-level multi-objective optimal operation of grid-connected microgrids. Electr. Power Syst. Res. **131**, 60–70 (2016)

12. Mula, J., Peidro, D., Díaz-Madroñero, M., Vicens, E.: Mathematical programming models for supply chain production and transport planning. Eur. J. Oper. Res. **204**(3), 377–390 (2010)

13. Said, A., Abbasi, R.A., Maqbool, O., Daud, A., Aljohani, N.R.: CC-GA: a clustering coefficient based genetic algorithm for detecting communities in social networks. Appl. Soft Comput. **63**, 59–70 (2018)

14. Wu, D.D., Luo, C., Wang, H., Birge, J.R.: Bi-level programing merger evaluation and application to banking operations. Prod. Oper. Manage. **25**(3), 498–515 (2016)

15. Wu, M., Kou, L., Hou, X., Ji, Y., Xu, B., Gao, H.: A bi-level robust planning model for active distribution networks considering uncertainties of renewable energies. Int. J. Electr. Power Energy Syst. **105**, 814–822 (2019)

16. Zhao, L., Wei, J.X.: A nested particle swarm algorithm based on sphere mutation to solve bi-level optimization. Soft Comput. **23**(21), 11331–11341 (2019). https://doi.org/10.1007/s00500-019-03888-6

Modifying the Symbolic Aggregate Approximation Method to Capture Segment Trend Information

Muhammad Marwan Muhammad Fuad$^{(\boxtimes)}$

Coventry University, Coventry CV1 5FB, UK
ad0263@coventry.ac.uk

Abstract. The Symbolic Aggregate approXimation (SAX) is a very popular symbolic dimensionality reduction technique of time series data, as it has several advantages over other dimensionality reduction techniques. One of its major advantages is its efficiency, as it uses precomputed distances. The other main advantage is that in SAX the distance measure defined on the reduced space lower bounds the distance measure defined on the original space. This enables SAX to return exact results in query-by-content tasks. Yet SAX has an inherent drawback, which is its inability to capture segment trend information. Several researchers have attempted to enhance SAX by proposing modifications to include trend information. However, this comes at the expense of giving up on one or more of the advantages of SAX. In this paper we investigate three modifications of SAX to add trend capturing ability to it. These modifications retain the same features of SAX in terms of simplicity, efficiency, as well as the exact results it returns. They are simple procedures based on a different segmentation of the time series than that used in classic-SAX. We test the performance of these three modifications on 45 time series datasets of different sizes, dimensions, and nature, on a classification task and we compare it to that of classic-SAX. The results we obtained show that one of these modifications manages to outperform classic-SAX and that another one slightly gives better results than classic-SAX.

Keywords: Classification · SAX · Time series mining

1 Introduction

Several medical, financial, and weather forecast activities produce data in the form of measurements recorded over a period of time. This type of data is known as time series. Time series data mining has witnessed substantial progress in the last two decades because of the variety of applications to this data type. It is estimated that much of today's data come in the form of time series [17].

There are a number of common time series data mining tasks, such as classification, clustering, query-by-content, anomaly detection, motif discovery, prediction, and others [7]. The key to performing these tasks effectively and efficiently is to have a high-quality representation of these data to capture their main characteristics.

© Springer Nature Switzerland AG 2020
V. Torra et al. (Eds.): MDAI 2020, LNAI 12256, pp. 230–239, 2020.
https://doi.org/10.1007/978-3-030-57524-3_19

Several time series representation methods have been proposed. The most common ones are *Discrete Fourier Transform* (DFT) [1, 2], *Discrete Wavelet Transform* (DWT) [5], *Singular Value Decomposition* (SVD) [12], *Adaptive Piecewise Constant Approximation* (APCA) [11], *Piecewise Aggregate Approximation* (PAA) [10, 23], *Piecewise Linear Approximation* (PLA) [18], and *Chebyshev Polynomials* (CP) [4].

Another very popular time series representation method, which is directly related to this paper, is the *Symbolic Aggregate approXimation* method (SAX) [13, 14]. The reason behind its popularity is its simplicity and efficiency, as it uses precomputed lookup tables. Another reason is its ability to return exact results in query-by-content tasks. The drawback of SAX is that during segmentation and symbolic representation, the trend information of the segments is lost, which results in lower-quality representation and less pruning power.

Several papers have spotted this drawback in SAX and there have been a few attempts to remedy it. All of them, however, had to sacrifice one, or even both, of the main advantages of SAX; its simplicity and its ability to return exact results in query-by-content tasks.

In this paper we propose three modifications of SAX that attempt to capture, to a certain degree, the trend information of segments. The particularity of our modifications is that they retain the two main advantages of the original SAX, which we call *classic-SAX* hereafter, as our modifications have the same simplicity and require exactly the same computational cost as classic-SAX. They also return exact results in query-by-content tasks.

We conduct classification experiments on a wide variety of time series datasets obtained from the time series archive to validate our method. The results were satisfying. It is important to mention here that we are not expecting the results to "drastically" outperform those of classic-SAX given that we kept exactly the same simplicity and efficiency of classic-SAX, which was the objective of our method while we were developing it.

The rest of this paper is organized as follows; Sect. 2 is a background section. The new method with its three versions is presented in Sect. 3, and is validated experimentally in Sect. 4. We conclude with Sect. 5.

2 Background

Time series data mining has witnessed increasing interest in the last two decades. The size of time series databases has also grown considerably. Because of the high-dimensionality and high feature correlation of time series data, representation methods, which are dimensionality reduction techniques, have been proposed as a means to perform data mining tasks on time series data.

The GEMINI framework [8] reduces the dimensionality of the time series from a point in an n-dimensional space into a point in an m-dimensional space (some methods use several low-dimensional spaces like [19]), where $m \ll n$. The similarity measure defined on the m-dimensional space is said to be *lower bounding* of the original similarity measure defined on the n-dimensional space if:

$$d^m(\dot{S}, \dot{T}) \leq d^n(S, T) \qquad (1)$$

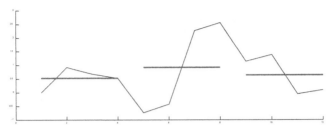

Fig. 1. PAA representation

where $\dot{S}$ and $\dot{T}$ are the representations of time series S, T, respectively, on the m-dimensional space. Applying the GEMINI framework guarantees that the similarity search queries will not produce false dismissals.

One of the most popular dimensionality reduction techniques of time series is the *Piecewise Aggregate Approximation* (PAA) [10, 23]. PAA divides a time series S of n-dimensions into m equal-sized segments (words) and maps each segment to a point in a lower m-dimensional space, where each point in the reduced space is the mean of values of the data points falling within this segment (Fig. 1). The similarity measure given in the following equation:

$$d^m(S, T) = \sqrt{\frac{n}{m}} \sqrt{\sum_{i=1}^{m} (\bar{s}_i - \bar{t}_i)^2} \qquad (2)$$

is defined on the m-dimensional space, where $\bar{s}_i$, $\bar{t}_i$ are the averages of the points in segment i in S, T, respectively.

PAA is the basis for another popular and very efficient time series dimensionality reduction technique, which is the *Symbolic Aggregate approXimation* – SAX [13, 14]. SAX is based on the assumption that normalized time series have a Gaussian distribution, so by determining the locations of the breakpoints that correspond to a particular alphabet size, chosen by the user, one can obtain equal-sized areas under the Gaussian curve. SAX is applied to normalized time series in three steps as follows:

1- The dimensionality of the time series is reduced using PAA.
2- The resulting PAA representation is discretized by determining the number and locations of the breakpoints. The number of the breakpoints *nrBreakPoints* is related to the alphabet size *alphabetSize*; i.e. *nrBreakPoints = alphabetSize* − 1. As for their locations, they are determined, as mentioned above, by using Gaussian lookup tables. The interval between two successive breakpoints is assigned to a symbol of the alphabet, and each segment of PAA that lies within that interval is discretized by that symbol.
3- The last step of SAX is using the following similarity measure:

Table 1. The lookup table of *MINDIST* for alphabet size = 3.

	a	b	c
a	0	0	0.86
b	0	0	0
c	0.86	0	0

$$MINDIST\left(\hat{S}, \hat{T}\right) = \sqrt{\frac{n}{m}} \sqrt{\sum_{i=1}^{m}\left(dist\left(\hat{s}_i, \hat{t}_i\right)\right)^2} \qquad (3)$$

Where n is the length of the original time series, m is the number of segments, $\hat{S}$ and $\hat{T}$ are the symbolic representations of the two time series S and T, respectively, and where the function $dist()$ is implemented by using the appropriate lookup table. For instance, the lookup table for an alphabet size of 3 is the one shown in Table 1. Figure 2 illustrates the different steps of SAX.

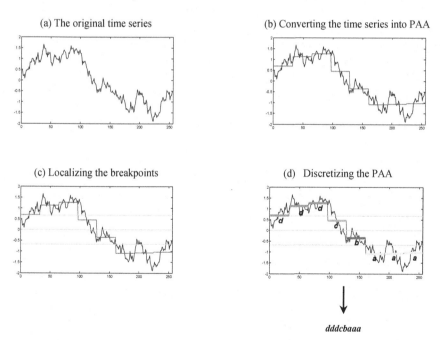

(a) The original time series

(b) Converting the time series into PAA

(c) Localizing the breakpoints

(d) Discretizing the PAA

dddcbaaa

Fig. 2. The different steps of SAX

It is proven in [13, 14] that the similarity distance defined in Eq. (3) is a lower bound of the Euclidean distance applied in the original space of time series.

Despite its popularity, SAX has a main drawback that it inherits from PAA, which is its inability to capture trend information during discretization. For example, the two segments:

$S_1 = [-6, -1, +7, +8]$ and $S_2 = [+9, +3, +1, -5]$ have the same PAA coefficient which is $+2$ so their SAX representation is the same, although, as we can clearly see, their trends are completely different.

Several researchers have reported this flaw in SAX and attempted to propose different solutions to handle it. In [16] the authors present 1d-SAX which incorporates trend information by applying linear regression to each segment of the time series and then discretizing it using a symbol that incorporates both trend and mean of the segment at the same time. The method applies a complex representation scheme. It also includes additional parameters that need training. In [24] another extension of SAX that takes into account both trend and value is proposed. This method first transforms each real value into a binary bit. This representation is used in a first pruning step. In the second step the trend is represented as a binary bit. This method requires two passes through the data which makes it inefficient for large datasets. [9] proposes to consider both value and trend information by combining PAA with an extended version of the clipping technique [21]. However, obtaining good results requires prior knowledge of the data.

In addition to all these drawbacks we mentioned above of each of these methods, which attempt to enable SAX to capture trend information, they all have two main disadvantages: a) whereas the main merit of SAX is its simplicity, all these methods are more/much more complex than classic-SAX. They require training of parameters and/or preprocessing steps. b) the main disadvantage of these methods is that the distance measure they propose on the low-dimension space does not lower bound the distance measure applied in the original space of the raw data. As a consequence, they do not return all the answers of a query-by-content task, unlike classic-SAX which returns exact results because it applies a lower bounding distance.

3 The Proposed Method

Our method aims to integrate trend information into classic-SAX all by fulfilling the following requirements:

i.   Conserving the simplicity of classic-SAX.
ii. Not adding any additional steps.
iii. Not adding any additional parameters.
iv. Keeping the lower bounding condition of classic-SAX.

Our method is based on the two following remarks:

a. Trend can be better captured by including points that are farther apart than if they were adjacent.
b. Very interestingly, Eq. (2) on which PAA is based, and consequently Eq. (3), does not stipulate that the points of a segment should be adjacent. It only stipulates that the points should be calculated once only in the summation, and of course, that each

point of one time series should be compared with its counter point in the other time series. In simple words, if, say we are using a compression ratio of 4:1, i.e. each segment of four points in the original space is represented by the mean of these four points, then these four points do not have to be adjacent for Eq. (2) to be valid.

Based on these two remarks, we propose the following three methods to segment the time series and calculate the means in PAA, and consequently in SAX:

1-Overlap-SAX: In this version, the end points of each segment are swapped with the end points of the two neighboring segments; i.e. the first point of segment m is swapped with the end point of segment $m - 1$, and the end point of segment m is swapped with the first point of segment $m + 1$. As for the first and last segments of the time series, only the last, first, respectively, points are swapped.

For example, for a time series of 16 points and for a compression ratio of 4:1, according to overlap-SAX this time series is divided into four segments as follows:
$\langle x_1, x_2, x_3, x_5 \rangle, \langle x_4, x_6, x_7, x_9 \rangle, \langle x_8, x_{10}, x_{11}, x_{13} \rangle, \langle x_{12}, x_{14}, x_{15}, x_{16} \rangle$
The next steps are identical to those of classic-SAX, i.e. the mean of each segment is calculated then discretized using the corresponding lookup table, and finally Eq. (3) is applied.

2-Intertwine-SAX: In this version, as the name suggests, each other point belongs to one segment, the next segment consists of the points we skipped when constructing the previous segment, and so on. For the same example as above, in intertwine-SAX this 16-point time series is divided into four segments as follows:
$\langle x_1, x_3, x_5, x_7 \rangle, \langle x_2, x_4, x_6, x_8 \rangle, \langle x_9, x_{11}, x_{13}, x_{15} \rangle, \langle x_{10}, x_{12}, x_{14}, x_{16} \rangle$
The rest of the steps are identical to those of classic-SAX.

3-Split-SAX: In this version, two successive points are assigned to a segment m, then two successive points are skipped, then the two successive points are assigned to m too, segment $m + 1$ consists of the skipped points when constructing m in addition to the two successive points following segment m, and so on. The 16-point time series in our example is divided into four segments in spilt-SAX as follows:
$\langle x_1, x_2, x_5, x_6 \rangle, \langle x_3, x_4, x_7, x_8 \rangle, \langle x_9, x_{10}, x_{13}, x_{14} \rangle, \langle x_{11}, x_{12}, x_{15}, x_{16} \rangle.$
This version attempts to capture trend on a wider range than the two previous ones.

As we can see, the three versions do not add any additional complexity to classic-SAX, they conserve all its main characteristics, mainly its simplicity. They also lower bound the original distance as we showed in remark (b) above. In fact, even coding each of these versions requires only a very simple modification of the original code of classic-SAX.

4   Experiments

Classification is one of the main tasks in data mining. In classification we have categorical variables which represent classes. The task is to assign class labels to the dataset according to a model learned during a learning stage on a training dataset, where the class labels are known. When given new data, the algorithm aims to classify these data based on the model acquired during the training stage, and later applied to a testing dataset.

There are a number of classification models, the most popular of which is *k-nearest-neighbor* (*k*NN). In this model the object is classified based on the *k* closest objects in its neighborhood. A special case of particular interest is when $k = 1$, which we use in this paper.

Time series classification has several real world applications such as health care [15], security [22], food safety [20], and many others.

The performance of classification algorithms can be evaluated using different methods. One of the widely used ones is *leave-one-out cross-validation* (LOOCV) (also known as *N-fold cross-validation*, or *jack-knifing*), where the dataset is divided into as many parts as there are instances, each instance effectively forming a test set of one. N classifiers are generated, each from N − 1 instances, and each is used to classify a single test instance. The classification error is then the total number of misclassified instances divided by the total number of instances [3].

We compared the performance of the three versions we presented in Sect. 3; overlap-SAX, intertwine-SAX, and split-SAX, to that of classic-SAX in a 1NN classification task on 45 time series datasets of different sizes and dimensions available at the *UCR Time Series Classification Archive* [6]. Each dataset in this archive is divided into a training dataset and a testing dataset. The dimension (length) of the time series on which we conducted our experiments varied between 24 (ItalyPowerDemand) and 1882 (InlineSkate). The size of the training sets varied between 16 instances (DiatomSizeReduction) and 560 instances (FaceAll). The size of the testing sets varied between 20 instances (BirdChicken), (BeetleFly) and 3840 instances (ChlorineConcentration). The number of classes varied between 2 (Gun-Point), (ECG200), (Coffee), (ECGFiveDays), (ItalyPowerDemand), (MoteStrain), (TwoLeadECG), (BeetleFly), (BirdChicken), (Strawberry), (Wine), and 37 (Adiac).

Applying the different versions of SAX comprises two stages. In the training stage we obtain the alphabet size that yields the minimum classification error for that version of SAX on the training dataset. Then in the testing stage we apply the investigated version of SAX (classic, overlap, intertwine, split) to the corresponding testing dataset, using the alphabet size obtained in the training stage, to obtain the classification error on the testing dataset.

In Table 2 we show the results of the experiments we conducted. The best result (the minimum classification error) for each dataset is shown in underlined, boldface printing. In the case where all versions give the same classification error for a certain dataset, the result is shown is italics.

There are several measures used to evaluate the performance of time series classification methods. In this paper we choose a simple and widely used one, which is to count how many datasets on which the method gave the best performance. Of the four versions tested, the performance of overlap-SAX is the best as it yielded the minimum classification error on 15 datasets. We have to say these were the results we were expecting. In fact, while we were developing the methods presented in this paper overlap-SAX was the first version we thought of as it mainly focusses on the principle information

Table 2. The 1NN classification error of classic-SAX, overlap-SAX, intertwine-SAX, and split-SAX The best result for each dataset is shown in underlined, boldface printing. Results shown in italics are those where the four versions gave the same classification error.

Dataset	Method			
	classic-SAX	overlap-SAX	intertwine-SAX	split-SAX
synthetic_control	**0.023333**	0.05	0.07	0.063333
Gun_Point	0.14667	0.14	0.19333	**0.13333**
CBF	0.075556	0.095556	**0.06**	0.085556
FaceAll	0.30473	**0.27811**	0.3	0.31953
OSULeaf	0.47521	0.4876	0.48347	**0.47107**
SwedishLeaf	**0.2528**	0.2656	0.2768	0.272
Trace	0.37	**0.28**	0.32	0.29
FaceFour	0.22727	0.20455	**0.17045**	0.20455
Lighting2	**0.19672**	0.21311	0.37705	0.40984
Lighting7	0.42466	0.39726	0.43836	**0.38356**
ECG200	0.12	**0.11**	0.12	0.13
Adiac	0.86701	0.86701	0.86445	**0.85678**
yoga	**0.18033**	0.18233	0.186	0.18433
FISH	0.26286	0.22286	**0.21714**	0.24571
Plane	*0.028571*	*0.028571*	*0.028571*	*0.028571*
Car	**0.26667**	0.28333	0.3	0.3
Beef	0.43333	**0.4**	0.43333	0.43333
Coffee	0.28571	**0.25**	0.28571	**0.25**
OliveOil	*0.83333*	*0.83333*	*0.83333*	*0.83333*
CinC_ECG_torso	**0.073188**	**0.073188**	0.075362	0.12246
ChlorineConcentration	0.58203	**0.54661**	0.57891	0.5625
DiatomSizeReduction	0.081699	0.088235	0.084967	**0.078431**
ECGFiveDays	0.14983	**0.13821**	0.18815	0.17886
FacesUCR	0.24244	**0.19366**	0.23024	0.24537
Haptics	0.64286	0.63636	**0.63312**	0.64935
InlineSkate	0.68	0.67818	0.67455	**0.67273**
ItalyPowerDemand	0.19242	**0.16035**	0.31487	0.25462
MALLAT	0.14328	0.14456	**0.14286**	0.15309
MoteStrain	0.21166	0.19249	0.1885	**0.17812**
SonyAIBORobotSurface	0.29784	**0.28785**	0.32612	0.2995
SonyAIBORobotSurfaceII	**0.14376**	0.19937	0.32529	0.15845
Symbols	**0.10251**	0.10452	0.10553	0.10553
TwoLeadECG	0.30904	0.34241	0.30729	**0.29939**
InsectWingbeatSound	0.44697	0.45253	0.44949	**0.44141**
ArrowHead	0.24571	0.22857	0.22857	**0.22286**
BeetleFly	*0.25*	*0.25*	*0.25*	*0.25*
BirdChicken	**0.35**	0.4	0.4	0.4
Ham	**0.34286**	0.39048	0.41905	0.41905
Herring	*0.40625*	*0.40625*	*0.40625*	*0.40625*
ToeSegmentation1	0.36404	**0.35088**	0.35965	0.39035
ToeSegmentation2	0.14615	0.20769	**0.13846**	0.18462
DistalPhalanxOutlineAgeGroup	0.2675	**0.235**	0.5325	0.4425
DistalPhalanxOutlineCorrect	0.34	0.38333	**0.28333**	0.29
DistalPhalanxTW	0.2925	**0.2725**	0.3175	0.3275
WordsSynonyms	0.37147	**0.3652**	0.37147	0.37147
	10	15	7	11

of each segment and only adds the "extra" required information to capture the trend by swapping the end points of each segment with the neighboring segments.

The second best version is split-SAX, which gave the minimum classification error on 11 datasets. The third best version is classic-SAX, whose performance is very close to that of split-SAX, as it gave the minimum classification error in 10 of the tested datasets. The last version is intertwine-SAX which gave the minimum classification error in only

7 of the tested datasets. The four versions gave the same classification error on three datasets (Plane), (OliveOil), and (BeetleFly).

It is interesting to notice that even the least performing version gave better results than the others on certain datasets. This can be a possible direction of further research for the work we are presenting in this paper - to investigate why a certain version works better with a certain dataset.

We have to add that although the versions we presented in Sect. 3 apply a compression ratio of 4:1, which is the compression ratio used in classic-SAX, the extension to a different compression ratio is straightforward, except for split-SAX which requires a simple modification in the case where the segment length is an odd number.

5   Conclusion and Future Work

In this paper we presented three modifications of classic-SAX, a powerful time series dimensionality reduction technique. These modifications were proposed to enhance the performance of classic-SAX by enabling it to capture trend information. These modifications; overlap-SAX, intertwine-SAX, and split-SAX, were designed so that they retain the exact same features of classic-SAX as they all have the same efficiency, simplicity, and they use a lower bounding distance. We compared their performance with that of classic-SAX, and we showed how one of them, overlap-SAX, gives better results than classic-SAX, and another one, split-SAX, gives slightly better results than classic-SAX. This improvement, although not substantial, is still interesting as it did not require any additional pre-processing/post-processing steps, or any supplementary storage requirement. It only required a very simply modification that is based solely on segmenting the time series differently compared to classic-SAX.

There are two directions of future work, the first is to relate the trend of each segment to the global trend of the time series, so the trend of each segment should actually only capture how it "deviates" from the global trend.

The other direction is to explore new segmentation methods, whether when applied to SAX or to other time series dimensionality reduction techniques. We believe another indirect outcome of this paper is that it opens the door to considering new schemes for segmenting time series that do not necessarily focus on grouping adjacent points.

References

1. Agrawal, R., Faloutsos, C., Swami, A.: Efficient similarity search in sequence databases. In: Lomet, D.B. (ed.) FODO 1993. LNCS, vol. 730, pp. 69–84. Springer, Heidelberg (1993). https://doi.org/10.1007/3-540-57301-1_5
2. Agrawal, R., Lin, K.I., Sawhney, H.S., Shim, K.: Fast similarity search in the presence of noise, scaling, and translation in time-series databases. In: Proceedings of the 21st International Conference on Very Large Databases. Zurich, Switzerland, pp. 490–501 (1995)
3. Bramer, M.: Principles of Data Mining. Springer, Heidelberg (2007)
4. Cai, Y., Ng, R.: Indexing spatio-temporal trajectories with Chebyshev polynomials. In: SIGMOD (2004)
5. Chan, K.P., Fu, A.W.-C.: Efficient time series matching by wavelets. In: Proceedings of 15th International Conference on Data Engineering (1999)

6. Chen,Y., Keogh, E., Hu, B., Begum, N., Bagnall, A., Mueen, A., Batista, G.: The UCR time series classification archive (2015). www.cs.ucr.edu/~eamonn/time_series_data
7. Esling, P., Agon, C.: Time-series data mining. ACM Comput. Surv. (CSUR) **45**(1), 12 (2012)
8. Faloutsos, C., Ranganathan, M., Manolopoulos, Y.: Fast subsequence matching in time-series databases. In: Proceedings of ACM SIGMOD Conference, Minneapolis (1994)
9. Kane,A.: Trend and value based time series representation for similarity search. In: 2017 IEEE Third International Conference Multimedia Big Data (BigMM), p. 252 (2017)
10. Keogh, E., Chakrabarti, K., Pazzani, M., Mehrotra, S.: Dimensionality reduction for fast similarity search in large time series databases. J. Knowl. Inform. Syst. **3**, 263–286 (2000)
11. Keogh, E., Chakrabarti, K., Pazzani, M., Mehrotra, S.: Locally adaptive dimensionality reduction for similarity search in large time series databases. In: SIGMOD, pp. 151–162 (2001)
12. Korn, F., Jagadish, H., Faloutsos, C.: Efficiently supporting ad hoc queries in large datasets of time sequences. In: Proceedings of SIGMOD 1997, Tucson, AZ, pp. 289–300 (1997)
13. Lin, J., Keogh, E., Lonardi, S., Chiu, B.Y.: A symbolic representation of time series, with implications for streaming algorithms. In: DMKD 2003, pp. 2–11 (2003)
14. Lin, J.E., Keogh, E., Wei, L., Lonardi, S.: Experiencing SAX: a novel symbolic representation of time series. Data Min. Knowl. Discov. **15**(2), 107–144 (2007)
15. Ma, T., Xiao, C., Wang, F.: Health-ATM: a deep architecture for multifaceted patient health record representation and risk prediction. In: SIAM International Conference on Data Mining (2018)
16. Malinowski, S., Guyet, T., Quiniou, R., Tavenard, R.: 1d-SAX: a novel symbolic representation for time series. In: Tucker, A., Höppner, F., Siebes, A., Swift, S. (eds.) IDA 2013. LNCS, vol. 8207, pp. 273–284. Springer, Heidelberg (2013). https://doi.org/10.1007/978-3-642-41398-8_24
17. Maimon, O., Rokach, L.: Data Mining and Knowledge Discovery Handbook. Springer, New York (2005)
18. Morinaka, Y., Yoshikawa, M., Amagasa, T., Uemura, S.: The L-index: an indexing structure for efficient subsequence matching in time sequence databases. In:Proceedings of 5th Pacific Asia Conference on Knowledge Discovery and Data Mining, pp. 51–60 (2001)
19. Muhammad Fuad, M.M., Marteau P.F.: Multi-resolution approach to time series retrieval. In: Fourteenth International Database Engineering & Applications Symposium– IDEAS 2010, Montreal, QC, Canada (2010)
20. Nawrocka, A., Lamorska, J.: Determination of food quality by using spectroscopic methods. In: Advances in Agrophysical Research (2013)
21. Ratanamahatana, C., Keogh, E., Bagnall, Anthony J., Lonardi, S.: A novel bit level time series representation with implication of similarity search and clustering. In: Ho, T.B., Cheung, D., Liu, H. (eds.) PAKDD 2005. LNCS (LNAI), vol. 3518, pp. 771–777. Springer, Heidelberg (2005). https://doi.org/10.1007/11430919_90
22. Tan, C.W., Webb, G.I., Petitjean, F.: Indexing and classifying gigabytes of time series under time warping. In: Proceedings of the 2017 SIAM International Conference on Data Mining, pp. 282–290. SIAM (2017)
23. Yi, B.K., Faloutsos, C.: Fast time sequence indexing for arbitrary Lp norms. In: Proceedings of the 26th International Conference on Very Large Databases, Cairo, Egypt (2000)
24. Zhang, T., Yue, D., Gu, Y., Wang, Y., Yu, G.: Adaptive correlation analysis in stream time series with sliding windows. Comput. Math Appl. **57**(6), 937–948 (2009)

Efficiently Mining Gapped and Window Constraint Frequent Sequential Patterns

Hugo Alatrista-Salas[1]([✉]) [ID], Agustin Guevara-Cogorno[2],
Yoshitomi Maehara[1] [ID], and Miguel Nunez-del-Prado[1] [ID]

[1] Universidad del Pacífico, Av. Salaverry 2020, Jesús María, Lima, Peru
{h.alatristas,ye.maeharaa,m.nunezdelpradoc}@up.edu.pe
[2] Pontificia Universidad Católica del Perú, Av. Universitaria 1801,
San Miguel, Lima, Peru
a20122661@pucp.pe

Abstract. Sequential pattern mining is one of the most widespread data mining tasks with several real-life decision-making applications. In this mining process, constraints were added to improve the mining efficiency for discovering patterns meeting specific user requirements. Therefore, the temporal constraints, in particular, those that arise from the implicit temporality of sequential patterns, will have the ability to efficiently apply temporary restrictions such as, window and gap constraints. In this paper, we propose a novel window and gap constrained algorithms based on the well-known PrefixSpan algorithm. For this purpose, we introduce the virtual multiplication operation aiming for a generalized window mining algorithm that preserves other constraints. We also extend the PrefixSpan Pseudo-Projection algorithm to mining patterns under the gap-constraint. Our performance study shows that these extensions have the same time complexity as PrefixSpan and good linear scalability.

Keywords: Sequential pattern mining · Gap constraint · Window constraint · Temporal constraints

1 Introduction

Data mining methods extract knowledge from vast amounts of data. Mainly, sequential pattern mining techniques provide a relevant solution to discover patterns - without *a priori* hypothesis - describing the temporal evolution of events characterizing a phenomenon. Nevertheless, extracted patterns can be misleading due to the strategy of appending an event after another, regardless of time constraints.

In this context, the temporal domain is fundamentally tied to the sequential nature of the patterns from the restriction that they be frequent in both itemset presence and the ordering. This temporal nature, however, has no bounds on the spacing between itemset occurrences. As such, a pattern's first and second

© Springer Nature Switzerland AG 2020
V. Torra et al. (Eds.): MDAI 2020, LNAI 12256, pp. 240–251, 2020.
https://doi.org/10.1007/978-3-030-57524-3_20

itemset could have a highly variable gap in each entry or be extremely big to the point where one might suspect it is a discretization artifact arising from the fact that a value range is likely to reoccur after long enough periods. Furthermore, the variable nature of gaps between itemsets means that sequential patterns cannot be used as predictors.

Constrained sequential pattern mining is a problem that precedes the development of the PrefixSpan algorithm. The paper on which PrefixSpan was first published [12] already listed the constraint satisfaction problem on its further works as an interesting extension to the algorithm.

In this paper, we propose the WinCopper algorithm, which extends the PrefixSpan-based Copper algorithm [4] by allowing mining *window constrained sequential patterns* that preserves any context-free constraint set. This extension is based on the Pseudo-Projection implementation of PrefixSpan that allows mining *gap constrained sequential pattern*.

Besides, we provide a proof of the correctness of these extensions. In detail, the preservation condition of the window constrained sequential patterns extension implicitly give an extension for the combined mining both window and gap constraints simultaneously.

To implement the extension, we perform a virtual multiplication operation, which naturally embeds the window condition into the database and by using the data structures used by the Pseudo-Projection implementation. In this manner, we satisfy the gap condition by restricting the items-for-appending search to candidates only. Indeed, the proposal follows the PrefixSpan philosophy of not generating candidates and testing but only extending values known to produce a sequential pattern.

The remainder of this article is organized as follows. Section 2 summarizes the state-of-the-art in the field of sequential pattern mining under temporal constraints. In Sect. 3, the objects necessary to construct the window and gap constraint were rigorously defined. Then, in Sect. 4, the experimental and performance results were presented. Finally, Sect. 5 summarized our study and point out possible future works.

2   Related Works

In the present section, we describe the works on the literature implementing gap, and Windows constrains.

Windows and gap constraints allow strict sequentiality, which is useful for interpreting sequential patterns as predictors. These constraints allow the validation of sequential pattern mining results using time series analysis. We accomplish this task by enforcing a *static distance* relationships between itemsets in a sequential pattern and using stochastic process modeling of data mining methods such as [7].

In particular, more relaxed temporal constraints allow modeling inter-itemset relationships, especially when paired with inclusion conditions for the elements in a pattern. For example, in an epidemiological dataset, it might be of interest to only mine patterns that have actual occurrences of a particular disease of interest. The temporal constraint ensures that the events surrounding the occurrence are temporally close and can be interpreted as either consequence, causes, or at least tentative indicators [5].

Several efforts were proposed in the literature to include the time constraints in the context of sequential patterns mining. For instance, in [8], the authors describe the Graph for Time Constraints (GTC) algorithm, which extracts sequential patterns by handling time constraints. The proposed algorithm considers time constraints directly in the mining process through apriori-like algorithms. Authors define the Frequent Generalized Sequential Pattern problem, which extracts frequent sequences taking into account a user-specified minimum time gap ($minGap$), a maximum time gap ($maxGap$), and a time window size ($windowSize$) as extra parameters (other than the minimal support). To measure results, the authors measure the performance of their proposal over three synthetic datasets and two real datasets.

Further, the algorithm DRL-Prefixspan is described in [3]. Unlike the algorithm proposed in [8], this algorithm uses the pattern-growth strategy to extract sequential patterns by integrating two restrictions. On the one hand, the Item constraint specifies the subset of items that should or should not be present in the patterns. On the other hand, the Adjacency constraint, in which sequential patterns must verify the timestamp of an item in the sequential pattern. Then, the variance of two timestamps must not be greater or lesser than a predefined threshold. The authors tested the performance of their proposal using a synthetic database.

In the same spirit, [2] extends the PrefixSpan algorithm to include the Gap, Recency, Compactness (GRC) constrains. The *gap* constraint verifies that each transaction in all sequence has a timestamp. Indeed, the timestamp variance (variance of days) among every two inline transactions in a revealed sequential pattern must not be larger than a given gap. Further, the *recency* constraint is stated by giving a minimum recency support, which is the number of days left from the starting date of the sequence database. Finally, the *compactness* constraint states that the sequential patterns must have the property such that the timestamp variance (variance of days) between the first and the former transactions in a discovered sequential pattern must not be greater than the given period. The drawback of this paper is the absence of experiments.

Moreover, Lin *et al.* [6] propose the Delimited Sequential Pattern Mining (DELISP) algorithm, which extracts the time-constrained sequential pattern through the pattern-growth strategy. For that, the authors handling all three-time constraints on sequential patterns, introduced in the context of GSP algorithm [13], within the pattern-growth framework, i.e., giving a minimum support, a minimum time gap, a maximum time gap, and a sliding time-window, DELISP discover the set of all time-constrained sequential patterns. Additionally, to accelerate the mining process by reducing the size of subsequences, the constraints are integrated into the projection (called bounded projection) to delimit the counting and growth of sequences. To measure the performance of the DELISP was tested on several datasets.

Unlike the articles cited above, WinCopper formally defines temporary restrictions (max gap, min gap, and time window) through virtual multiplication procedures preserving other context-free constraints. Additionally, given that WinCopper is an extension of Copper [4], the algorithm also integrates the restrictions of Soft Inclusion, Sequence Size Constraint, and Itemset Size Constraint. This set of restrictions allows to include knowledge of the experts in the mining process, reducing the number of patterns that experts must validate. Finally, we test our WinCopper implementation on four real datasets and a synthetic dataset, showing its efficiency. In the next section, we detail our approach.

3   PrefixSpan Under Constraints

Suppose we have two sequences $S_1 = <abcd>$ and $S_2 = <axyzb>$. In these sequences, a week has passed between each itemset, *i.e.*, b appeared one week after a in S_1. If we apply the PrefixSpan algorithm with minimal support of 2, one of the extracted patterns is $<a\ b>$ because it appears in 2 sequences. The pattern is valid; however, in sequence S_2, the gap between a and b is four weeks, while in sequence S_1 a and b are consecutive. Our proposal tackled these temporal configurations. In this regard, the goal of the present section is twofold. On the one hand, we define the relevant terms and prove the necessary lemmas for the formal verification of our approach. On the other hand, we describe the WinCopper Algorithm.

3.1   Formal Definitions

Definitions (Itemset, Pattern, Prefix, Database). Given $\Omega = \{i_1, i_2, \ldots, i_N\}$ the set of all items. An **Itemset** is a subset of Ω with a timestamp (s_t) denoted $I = < s_t, s_0, s_1, \ldots, s_{n-1} >$ where $s_i \in \Omega, \forall\ 0 \leq i \leq n-1$. Where n is the number of items in an itemset the **itemset's length**. A **Pattern** is a sequence of Itemsets without timestamps denoted $P = [I_0, I_1, \ldots, I_{N-1}]$, the number of itemsets in a pattern is called the **pattern's span**. Additionally, the sum of the pattern's span is the **pattern's length**. We say an itemset I_0 is **contained** by I_1 denoted $I_0 \subset I_1$ when $\forall\ s \in I_0 \cap \Omega \implies s \in I_1$ (contention ignores the timestamp). A pattern $P^0 = [I_0^0, I_1^0, \ldots, I_{N-1}^0]$ is called a **prefix** of another pattern $P^1 = [I_0^1, I_1^1, \ldots, I_{J-1}^1]$ when $N \leq J, I_i^0 = I_i^1, \forall\ 0 \leq i \leq N-2$, $I_{N-1}^0 \subset I_{N-1}^1$ and $i_k \in I_{N-1}^1 \setminus I_{N-1}^0 \implies k > s, \forall\ i_s \in I_{N-1}^0$. A **Database Entry** consists of an ordered sequence of itemsets with a total ordering induced by the temporal projection $\Pi_t(I) = s_t$. A **Database** is a set of database entries.

Definitions (Occurrence, Constraints and Virtual Multiplication). Given pattern $P = [I_0, I_1, \ldots, I_N]$ and a database entry $D = [D_0, D_1, \ldots, D_M]$, an ordered index set $\lambda = [\lambda_0, \lambda_1, \ldots, \lambda_N]$ with $\lambda_i \leq \lambda_j, \forall i, j$ such that $i \leq j \leq N$ is an **occurrence** of P in D if $I_i \subset DI_{\lambda_i}, \forall 0 \leq i \leq N$.

A **constraint** is a boolean function $f(D, P, \lambda)$ which takes D a database entry, P a pattern and λ an occurrence of P in D. Additionally we call a constraint **context free** when $f = g(P, \delta_\lambda)$ where δ_λ is the set of all finite differences of λ (the values in λ are not relevant, only their difference with other values of λ). From this point forward all constraints are considered context free unless otherwise noted.

A **window constraint** ω is given by a scalar in the units of the temporal dimension of the database, we say an occurrence λ of P in D satisfies a window constraint if $\Pi_t(D_{\lambda_1}) - \Pi_t(D_{\lambda_N}) \leq \omega$ where Π_t is the projection in the temporal dimension.

A **gap constraint** γ is given by a scalar in the units of the temporal dimension of the database, we say an occurrence λ of P in D satisfies a window constraint if $\Pi_t(D_{\lambda_i}) - \Pi_t(D_{\lambda_i+1}) \leq \gamma, \forall i \in [0, N)_\mathbb{N}$.

If $\exists c$ such that $\forall 0 \leq i < M, D_i - D_{i-1} = c$ for all database entries (the database entries' frequency is constant and there are no missing data points) we can write both window and gap constraints as integers and express the constraints as $\lambda_1 - \lambda_N \leq \omega$ and $\lambda_i - \lambda_{i+1} \leq \gamma, \forall i \in [0, N)_\mathbb{N}$ respectively. From this point onwards we will assume this hypothesis is satisfied by all databases unless otherwise specified.

The **support of a pattern** P in a database $D = [D^0, D^1, \ldots, D^K]$, where $D^i = [D_0^i, D_1^i, \ldots, D_{M_i}^i]$, under a set of constraints $C = \{c_0, c_1, \ldots, c_n\}$ is the number of entries D_k such that $\exists \lambda_k$ occurrence of P in D_k such that λ_k satisfies $c_i, \forall i \in [0, n]_\mathbb{N}$ (a single occurrence must satisfy all constraints).

A **virtual multiplication** of a database D given a constraint set is a database D' such that every entry in D' is associated with an entry in D by a function $H : D' \mapsto D$ and $\forall P$ frequent pattern of D satisfying the constraint set and with support t, P is frequent in D' with a modified support function $\#([D_i]_H | D_i \in D', P \subseteq D_i)$ where $[D_i]_H = [D_j]_H \iff H(D_i) = H(D_j)$.

A constraint S is said to have the **Prefix Downward Closure** property if for every pattern P, for every occurrence λ that satisfies S, for every prefix P' of P, $\exists \lambda'$ such that $\lambda' \subset \lambda$ and $S(P', \lambda') = S(P, \lambda)$.

Lemma 1 (Window compatible Virtual Multiplication). *Given a set of constraints* $S = \{s_1, s_2, \ldots, s_{n-1}, \omega\}$ *where* ω *is a window constraint and a database* D*; the constraint satisfying pattern search space partition given by all patterns prefixed by the unitary span pattern* $[I_0]$*, denoted* P_{I_0}*; and the set of occurrences of* I_0 *for each database entry* D^i*, denoted* λ_{D^i}*. The pair* (D', H)*,* $D' = \{D_{\lambda, \lambda+\omega}^i | \lambda \in \lambda_{D^i}\}$*,* $H(D_{\lambda, \lambda+\omega}^i) = D^i$ *is a virtual multiplication of* D *for all elements of* P_{I_0}*.*

Proof. Let P be a frequent pattern prefixed by I_0 that satisfies S. Let λ be an occurrence of P in D^i of length N satisfying S which implies $I_0 \subset D_{\lambda_0}^i$, $\lambda_{N-1} - \lambda_0 \leq \omega$. $I_0 \subset D_{\lambda_0}^i \implies D_{\lambda_0, \lambda_0+\omega}^i \in D'$ by construction. And $P \subset D_{\lambda_0, \lambda_0+\omega}^i$ due to $\lambda_{N-1} \leq \lambda_0 + \omega$.

Conversely, let P be a frequent pattern in D' under the modified support induced by H, prefixed by I_0 and which satisfies S. P supported by $[D^i] \implies \exists \epsilon_0$

Algorithm 1. WinCopper Algorithm

Require: ω a window constraint, γ a gap constraing, t the support threshold, D a database

1: **procedure** PRELIMINARYROUTINE(ω, γ, t, D)
2:     *unit_patterns* $\leftarrow$ frequent items in D
3:     **for** *unit_pattern* in *unit_patterns* **do**
4:         *positions* $\leftarrow$ PositionsOfIn(*unit_pattern*, D)
5:         *VM_Proj* $\leftarrow$ VirtualMultiplication(ω, *positions*, D)
6:         $D'|\alpha \leftarrow$ PseudoProject(VM_Proj, $[< unit\_pattern >]$)
7:         **WindowPrefixspan**($[< unit\_pattern >]$, γ, t, $D'|\alpha$)
8:     **end for**
9: **end procedure**

Require: α a frequent sequential pattern, γ a gap constrain, t the support threshold, $D'|\alpha$ a virtual multiplication of a database Pseudo-Projected by α

10: **procedure** WINDOWPREFIXSPAN(α, γ, t, $D'|\alpha$)
11:     Output(α)
12:     *asc_unit_patterns* $\leftarrow$ FindAssemblyUnit_pattern($D'|\alpha$,t)
13:     **for** *unit_pattern* in *asc_unit_patterns* **do**
14:         $\alpha' \leftarrow \alpha$.assemble(*unit_pattern*)
15:         $D'|\alpha' \leftarrow$ PseudoProject(VM_Proj, α')
16:         **WindowPrefixspan**(α', γ, t, $D'|\alpha'$)
17:     **end for**
18:     *app_unit_patterns* $\leftarrow$ FindAppendingCandidates($D'|\alpha$,t,γ)
19:     **for** *unit_pattern* in *app_unit_patterns* **do**
20:         $\alpha' \leftarrow \alpha$.append(*unit_pattern*)
21:         $D'|\alpha' \leftarrow$ PseudoProject(VM_Proj, α')
22:         **WindowPrefixspan**(α', γ, t, $D'|\alpha'$)
23:     **end for**
24: **end procedure**

---

such that $P \subset D^i_{\epsilon_0, \epsilon_0 + \omega} \implies \exists \lambda$ occurrence of P in $D^i_{\epsilon_0, \epsilon_0 + \omega}$, let $\lambda' = \lambda + \epsilon_0$ (we add ϵ_0 to each entry of λ), λ' is then an occurrence of P in D^i that satisfies S.

We conclude that mining D for frequent patterns satisfying S prefixed by P is equivalent to mining D' for frequent patterns satisfying $S \setminus \omega$ with the modified support function.

Lemma 2 (Gap Constraint Partial Downward Closure). *The Gap constraint has partial downward closure.*

Proof. Let P pattern that satisfies γ a gap constraint with occurrence $\lambda = [\lambda_0, \lambda_1, \ldots, \lambda_{N-1}]$. Let P' prefix of P with span $M - 1$, then $\lambda' = [\lambda_0, \lambda_1, \ldots, \lambda_{M-1}]$ is an occurrence of P' (observe that $P'_i = P_i \forall i < M - 1$ and $P'_{M-1} \subset P_{M-1}$. Since $\lambda_{i+1} - \lambda_i < \gamma \ \forall 0 \leq i < N - 1$ in particular $0 \leq i < M - 1$, λ' also satisfies γ. Hence we conclude the Gap constraint has partial downward closure.

3.2 WinCopper Algorithm

In this section, we presents WinCopper, an extension of the Copper algorithm [4], which already includes Soft-Inclusion, Sequence Size, and Itemset Size constraints.

We add both Window and Gap extensions in Algorithm 1. The WinCopper extension takes as input a Window Constraint ω (*c.f.*, Lemma 1), a Gap Constraint γ (*c.f.*, Lemma 2), a support threshold t and a database D. The algorithm computes the unitary patterns with an occurrence grater than t (*c.f.*, line 2). Then, the algorithm maps the positions of the unitary patterns to perform a Window Compatible Virtual Multiplication (*c.f.*, line 4–5). In this step, the algorithm couple the resulting database with a modified support function. Finally, mining the new database with the pattern as the starting point would yield the desired frequent patterns for the projection and the constraint set. Finally, a database projection is built to perform the WindowsPrefixspan routine. (*c.f.*, line 6–7).

The routine WindowsPrefixspan receives as parameters a frequent sequential pattern α, a gap constrain γ, the support threshold t and the virtual multiplication of a database Pseudo-Projected by α, $D'|\alpha$. Line 11 of Algorithm 1 output the frequent patterns. Later, we perform two kinds of extended patterns: (i) by assembling, *i.e.*, adding an item to the same itemset (lines 12 to 16); and, (ii) by appending, *i.e.*, adding a new itemset (lines 18 to 22).[1] Then, for each extension, we add valid items, *i.e.*, with a support greater than t and a temporal gap less than *gamma*. It is worth noting that, for both extensions, α' is constructed by assembling (line 14) or appending (line 20), a new candidate. Later, for each α', the pseudo projection is computed once again over *VM_Proj* (lines 15 and 20). Finally, the routine WindowsPrefixspan is called attractively until no more patterns can be extended.

4 Experiments

In the current section, we present the synthetic and four real datasets used to test the performance and efficiency of our proposal (see Table 1).

The first dataset was artificially generated by IBM Quest named *Synthetic Generated (GA)* dataset. This dataset, used in Pei *et al.* [11], comprises 9219 sequences (*c.f.*, Table 1). Regarding the real datasets, four different datasets were employed. The *Smartphone Apps Usage (SAU)* contains itemsets of 26 different apps, such as Facebook, Twitter, Instagram, Netflix, *etc.* during July and August 2018. This dataset contains more than 5M of sequences. The *German Credit (GC)* dataset [10] posses 1000 sequences about clients requesting credits to a German bank. Each sequence contains the bank answer (credit approved or denied) and 89 different features about the client's socio-economical status.

---

[1] Specifically, we modify the appending candidate search function range coupled to the union of ranges generated from the intervals starting on the pointers used by the Pseudo-projections and ending γ positions away.

Table 1. Summary of characteristics of datasets

Name	GA	SAU	GC	CC	AA
Sequences	9219	5820844	1000	175118	733
Minimum itemsets	1	3	21	1	2
Maximum itemsets	18	13	21	76	15
Mean itemsets	6	6	21	7	4
Minimum items	1	1	1	1	1
Maximum items	27	1	1	1	2
Mean items	8	1	1	1	1
Different items	915	26	90	162	203

The *Clusters of Consumption (CC)* dataset consists of around 175 K sequences with 162 different items (clusters) of weekly clustered debit/credit card payments. Finally, the *Admission Applicants (AA)* dataset has 203 different items belonging to 733 sequences representing the participation in newcomers' admission activities while they are university candidates. Concerning the results, we compared our *WinCopper* algorithm (*c.f.*, Algorithm 1) to *PrefixSpan* [11] and *Copper* [4] algorithms.

First, we compare the performance of the four algorithms under the same scenario setup without applying any constraint. It is worth noting that each algorithm was executed in a similar scenario, *i.e.*, without applying any constraint, and we obtain the same number of patterns (completeness). Figure 1a depicts a similar behavior for the four algorithms using the synthetic dataset. While for the real datasets in Figs. 1b–e, the WinCopper algorithm is more time-consuming than the other two algorithms. This effect is due to the virtual multiplication step in the projection dataset process, which needs additional time to check whether the frequent patterns founded fulfills with the temporal constraints. On the contrary, in most cases, the Copper and PrefixSpan algorithms have a similar time consumption. This effect has widely discussed in [4].

Concerning the memory usage, the behavior of curves is different. Indeed, memory usage is proportional to the number of patterns founded for a given minimal support (*c.f.*, Figs. 1f, h, and i). It is important to notice that, for the Smartphone Apps Usage dataset (*c.f.*, Fig. 1g) and the Admission Applicants (*c.f.*, Fig. 1j), the behavior is quite different due to the shape of the dataset. Both dataset have itemsets of size one. This feature impacts the mining process of the WinCopper algorithm negatively.

Figure 2 illustrates the comparison between Prefixspan and WinCopper, integrating two constraints. Therefore, the window gap constraint of the WinCopper algorithm was fixed to the minimum value (*i.e.*, sequences with consecutive itemsets were formed). Also, an item was randomly selected to test the performance of the soft-inclusion constraint.

As expected, the WinCopper algorithm obtains fewer patterns due to the Windows Gap and Soft-inclusion constraints applied in the mining process.

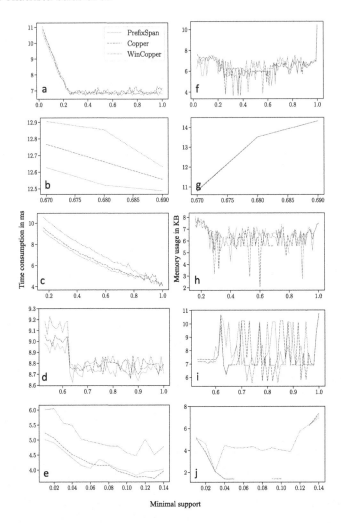

Fig. 1. Processing time (first column) and memory usage (second column) for different Support thresholds using Synthetic Generated *(a,f)*, Applications Usage *(b,g)*, German Credits *(c,h)*, Consumption Cluster *(d,i)*, and Admission Applicants *(e,j)*.

Nevertheless, three particular behavior should be highlighted. For the synthetic dataset in Fig. 2a, no patterns were found due to the dataset shape. Indeed, no sequences with consecutive itemsets were found. Concerning the Smartphone Apps Usage dataset, our approach extracts only one pattern due to the Windows Gap constraint in Fig. 2b. This dataset contains sequences of 1-size itemset, and users frequently use the same apps, making the patterns extraction difficult under constraints difficult. For the Consumption Clusters, dataset, both algorithms obtain the same number of sequences, and the WinCopper constraints do not impact the selection of sequences (see Fig. 2d).

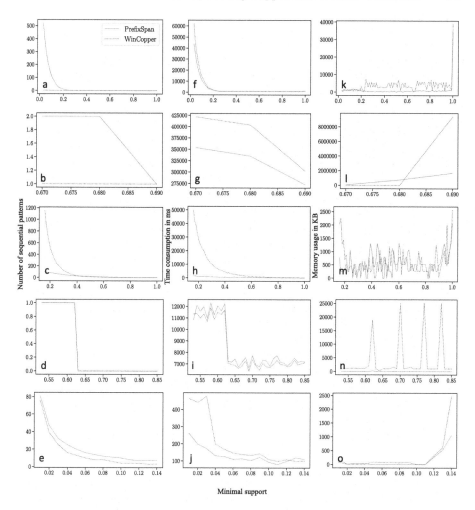

Fig. 2. Number of sequential patterns (first column), Processing time (second column), and Memory usage (third column) for PrefixSpan and WinCopper algorithms over Synthetic Generated *(a,f,k)*, Applications Usage *(b,g,l)*, German Credits *(c,h,m)*, Consumption Cluster *(d,i,n)*, and Admission Applicants *(e,j,o)*.

Concerning time consumption, we observe the same behavior of processing time in Figs. 2f and i. While in Figs. 2g, h and j, the WinCopper algorithm shows more processing time when the support is small and converging when the support tends to one. About the memory usage, Figs. 2l, m and o evidence that both WinCopper and PrefixSpan memory usage differs in a small scale. Whereas, in Figs. 2k and n, PrefixSpan uses more memory that our proposed algorithm.

To summarize, WinCopper can extract useful patterns, including expert-knowledge in the mining process, in the form of constraints without severely degrading the computational resources consumption. Indeed, WinCopper incorporates several constraints, namely gap constraint, and windows gap constraint, paired to constraint integrated into Copper, which is the preliminary version of our proposal. Due to the way these constraints were integrated into the process mining, our proposal has the same time complexity as PrefixSpan and good linear scalability.

5   Conclusions

In this paper, we propose a scalable and efficient temporal constraint implementation for sequential pattern mining over the PrefixSpan algorithm. Our proposal implements a virtual multiplication procedure that allows mining a dataset that has the window constraint naturally embedded while modifying the support function to preserve the original support values in the mining process. Besides, an extension for gap constraint mining involves the restriction of the search space for candidates when performing an appending operation to those that would satisfy the gap constraint. This restriction was performed on top of a previous PrefixSpan optimization known as Pseudo-Projection for its implementation. We show the windowed extension has the same time complexity as PrefixSpan and good linear scalability.

In detail, as the virtual multiplication procedure used for enforcing the window constraint preserves other context-free constraints, it is possible to combine it with any other such constraint directly. Besides, as the gap constraint depends only on the difference in timing between occurrences, the mining procedure can also be directly combined with any constraint that only limits pattern properties and not its occurrence vector. Properties such as max/minimum length for pattern and itemsets, taxonomical conditions, and certain regular expression conditions can be directly integrated. The mining of some of these constraints is described in a previous paper [4].

Window and gap mining are promising tools for reducing result spaces when mining databases where the number of itemsets per entry is high such as those used in propagation analysis through sequential pattern mining [1], and economics [9]. It would be interesting to generalize the gap algorithm presented to other PrefixSpan-like implementations, especially Bi-level projection, as it can also be implemented in terms of pseudo-projections. Further temporal constraints such as minimum gaps or allowing the timestamps in the databases to be non-uniform can be similarly developed naturally from the ideas here presented. Finally, introducing geographical elements to the mining process and propagation would provide a valuable complement to the temporal constraints formulated in this paper.

References

1. Alatrista-Salas, H., Bringay, S., Flouvat, F., Selmaoui-Folcher, N., Teisseire, M.: Spatio-sequential patterns mining: beyond the boundaries. Intell. Data Anal. **20**(2), 293–316 (2016)

2. Chaudhari, M., Mehta, C.: Extension of prefix span approach with grc constraints for sequential pattern mining. In: 2016 International Conference on Electrical, Electronics, and Optimization Techniques (ICEEOT), pp. 2496–2498 (2016)

3. George, A., Binu, D.: DRL-prefixspan: a novel pattern growth algorithm for discovering downturn, revision and launch (DRL) sequential patterns. Cent. Eur. J. Comput. Sci. **2**, 426–439 (2012)

4. Guevara-Cogorno, A., Flamand, C., Alatrista-Salas, H.: Copper - constraint optimized prefixspan for epidemiological research. Procedia Comput. Sci. **63**, 433–438 (2015). https://doi.org/10.1016/j.procs.2015.08.364. The 6th International Conference on Emerging Ubiquitous Systems and Pervasive Networks (EUSPN 2015)/ The 5th International Conference on Current and Future Trends of Information and Communication Technologies in Healthcare (ICTH-2015)

5. He, D., Stone, L.: Spatio-temporal synchronization of recurrent epidemics. Proc. Biol. Sci. R. Soc. **270**(1523), 1519–1526 (2003). https://doi.org/10.1098/rspb.2003. 2366

6. Lin, M.-Y., Lee, S.-Y.: Efficient mining of sequential patterns with time constraints by delimited pattern growth. Knowl. Inf. Syst. **7**(4), 499–514 (2004). https://doi. org/10.1007/s10115-004-0182-5

7. Mari, J.F., Ber, F.L.: Temporal and spatial data mining with second-order hidden markov models. Soft Comput. Fus. Found. Methodol. Appl. **10**(5), 406–414 (2006)

8. Masseglia, F., Poncelet, P., Teisseire, M.: Efficient mining of sequential patterns with time constraints: reducing the combinations. Exp. Syst. Appl. **36**(2), 2677–2690 (2009). https://doi.org/10.1016/j.eswa.2008.01.021

9. Mennis, J., Liu, J.W.: Mining association rules in spatio-temporal data: an analysis of urban socioeconomic and land cover change. Trans. GIS **9**(1), 5–17 (2005). https://doi.org/10.1111/j.1467-9671.2005.00202.x

10. Pedreshi, D., Ruggieri, S., Turini, F.: Discrimination-aware data mining. In: Proceedings of the 14th ACM SIGKDD International Conference on Knowledge Discovery and Data Mining, KDD 2008, pp. 560–568. Association for Computing Machinery, New York (2008). https://doi.org/10.1145/1401890.1401959

11. Pei, J., et al.: Mining sequential patterns by pattern-growth: the prefixspan approach. IEEE Trans. Knowl. Data Eng. **6**(11), 1424–1440 (2004). https://doi.org/10.1109/TKDE.2004.77

12. Pei, J., et al.: Mining sequential patterns by pattern-growth: the prefixspan approach. Proc. IEEE TKDE **16**(11), 1424–1440 (2004)

13. Srikant, R., Agrawal, R.: Mining sequential patterns: generalizations and performance improvements. In: Apers, P., Bouzeghoub, M., Gardarin, G. (eds.) EDBT 1996. LNCS, vol. 1057, pp. 1–17. Springer, Heidelberg (1996). https://doi.org/10.1007/BFb0014140

Aggregating News Reporting Sentiment by Means of Hesitant Linguistic Terms

Jennifer Nguyen[1], Albert Armisen[2], Núria Agell[1]($\boxtimes$), and Ángel Saz[1]

[1] ESADE Business School, Ramon Llull University, Barcelona, Spain
nuria.agell@esade.edu
[2] Faculty of Business and Communication, Vic University - Central University
of Catalonia, Vic, Spain

Abstract. This paper focuses on analyzing the underlying sentiment of news articles, taken to be factual rather than comprised of opinions. The sentiment of each article towards a specific theme can be expressed in fuzzy linguistic terms and aggregated into a centralized sentiment which can be trended. This allows the interpretation of sentiments without conversion to numerical values. The methodology, as defined, maintains the range of sentiment articulated in each news article. In addition, a measure of consensus is defined for each day as the degree to which the articles published agree in terms of the sentiment presented. A real case example is presented for a controversial event in recent history with the analysis of 82,054 articles over a three day period. The results show that considering linguistic terms obtain compatible values to numerical values, however in a more humanistic expression. In addition, the methodology returns an internal consensus among all the articles written each day for a specific country. Therefore, hesitant linguistic terms can be considered well suited for expressing the tone of articles.

Keywords: Hesitant fuzzy linguistic terms · Sentiment analysis · Consensus measurement

1 Introduction

Exploiting sentiment analysis for stock predictions [3], elections [8], sentiment tracking [6], has become common. Different sentiment techniques have been developed to identify sentiment in data such as blogs, online reviews, and microblogs [18]. Methodologies have been applied to trending sentiments towards particular topics such as election candidates, monitoring customer sentiment towards a product or business, aggregating customer reviews and so on. Many of these methods begin by assessing sentiment of written text on a n-point scale or as positive, negative or neutral labels [16]. However, this initial assessment

This project has received funding from the European Union's Horizon 2020 research and innovation programme (grant agreement n° 822654).

V. Torra et al. (Eds.): MDAI 2020, LNAI 12256, pp. 252–260, 2020.
https://doi.org/10.1007/978-3-030-57524-3_21

can loose the original sentiment contained within the text if the positive and negative values assigned to each word in a text are summed together for an overall score. In addition, the sentiment can be further removed when the sentiment of groups of texts are aggregated. As sentiment can be both positive and negative in a group of texts, the average sentiment may appear neutral. Furthermore, the extent to which an individual text may be positive or negative is no longer evident in the mean. To this end, the methodology presented in this paper proposes to represent sentiment as fuzzy linguistic terms in order to capture the range of positiveness and negativeness in an individual written text.

As the consumption of fake news becomes more and more of a concern, the professional role of media is at the forefront. Sensationalism contradicts the role of the news to report accurate information about events [7]. Journalists control what is explained and how the story is framed [1,11]. This framing shapes how an audience discusses a story. However, its influence can diminish with an audience's awareness of issues or events in their communities or the world [4]. In addition, research has demonstrated that newspapers have the ability to alter the perception of the importance and dominant opinion a community associates with specific issues [15]. As such, newspapers may be able to indirectly change public opinion [15].

Public opinion can result in different actions. Therefore, there has been interest in studying investor sentiment [3], voter sentiment [8,19], sentiment tracking [6]. These sentiments have been studied to predict stock forecasts [3], election outcomes [8], voter social media behavior [19], and detecting events related to sentiment change [6]. Some methods applied a mean sentiment score [3,19], others analyzed the frequency with which texts scored a specific sentiment [6,8], others a moving average [6]. For some papers, the overall sentiment scores were defined for each tweet as the sum of the positive and negative sentiments assigned [6,19]. Another paper, traced the divergence of opinions over time [2]. In order to study public sentiment, each of these studies focused on a particular theme such as political candidates, specific stocks or event. Lastly, these papers analyzed sentiment from Twitter posts rather than news articles, a platform where expressions of opinion are more accessible and possibly more obvious.

In this paper, we propose a methodology which highlights central sentiment and degree of consensus among a group of news articles. The sentiment of each article is expressed in fuzzy linguistic terms. As centralized sentiment is computed to reflect country specific opinion and a comparison is made among the positions taken by each country. The approach is novel, to our knowledge, in considering article sentiment as intervals rather than separately, or as an average or substraction of positive and negative sentiment preserving the original opinion. In addition, sentiment is described in linguistic terms to better reflect the natural manner in which humans discuss articles.

The rest of the paper is organized as follow. First, the preliminaries necessary to perform the methodology are presented in Sect. 2. Next, the proposed methodology is introduced in Sect. 3. It is followed by a real case example and a discussion of the results from its implementation in Sect. 4. Lastly, the conclusion and future work are presented in Sect. 5.

2   Preliminaries

In this section, a summary of basic concepts related to Hesitant Fuzzy Linguistic Term Sets (HFLTS) which will be referenced in the methodology are presented.

Let S denote a finite totally ordered set of linguistic terms, $S = \{a_1, \ldots, a_n\}$ with $a_1 < \cdots < a_n$. In this article, a HFLTS [17] is defined as a set $\{x \in S | a_i \leq x \leq a_j\}$ that is denoted as $[a_i, a_j]$ if $i < j$ or $\{a_i\}$ if $j = i$. Then, $\mathcal{H}_S$ is defined as the set of all possible HFLTS over S including the empty HFLTS, $\{\emptyset\}$, such that $\mathcal{H}_S^* = \mathcal{H}_S - \{\emptyset\}$, according to Montserrat et al. [13].

The set $\mathcal{H}_S$ is extended to $\overline{\mathcal{H}_S}$, to include the concepts of *positive HFLTS*, *negative HFLTS* and *zero HFLTS*. *Positive HFLTS* come from two HFLTS with some linguistic terms in common, *zero HFLTS* are the result of two consecutive HFLTS, while *negative HFLTS* are the result of two HFLTS with no common or consecutive linguistic terms.

In addition, the extended connected union and extended intersection operators are considered in this context as explained in [13].

1. The *extended intersection* of H_1 and H_2, $H_1 \sqcap H_2$, is the largest element in $\overline{\mathcal{H}_S}$ that is contained in H_1 and H_2.
2. The *extended connected union* of H_1 and H_2, $H_1 \sqcup H_2$, is the smallest element in $\overline{\mathcal{H}_S}$ that contains H_1 and H_2.

The extended intersection and extended connected union can be used to compute the distance between two HFLTS as defined in [13]. Given H_1 and $H_2 \in \overline{\mathcal{H}_S}$, the *width* of H, $\mathcal{W}(H)$, is defined as the number of linguistic terms contained in H, or cardinality, $card(H)$, if $H \in \mathcal{H}_S$ or $-card(-H)$ if H is a negative HFLTS. Then the distance between HFLTS in $\overline{\mathcal{H}_S}$ is computed between H_1 and H_2, as:

$$D(H_1, H_2) := \mathcal{W}(H_1 \sqcup H_2) - \mathcal{W}(H_1 \sqcap H_2). \tag{1}$$

To obtain the central sentiment (or centroid) of a set of articles about a specific theme λ, the distance D is applied as follows:

Definition 1 ([13]). *Let λ be a theme, G a set of r articles and $H_1, \ldots, H_r$ the HFLTS expressed by the articles in G with respect to the theme, λ. Then, the centroid of the set is:*

$$C_o = \arg \min_{H \in \mathcal{H}_S^*} \sum_{i=1}^{r} D(H, H_i). \tag{2}$$

The centroid is a central measure for ordinal scales with hesitancy. In addition to the centroid, we consider the consensus degree proposed by [12] to quantify the sentiment agreement among a set of articles.

Definition 2 ([14]). *Let G be a set of r articles of a theme* λ, *and* $H_1, \ldots, H_r$ *be their respective sentiments in HFLTS. Let* C_o *be the central sentiment of the set. Then, the degree of consensus of G on* λ *is defined as:*

$$\delta_\lambda(G) = 1 - \frac{\sum\limits_{i=1}^{r} D(C_o, H_i)}{r \cdot (n-1)}. \tag{3}$$

Note that $0 \leq \delta_\lambda(G) \leq 1$ as $r \cdot (n-1)$ is an upper bound of the addition of distances between the centroid and sentiment expressed as HFLTS [14].

3   The Proposed Approach to Detecting Contrasting Sentiment

In this section, we present the formal framework to determine the centroid and consensus among the sentiments of articles. Generally, articles published on the same day about the same theme do not have to reflect the same sentiments. However when it does, it could indicate that the sources of the articles are emotivated in the same direction. Further analysis could be performed by evaluating the consensus of articles from neighboring or allied countries. Likewise, articles from a specific country could be trended by representing the aggregate articles for each day in terms of their centroid. A spike in the any direction different from the trend can draw attention.

The methodology requires as input a set of articles previously identified with positive and negative sentiment, and themes. The process of identifying the sentiment and theme are considered out of scope of this methodology as we are focused on identifying the centroid and measuring the consensus. The process has four steps: 1) Select theme for analysis and the corresponding articles, 2) Represent sentiment in linguistic terms, 3) Identify the centroid, and 4) Measure the consensus as shown in Fig. 1.

1. *Select theme for analysis and the corresponding articles:* In this step, a theme is selected in order to focus the analysis. The data set is filtered for only those articles which reference a particular theme regardless of the degree to which a theme is mentioned.
2. *Represent sentiment in linguistic terms:* Next, for each article there are positive and negative sentiments. If the article is associated with positive and negative sentiment for each word in the article, the percent of negative and percent of positive words needs to be computed. We will refer to these percentages as positive and negative scores going forward.

Different from the methodologies previously discussed in Sect. 1, the methodology presented in this paper proposes to utilize intervals to represent the article sentiments. Intervals assist with distinguishing cases in which you have a polarization in sentiment. For example, an article with a positive score, $+12$, and negative score, -11, could be summarized by its average sentiment, 0.5. Similarly,

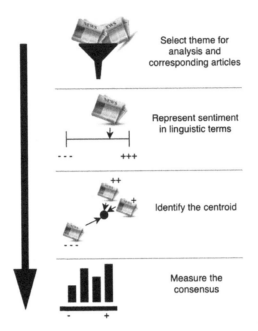

Fig. 1. Framework to determine the centroid and consensus among article sentiment

an article with a positive score, +5, and negative score, −3.5, would be summarized by its average sentiment, 0.75. Both of these examples would appear neutral. However, an interval would highlight that the first article expressed extreme sentiment in both directions. Whereas, the second article communicated with lesser sentiment. In addition, each interval is converted to linguistic terms to better represent how humans would describe an article.

Given an article with negative and positive scores, A^- and A^+, respectively, the sentiment can be represented in linguistic terms as:

$$H_A = min\{H \in \mathcal{H}_\mathcal{S}/[A^-, A^+] \subset H\},\qquad(4)$$

when $A^- \neq 0$ and $A^+ \neq 0$. In the case that $A^- = 0$ or $A^+ = 0$, then H_A is $min\{H \in \mathcal{S}/A^- \in \mathcal{S}\}$ or $min\{H \in \mathcal{S}/A^+ \in \mathcal{S}\}$, respectively.

Example 1. Let us consider a set of possible sentiments in linguistic terms: $\mathcal{S} = \{very\ negative,\ negative,\ somewhat\ negative,\ somewhat\ positive,\ positive,\ very\ positive\}$ where *very negative* (VN) $= [-100, -10]$, *negative* (N) $= (-10, -5]$, *somewhat negative* (SN) $= (-5, 0]$, *somewhat positive* (SP) $= (0, 5]$, *positive* (P) $= (5, 10]$, *very positive* (VP) $= (10, 100]$ from which an article's sentiment may be described. Given an article with positive score $A^+ = 3$ and negative score $A^- = -8$, the representation of the sentiment of the article in linguistic terms would be $[N, SP]$. Similarly, given an article with positive score $A^+ = 12$ and negative score $A^- = 0$, the representation of the sentiment of the article in linguistic terms would be $[VP]$.

3. *Identify the centroid:* Once all the pairs of positive and negative scores for each article in the set have been translated into linguistic terms, the centroid can be computed according to Eq. 2 and distance D from Eq. 1. This represents the central sentiment of the set of articles.

Example 2. Let us consider G to be a set of 5 articles written about a theme λ. The sentiment of each article is expressed in HFLTS over the set S from Example 1. If H_1, H_2, H_3, H_4, H_5 are the HFLTS of the sentiment communicated in the 5 articles, then the centroid of the set of articles, C_o, can be identified as shown in Table 1.

Table 1. Centroid of the set of articles G related to theme λ.

	H_1	H_2	H_3	H_4	H_5	C_o
λ	$[N, SN]$	$\{N\}$	$[SP, P]$	$[VN, N]$	$[N, SP]$	$\boldsymbol{[N, SN]}$

4. *Measure the consensus:* To understand to what extent articles in the set share similar sentiment, we compute the distance of each one of them to the central sentiment and determine its consensus δ from Eq. 3.

Example 3. Continuing with Example 2, the distances, D, between the centroid C_o and the sentiment of each article are computed using Eq. 1. The distances are shown in Table 2 along with their associated degrees of consensus.

Table 2. Consensus of sentiments for articles in set G related to theme λ

	D_1	D_2	D_3	D_4	D_5	$\sum_{i=1}^{5} D_i$	$\delta_\lambda(G)$
λ	0	1	4	2	2	9	0.45

4   A Real Case Example

We demonstrate the presented methodology works with a subset of data from GDELT [10], an open news platform. "GDELT monitors print, broadcast, and web news media in over 100 languages from across every country in the world..."[1]. Its archives are continuously updated every 15 min providing information on the people, locations, organizations, themes, sources, emotions, counts, quotes, images and events discussed in each article. GDELT has been used in previous

---

[1] https://www.gdeltproject.org.

research studies related to global news coverage of disasters [9], effects of political conflict [20], and predicting social unrest [5].

The GDELT data set is publicly available[2]. For each article, a list of all the themes found in the document are provided. There are 2589 possible themes from which a document can be labeled. The data set includes the tone or sentiment for each of the articles. The tone for an article is described in terms of six emotional dimensions: the average tone of the article, the positive score, negative score, percentage of words found in the tonal dictionary, percentage of active words, and percentage of self/group reference. The average tone is the positive score minus the negative score. The positive and negative scores are the percentage of all words in the article found to have positive and negative emotion, respectively.

To illustrate the viability of the methodology, we selected data before, during, and after a controversial news event. Specifically, we selected January 7 through January 9, 2020 during which time the crash of a Ukrainian airplane was being questioned. During this time frame, information for 589,815 articles were collected from GDELT's database. The articles were filtered for those labeled as "Conflict and Violence" to narrow the articles to those most related to the event. This reduced the data set to 82,054 articles. The positive and negative scores for each of the articles were selected to represent the tone. These were selected as they represented the range and quantity of each type of sentiment present in an article, making them a better descriptor than the average tone, as previously mentioned.

Next, we represented each article's tone in linguistic terms following Eq. 4 and Example 1. Table 3 provides an example of five articles to show the comparison of the tone obtained from GDELT's platform (a numerical value) and the value we obtained in linguistic terms.

Table 3. Comparison of the tones given by GDELT and that of the proposed methodology

Article	Country	Date	Tone (numerical)	Tone (linguistic)
1	US	07/01/2020	−5.74	[N, SP]
2	US	07/01/2020	3.12	[SN, P]
3	US	07/01/2020	−6.12	[N, SP]
4	US	07/01/2020	−4.61	SN
5	US	07/01/2020	−6.75	[N, SP]

Then, we analyzed the consensus on the tone of the articles to each country level. Therefore, the centroid was computed as the central sentiment for each of the three days selected for all the articles published. At this level we were able to analyze changes in sentiment by a given country during the selected

[2] http://data.gdeltproject.org/gdeltv2/masterfilelist.txt.

event period. Table 4 depicts the linguistic term that are consistent across the different days, but a different consensus measure for US.

Table 4. Linguistic terms & consensus of the US for the three different days

Country	Date	Linguistic term	Consensus
US	07/01/2020	[N,SP]	0.894
US	08/01/2020	[N,SP]	0.900
US	09/01/2020	[N,SP]	0.890

5   Conclusion and Future Research

The proposed methodology translates positive and negative sentiment scores into linguistic terms. These terms are expressed as elements of the lattice of HFLTS. This allows on the one hand the computation of distances among article sentiments and on the other hand the identification of the central representation together with the consensus of the sentiment. By using these linguistic terms, the methodology enables more explainable results compared to the results obtained when using numerical values. Intelligent systems requiring user machine interaction can benefit from expressing opinions in a human-like manner.

A limitation of the current methodology is that the analysis of the positive and negative sentiment of an individual article can be misinterpreted when coming from an individual theme. This is due to the sentiment scores of an article being associated with the entire article. Therefore, they cannot be separated into the different themes discussed. As such, the degree of positiveness or negativeness attributed to the theme analyzed cannot be certain. In this direction, as a future work, we will consider to use of different aggregation functions to take into account the imbalance among the themes represented in the articles. A second direction for our future work is to track sentiment trends to observe how different types of events affect the news reported.

References

1. Cain-Arzu, D.L.: Sensationalism in newspapers: a look at the reporter Andaman-dala in Belize 2010–2014 (2016)
2. Cao, N., Lu, L., Lin, Y.R., Wang, F., Wen, Z.: SocialHelix: visual analysis of sentiment divergence in social media. J. vis. **18**(2), 221–235 (2015)
3. Corea, F.: Can Twitter proxy the investors' sentiment? The case for the technology sector. Big Data Res. **4**, 70–74 (2016)
4. Ducat, L., Thomas, S., Blood, W.: Sensationalising sex offenders and sexual recidivism: impact of the serious sex offender monitoring act 2005 on media reportage. Aust. Psychol. **44**(3), 156–165 (2009)

5. Galla, D., Burke, J.: Predicting social unrest using GDELT. In: Perner, P. (ed.) MLDM 2018. LNCS (LNAI), vol. 10935, pp. 103–116. Springer, Cham (2018). https://doi.org/10.1007/978-3-319-96133-0_8

6. Giachanou, A., Crestani, F.: Tracking sentiment by time series analysis. In: Proceedings of the 39th International ACM SIGIR conference on Research and Development in Information Retrieval, pp. 1037–1040 (2016)

7. Hendriks Vettehen, P., Nuijten, K., Beentjes, J.: News in an age of competition: the case of sensationalism in Dutch television news, 1995–2001. J. Broadcast. Electron. Media **49**(3), 282–295 (2005)

8. Kolagani, S.H.D., Negahban, A., Witt, C.: Identifying trending sentiments in the 2016 us presidential election: a case study of Twitter analytics. Issues Inf. Syst. **18**(2), 80–86 (2017)

9. Kwak, H., An, J.: A first look at global news coverage of disasters by using the GDELT dataset. In: Aiello, L.M., McFarland, D. (eds.) SocInfo 2014. LNCS, vol. 8851, pp. 300–308. Springer, Cham (2014). https://doi.org/10.1007/978-3-319-13734-6_22

10. Leetaru, K., Schrodt, P.A.: GDELT: global data on events, location, and tone, 1979–2012. In: ISA Annual Convention, vol. 2, pp. 1–49. Citeseer (2013)

11. McCombs, M.E., Shaw, D.L.: The agenda-setting function of mass media. Public Opin. Q. **36**(2), 176–187 (1972)

12. Montserrat-Adell, J., Agell, N., Sánchez, M., Ruiz, F.J.: A representative in group decision by means of the extended set of hesitant fuzzy linguistic term sets. In: Torra, V., Narukawa, Y., Navarro-Arribas, G., Yañez, C. (eds.) Modeling Decisions for Artificial Intelligence, pp. 56–67. Springer, Cham (2016). https://doi.org/10.1007/978-3-319-45656-0_5

13. Montserrat-Adell, J., Agell, N., Sánchez, M., Prats, F., Ruiz, F.J.: Modeling group assessments by means of hesitant fuzzy linguistic term sets. J. Appl. Logic **23**, 40–50 (2017)

14. Montserrat-Adell, J., Agell, N., Sánchez, M., Ruiz, F.J.: Consensus, dissension and precision in group decision making by means of an algebraic extension of hesitant fuzzy linguistic term sets. Inf. Fusion **42**, 1–11 (2018)

15. Mutz, D.C., Soss, J.: Reading public opinion: the influence of news coverage on perceptions of public sentiment. Public Opin. Q. **61**, 431–451 (1997)

16. Prabowo, R., Thelwall, M.: Sentiment analysis: a combined approach. J. Inf. **3**(2), 143–157 (2009)

17. Rodriguez, R.M., Martinez, L., Herrera, F.: Hesitant fuzzy linguistic term sets for decision making. IEEE Trans. Fuzzy Syst. **20**(1), 109–119 (2011)

18. Vinodhini, G., Chandrasekaran, R.: Sentiment analysis and opinion mining: a survey. Int. J. **2**(6), 282–292 (2012)

19. Yaqub, U., Chun, S.A., Atluri, V., Vaidya, J.: Sentiment based analysis of tweets during the us presidential elections. In: Proceedings of the 18th Annual International Conference on Digital Government Research, pp. 1–10 (2017)

20. Yonamine, J.E.: A nuanced study of political conflict using the global datasets of events location and tone (GDELT) dataset (2013)

Decision Trees as a Tool for Data Analysis. Elections in Barcelona: A Case Study

E. Armengol[(✉)] and À. García-Cerdaña

Artificial Intelligence Research Institute, (IIIA-CSIC), Campus UAB,
Camí de Can Planes, s/n, 08193 Bellaterra, Barcelona, Spain
`eva@iiia.csic.es`

Abstract. Decision trees are inductive learning methods that construct a domain model easy to understand from domain experts. For this reason, we claim that the description of a given data set using decision trees is an easy way to both discover patterns and compare the classes that form the domain at hand. It is also an easy way to compare different models of the same domain. In the current paper, we have used decision trees to analyze the vote of the Barcelona citizens in several electoral convocations. Thus, the comparison of the models we have obtained has let us know that the percentage of people with a university degree is the most important aspect to separate the neighbourhoods of Barcelona according to the most voted party in a neighbourhood. We also show that in some neighbourhoods has always won the same party independently of the kind of convocation (local or general).

Keywords: Inductive learning methods · Decision trees · Analysis of electoral results

1 Introduction

Decision trees are inductive machine learning algorithms useful to construct domain models. Commonly, such models are built having prediction in mind. The advantage of decision trees in front of other machine learning algorithms (i.e., support vector machines, neural networks, etc.) is that they are easily understandable by experts [8]. The most of our previous work has focused on building predictive models [2,4] but, during the interaction with the experts, we observed that sometimes the expert was more interested on the attributes taken into account during the construction of the tree than in the predictivity of the final model.

In [1] we pointed out that, given a decision tree, the path from the root to a leaf can be interpreted as an explanation of the classification since it contains the pairs attribute-value relevant for the classification. In addition, in [3] we argued that the tree can be used to analyze a database. For instance, when a tree has a high depth, this means that all classes are very similar. In that case, we could conclude that the attributes used to describe the domain objects

© Springer Nature Switzerland AG 2020
V. Torra et al. (Eds.): MDAI 2020, LNAI 12256, pp. 261–272, 2020.
https://doi.org/10.1007/978-3-030-57524-3_22

are not appropriated. Conversely, when two classes are separable using a few attributes, this means that the classes are very different. We have used this kind of analysis to assess the life quality of people with intelectual disabilities [3] and caracterization of melanomas [3] [A2 = v2] and also to classify cows according their milk production [6].

In the present paper, we propose the use of decision trees to compare models of a domain or also to compare models of different domains. By the way they are built, decision trees let us know which attribute is the most important. Therefore, this allows making a first comparison to easily detect what is important for each model. As the tree is growing, the domain objects are distributed in the leaves and this also gives an idea of the similarities and differences between the classes.

To prove the feasibility of performing the comparison of models using decision trees, we analyzed several electoral results of the city of Barcelona. In particular, we have analyzed results of the elections of four convocations: Catalan Parliament held in 2017; Spanish Parliament held in April 2019; Council Hall held in May 2019; and Spanish Parliament held in November 2019. Our goal is to compare the results of these convocations and to check if the electoral behaviour of the voters has changed.

The paper is organized as follows. In Sect. 2 there is a brief explanation of decision trees. In Sect. 3 there is the description of the database used in the experiments. Section 4 contains a description and a discussion of the experiments carried on. Finally, Sect. 5 is devoted to conclusions and future work.

2   Decision Trees

A *Decision Tree* (DT) is a directed acyclic graph in the form of a tree. The root of the tree has not incoming edges and the remaining ones have exactly one incoming edge. Nodes without outgoing edges are called *leaf* nodes and the others are *internal* nodes. A DT is a classifier expressed as a recursive partition of the set of known examples of a domain [7]. The goal is to create a domain model predictive enough to classify future unseen domain objects.

Each node of a tree has associated a set of examples that are those satisfying the path from the root to that node. The leaves determine a partition of the original set of examples since each domain object only can be classified following one of the paths of the tree. The construction of a decision tree is performed by splitting the source set of examples into subsets based on an attribute-value test. This process is repeated on each derived subset in a recursive manner. Algorithm 1 shows the ID3 procedure [9,10] commonly used to grow decision trees. From a decision tree we can extract rules (i.e., patterns) giving descriptions of classes, since each path from the root to a leaf forms a classification rule. When all the examples of a leaf belong to the same class such description is *discriminant*. Otherwise, the description is *no discriminant*.

A key issue of the construction of decision trees is the selection of *the most relevant attribute* to split a node. There are different criteria to split a node and therefore, the selected attribute could be different depending on it and thus

---

Algorithm 1. ID3 algorithm for growing a decision tree.

---

procedure ID3(E, A)   ▷ E: Set of Examples; A: Set of attributes
    Create *node*
    if all $e \in E$ belong to the same *class* **then**
        return *class* as the label for *node*
    else
        $a_t \leftarrow$ best attribute
        for each value v_j of a_i **do**
            add a new tree branch below *node*
            $E_{a_i} \leftarrow$ subset of examples of E such that $a_i = v_j$
            ID3($E_{a_i}, A - \{a_i\}$)
        end for
    end if
    return *node*
end procedure

---

the whole tree could also be different. In our experiments we used the López de Mántaras' distance [5], which is an entropy-based normalized metric defined in the set of partitions of a finite set. It compares the partition induced by an attribute, say a_i, with the *correct partition*, i.e., the partition that classifies correctly all the known examples. The best attribute is the one inducing the partition which is closest to the correct partition. Given a finite set X and a partition $\mathcal{P} = \{P_1, \ldots, P_n\}$ of X in n sets, the entropy of $\mathcal{P}$ is defined as ($|\cdot|$ is the cardinality function):

$$H(\mathcal{P}) = -\sum_{i=1}^{n} p_i \cdot \log_2 p_i, \text{ where } p_i = \frac{|P_i|}{|X|}$$

and where the function $x \cdot \log_2 x$ is defined to be 0 when $x = 0$. The *López de Mántaras'* distance (LM) between two partitions $\mathcal{P} = \{P_1, \ldots, P_n\}$ and $\mathcal{Q} = \{Q_1, \ldots, Q_m\}$ is defined as:

$$\text{LM}(\mathcal{P}, \mathcal{Q}) = \frac{H(\mathcal{P}|\mathcal{Q}) + H(\mathcal{Q}|\mathcal{P})}{H(\mathcal{P} \cap \mathcal{Q})}, \tag{1}$$

where

$$H(\mathcal{P}|\mathcal{Q}) = -\sum_{i=1}^{n}\sum_{j=1}^{m} r_{ij} \cdot \log_2 \frac{r_{ij}}{q_j}, \quad H(\mathcal{Q}|\mathcal{P}) = -\sum_{j=1}^{m}\sum_{i=1}^{n} r_{ij} \cdot \log_2 \frac{r_{ij}}{p_i},$$

$$H(\mathcal{P} \cap \mathcal{Q}) = -\sum_{i=1}^{n}\sum_{j=1}^{m} r_{ij} \cdot \log_2 r_{ij},$$

$$\text{with } q_j = \frac{|Q_j|}{|X|}, \text{ and } r_{ij} = \frac{|P_i \cap Q_j|}{|X|}.$$

Decision trees can be useful for our purpose because their paths give us *patterns* describing classes of objects (electoral sections in our approach) in a

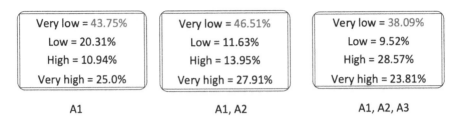

                 A1   A1, A2   A1, A2, A3

Fig. 1. Example of the stopping condition we used when growing a decision tree.

user-friendly manner. One shortcoming of decision trees is *overfitting*, meaning that there are few objects in most of the leaves of the tree. In other words, paths are actually descriptions that poorly represent the domain. The responsible of overfitting is the stoping condition of the algorithm: the set of examples has to be partitioned until all the examples of a node belong to the same solution class.

A way to either avoid or reduce overfitting is by pruning the tree, i.e., to expand all the nodes and then, with a post-process to merge two o more nodes; or, under some conditions, a node is no longer expanded. However, in both cases, this means that leaves can contain objects belonging to several classes and, therefore, paths do not represent discriminatory descriptions of classes. In other words, the descriptions or patterns represented by the branches of the tree are satisfied by objects of more than one class.

In our approach, we managed overfitting by controlling the percentage of elements of each class. Let S_N be the set of objects associated with an internal node N. The stopping condition in expanding N (the *if* of the ID3 algorithm) holds when the percentage of objects in S_N that belong to the majority class decreases in one of the children nodes. In such a situation, the node N is considered as a leaf.

As an example, let us suppose that the examples of a database can be classified in one of the following classes: *Very low*, *Low*, *High*, and *Very high*. Let us suppose now that when we grow a decision tree, we find that the most relevant attribute is A1. For the value $v1$ of such attribute we have the tree path (description) $D1 : [A1 = v1]$. The left hand side of Fig. 1 shows the distribution of the objects satifying $D1$ in each class. We see that the majority class with the 43.75% of examples is Very low. The next most relevant attribute is A2 and, for a value $v2$ we have the description $D2 : [[A1 = v1], [A2 = v2]]$. The centre of Fig. 1 shows the distribution of the objects satifying $D2$ in each class. Here the majority class is again Very low and the percentage is 46.51%; therefore the addition of A2 has improved the classification. Let us suppose that the next most relevant attribute is A3 and, for a value $v3$, we have the description $D3 : [[A1 = v1], [A2 = v2], [A3 = v3]]$. The right side of Fig. 1 shows that now the percentage of the majority class is 38.09%, i.e., lower than the one of $D2$. Therefore, now the procedure stops and the tree path we can use as description is $D2$.

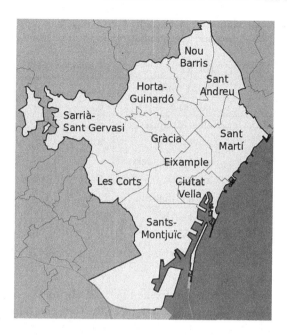

Fig. 2. Administrative division of Barcelona in 10 districts.

3   A Case Study: An Analysis of the Electoral Results in Barcelona

In Catalonia, there are four different kind of elections: Municipalities, Catalan Parliament, Spanish Parliament and European Parliament. In this study we want to use decision trees to compare the results of four electoral convocations: Catalan Parliament 2017, Spanish Parliament April 2019, Municipal elections 2019, and Spanish Parliament November 2019.

Previously to describe the database we have used in our experiments, we briefly explain the administrative organization in neighbourhoods of Barcelona and the political context.

3.1   Administrative Organization of Barcelona

From the administrative point of view, Barcelona is composed of 10 districts (Fig. 2) each one in turn, composed of neighbourhoods. Thus, Barcelona is composed of 73 neighbourhoods. We focus our study in the political party that had won in each neighbourhood.

3.2   Electoral Organization of Barcelona and Political Context

Electoral landscape of Catalonia is formed by 5048 *electoral sections* each one of them composed of a minimum of 500 potential voters and a maximum of 2000.

Table 1. Division of Barcelona city in districts. For each district it is shown the number of neighbourhoods (#Neighb) and the number of electoral sections (#ES).

Number	District	#Neighb.	#ES
1	Ciutat Vella	4	55
2	Eixample	6	173
3	Sants-Monjuïc	8	117
4	Les Corts	3	57
5	Sarrià-Sant Gervasi	6	98
6	Gràcia	5	88
7	Horta-Guinardó	11	123
8	Nou Barris	13	117
9	Sant Andreu	7	96
10	Sant Martí	10	147

Following such criteria, Barcelona is formed by 1071 electoral sections distributed between the 10 districts as Table 1 shows. Notice that the number of electoral sections of each district gives an idea of its density of population.

From 2010 there is a complex political framework in Catalonia. In addition to the traditional ideologies left-right a new issue appears: the independence of Catalonia from Spain. Some of the historical Catalan parties already had the independence in their program, however it was not a main objective. Nevertheless, from a set of reasons that are out of the scope of this paper, the independence of Catalonia has become a priority for many population and for some parties, to the point that the choice independence/no independence has put the left-right dichotomy in a second term. It would be interesting to know how this factor has influenced the behaviour of the voters. Barcelona is a very populated city with many people having his origins in other Spanish regions or in other countries. For this reason we have considered interesting to study if the independence issue has some influence in the vote and which are the most reluctant neighbourhoods to independence.

Table 2 shows the political parties that concurred to the electoral convocations we analyzed and their ideology. For the sake of simplicity we call *ECP* a party that has concurred with different names in all the elections: *Barcelona en Comú, En Comú Podem, Unidas Podemos*, among others. In that table we only show those parties that have been winners in some of the neigborhoods, therefore it is not an exhaustive table of all the parties that were eligible.

3.3   The Database

The database we have is composed of 73 records, each one of them corresponds to one neighbourhood of Barcelona. Each record has socio-demographic information and the party that had won in each one of the electoral convocations. Socio-demographic data has been obtained from the files of the official web of the Barcelona City Hall (https://www.bcn.cat/estadistica/angles/dades/inf/barris/a2018/index.htm) that contains socio-demographic information about each neighbourhood of Barcelona. The most voted party for each neighborhood has been obtained from the public results in https://www.bcn.cat/estadistica/angles/dades/inf/ele/index.htm.

Socio-demographic attributes are the following: density, women, men, age0–14, age15–25, age25–64, age+65, Barcelona, Catalunya, Spain, other, university degree, birth rate, alone+65, over-aging rate, unemployed and income. The over-aging rate has been calculated as the rate of $\frac{people+75}{people+65} * 100$ and the income is a percentage that has been calculated taking 100 as the index of the whole city.

The majority of the data above are percentages. We discretized them by dividing the whole range of an attribute in intervals of equal length. We have used the elbow method to determine the better number of intervals for each attribute. For the attributes discretized in three intervals, we have associated the labels *L (low)*, *M (medium)*, and *H (high)*. For the attributes discretized in four intervals, we have associated the labels *VL (very low)*, *L (low)*, *H (high)*, and *VH (very high)*. The attribute income has been discretized in four intervals, but we have used the labels *L (low)*, *M (medium)*, *H (high)*, and *VVH (very very high)*, where *VVH* corresponds to those neighbourhood having an income greater than 100%, the other intervals have been calculated using the equal length width and the elbow method.

4 Experiments

Our goal is to analyze how the most voted party changes for each electoral convocation at each neighborhood of Barcelona. We focused on the results of four electoral convocations: Catalan Parliament 2017, Spanish Parliament April 2019, Municipal elections 2019, and Spanish Parliament November 2019. In all the experiments we have considered all the socio-demographic attributes and, as solution class, the winner of each neighbourhood. We performed four independent experiments, one for each electoral convocation. Figure 3 shows the decision trees we have obtained.

Table 2. Political parties that have won in some neighbourhoods in some of the electoral convocations we analyzed. For each party we show its ideology in terms of right.left and independentist-no independentist.

Party	Ideology	Independentist?
Cs	Right	No
ERC	Center-left	Yes
JxC	Center-right	Yes
ECP	Left	No defined
PP	Right	No
PSC	Cener-left	No

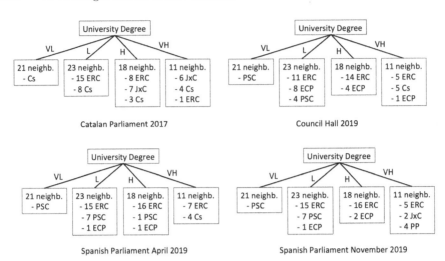

Fig. 3. Models of each one of the electoral convocations. Each node shows the number of neighbourhoods (*neigb.*) satisfying it, and also how many of them have voted to the parties.

Notice that, for all of them, the relevant attribute is the percentage of people having a university degree. Concerning the results of the convocation of 2017, seems clear that the voters were polarized according independentist/no independentist options since all the votes went to ERC and JxC (both independentists) and Cs (no independentist). The majority for the independentist option is clear, however notice that all the neighbourhoods having very low percentage of people with university degree had voted to Cs. As the percentage of university degrees increases, the percentage of votes to Cs decreases. In the rest of convocations these neighbourhoods changed the vote to PSC (center left). In any case, no voters had moved to independentist options. Also, it is interesting to remark that elections of 2017 were specially polarized by the independence/no independence option, and in this particular aspect, the political party Cs was more beligerant against independence than PSC. This can also be shown in the 18 neighbourhoods with high percentage of people with university degree. In 2017, 15 of them had voted independentist (8 to ERC and 7 to JxC) whereas the remaining 3 voted to Cs. In April 2019 the no independentist vote was divided between PSC and ECP; and in the remaining two convocations the vote went to ECP (this party has not a defined position about Catalonia independence). For the other percentages of university degrees there is not a clear separation using only that attribute. It is also interesting to see that the 21 neighbourhoods with low percentage of university degree that in 2017 had voted to Cs, in the rest of convocations have changed their vote in favour to PSC (a more moderated option). Also notice that the votes to Cs of the neighbourhoods with very high percentage of university degrees had change to PP (right, no independentist) in the last elections.

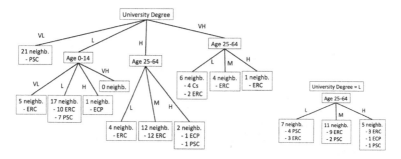

Fig. 4. Model for the Spanish elections held in April 2019. Each node shows the number of neighbourhoods (*neigb.*) satisfying it, and also how many of them have voted to the parties.

The complete model for the convocation held in April 2019 can be seen at the left hand side of Fig. 4. Notice that for neighbourhoods with high or very high percentage of people with university degree, the most relevant attribute is the range of age between 25 and 64, whereas when the percentage is low, the most relevant attribute is the range of age between 0 and 14. When the percentage is very low no additional attributes are necessary. Notice that neighbourhoods with low percentage of young people (from 0 to 14) could be interpreted as neighbourhoods with adult people who almost all of them can vote (in Spain the minimum age to vote is 18). In fact, this node could be considered as expressing the same as the one corresponding to [age25–64=H], since this last means that are neighbourhoods with low people under 24. For this reason, we have forced to use the attribute age25–64 for all the values of university degree. The result is the subtree show at the right hand side of Fig. 4.

The analysis of this result shows that for those neighbourhoods with either low or high percentage of university degrees, the vote is divided between independentist (ERC) and moderately no independentist (PSC) and ECP (undefined about independence). In neighbourhoods where the percentage of university degrees is high and the population with age between 25 and 64 is high, the vote goes to no independentist or undefined options (PSC or ECP) whereas in other situations, this means, in neighbourhoods with low or medium percentage of people between 24 and 64 years (i.e., mostly young people between 18 and 24 or people over 65) the vote goes to independentist options. Notice that the neighbourhoods with very high percentage of university degrees and low percentage of people between 25 and 64 years are the only ones that vote strong no independentist options (Cs).

If we let expand the tree, for the elections to the Council Hall (see Fig. 5) the next most relevant attribute is Other, i.e., the percentage of people of a neighbourhood that has born in a country different of Spain. In other words, this attribute represents the immigration percentage. We can see that in neighbourhoods with very high percentage of university degrees there are not neighbourhoods having high or very high percentage of immigration. The majority of the

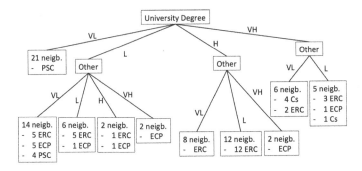

Fig. 5. Model for the Council Hall elections held in May 2019. Each node shows the number of neighbourhoods (*neigb.*) satisfying it, and also how many of them have voted to the parties.

neighbourhoods with very low percentage of immigration vote to Cs (strongly no independentist) and the vote of the neighbourhoods with low percentage of immigration is divided between ERC, ECP and Cs. In fact, the party Cs has win only in some neighbourhoods with [university degree=VH]. When the percentage of university degrees is high, only in the neighbourhoods with very high percentage of immigration ([Other=VH]) have voted the no independentist option of ECP. When the percentage of university degree is low, in the two neighbourhoods with very high immigration, the winner was ECP and in the 6 neighbourhoods with low percentage of immigration in 5 of them had won ERC. For other percentages there is not a clear winner and there is no way to expand the tree to separate the neighbourhoods.

Finally, the model for the elections held in November 2019 (see Fig. 6) show that the most relevant attribute is also age25–64 as in the elections of April 2019. Again, for neighbourhoods where the percentage of university degrees is low, the most relevant attribute is age0–14. As we explained, such atribute could be seen as complementary to age25–64, for this reason in Fig. 6 we used this last attribute for all the values of university-degree.

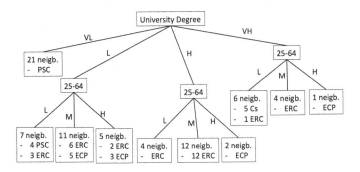

Fig. 6. Model for the Spanish elections held in November 2019. Each node shows the number of neighbourhoods (*neigb.*) satisfying it, and also how many of them have voted to the parties.

Comparing the subtrees of [university-degree=L] of the models of the Spanish Elections (Figs. 4 and 6), we can see that in essence they are not so different. For [age25–64=L] the distribution of votes is the same; for [age25–64=M] three neighbours of ERC (independentist) in April have changed to ECP (undefined) in November, and all the votes of PSC have went to ECP; and when [age25–64=H], ERC has lost one neighbour with respect to the results of April, and ECP has won all the no independentist neighbour that in April had voted to PSC. For [university-degree=H] the only change is that when [age25–64 = H] in both neighbourhoods the winner has been ECP. For [university-degree = VH] Cs has won in one more neighbourhood in November than in April; and also when [age25-64 = H] in the neighbourhood that in April had won ERC, in November has won ECP.

5   Conclusions

In this paper we have introduced a new approach to analyze electoral data: the decision trees. This kind of methods are commonly used to construct domain models useful for prediction. Our focus has been the political party that has won in each Barcelona neighbourhood.

Thanks to the representation as a tree we can see that: (1) the most relevant attribute to characterize the neighbourhoods of Barcelona according the winner party is university-degree; (2) from the point of view independentist/no independentist, the results are not substantially different in the four analyzed convocations; (3) the 21 neighbourhoods having a very low percentage of people with university degree, always have voted no independentist options, although in 2017 the winner was Cs (strongly against independence) and in the next convocations changed to PSC (more moderated); (4) The Cs party has had and important decrement of votes, however, the neighbourhoods with [university-degreeVH] and [age25-64=L] have a great fidelity to this party since there are the ones in which this party has won in all the convocations; and (5) the percentage of immigration is only important for the elections to the Council Hall, and the ECP party has won in more neighbourhoods than in other convocations.

The conclusion is that the citizens of Barcelona have not substantially changed their vote according to the kind of elections, since we have found very similar models for each one of them. In addition, a very important conclusion is that the percentage of university degree is the most important factor influencing the electoral result of a neighbourhood. Clearly, the independentist/no independentist dichotomy has had an influence in the result since in convocations previous to 2010 the sense of vote changed according to the elections: in convocations to Spanish Parliament tend to win parties that are delegations of national parties (for instance PSC) whereas in Catalan elections tend to win Catalan parties.

In the future we plan to analyze in the same way other Catalan cities as, for instance Girona, Lleida and Tarragona (the main cities of each one of the Catalonia concurrencies) and compare the similarities and differences.

Acknowledgments. This research is funded by the project RPREF (CSIC Intramural 201650E044); and the grant 2014-SGR-118 from the Generalitat de Catalunya.

References

1. Armengol, E.: Usages of generalization in case-based reasoning. In: Weber, R.O., Richter, M.M. (eds.) ICCBR 2007. LNCS (LNAI), vol. 4626, pp. 31–45. Springer, Heidelberg (2007). https://doi.org/10.1007/978-3-540-74141-1_3
2. Armengol, E.: Building partial domain theories from explanations. Knowl. Intell. **2**(08), 19–24 (2008)
3. Armengol, E., García-Cerdaña, À., Dellunde, P.: Experiences using decision trees for knowledge discovery. In: Torra, V., Dahlbom, A., Narukawa, Y. (eds.) Fuzzy Sets, Rough Sets, Multisets and Clustering. SCI, vol. 671, pp. 169–191. Springer, Cham (2017). https://doi.org/10.1007/978-3-319-47557-8_11
4. Armengol, E., Plaza, E.: Discovery of toxicological patterns with lazy learning. In: Palade, V., Howlett, R.J., Jain, L. (eds.) KES 2003. LNCS (LNAI), vol. 2774, pp. 919–926. Springer, Heidelberg (2003). https://doi.org/10.1007/978-3-540-45226-3_126
5. López de Mántaras, R.: A distance-based attribute selection measure for decision tree induction. Mach. Learn. **6**, 81–92 (1991)
6. Lopez-Suarez, M., Armengol, E., Calsamiglia, S., Castillejos, L.: Using decision trees to extract patterns for dairy culling management. In: Iliadis, L., Maglogiannis, I., Plagianakos, V. (eds.) AIAI 2018. IAICT, vol. 519, pp. 231–239. Springer, Cham (2018). https://doi.org/10.1007/978-3-319-92007-8_20
7. Maimon, O., Rokach, L. (eds.): Data Mining and Knowledge Discovery Handbook, 2nd edn. Springer, Boston (2010)
8. Pazzani, M.J.: Knowledge discovery from data? IEEE Intell. Syst. **15**(2), 10–13 (2000)
9. Quinlan, J.R.: Discovering rules by induction from large collection of examples. In: Michie, D. (ed.) Expert Systems in the Microelectronic Age, pp. 168–201. Edinburg University Press, Edinburg (1979)
10. Quinlan, J.R.: Induction of decision trees. Mach. Learn. **1**(1), 81–106 (1986)

Explaining Misclassification and Attacks in Deep Learning via Random Forests

Rami Haffar, Josep Domingo-Ferrer$^{(\boxtimes)}$, and David Sánchez

UNESCO Chair in Data Privacy, CYBERCAT-Center for Cybersecurity Research of Catalonia, Department of Computer Engineering and Mathematics, Universitat Rovira i Virgili, Av. Països Catalans 26, 43007 Tarragona, Catalonia, Spain
{rami.haffar,josep.domingo,david.sanchez}@urv.cat

Abstract. Artificial intelligence, and machine learning (ML) in particular, is being used for different purposes that are critical for human life. To avoid an algorithm-based authoritarian society, AI-based decisions should generate trust by being explainable. Explainability is not only a moral requirement, but also a legal requirement stated in the European General Data Protection Regulation (GDPR). Additionally, it is also beneficial for researchers and practitioners relying on AI methods, who need to know whether the decisions made by the algorithms they use are rational, lack bias and have not been subjected to learning attacks. To achieve AI explainability, it must be possible to derive explanations in a systematic and automatic way. A common approach is to use a simpler, more understandable decision algorithm to build a surrogate model of the unexplainable, a.k.a. black-box model (typically a deep learning algorithm). To this end, we should avoid surrogate models that are too large to be understood by humans. In this work we focus on explaining the behavior of black-box models by using random forests containing a fixed number of decision trees of limited depth as surrogates. In particular, our aim is to determine the causes underlying misclassification by the black box model. Our approach is to leverage partial decision trees in the forest to calculate the importance of the features involved in the wrong decisions. We achieve great accuracy in detecting and explaining misclassification by deep learning models constructed via federated learning that have suffered attacks.

Keywords: Explainability · Machine learning · Deep learning · Random forest

1 Introduction

The last two decades have witnessed major progress in artificial intelligence (AI). Deep learning, based on deep neural networks, has ushered in a breakthrough in the capability of AI systems. This forward leap has been made possible by the increase of computing power and the availability of big data, which together

© Springer Nature Switzerland AG 2020
V. Torra et al. (Eds.): MDAI 2020, LNAI 12256, pp. 273–285, 2020.
https://doi.org/10.1007/978-3-030-57524-3_23

allow training deep learning models. However, not all big data sources are public-domain. In order to leverage private data for learning, in [5] a new concept called federated learning was proposed. In federated learning the model is learned in a decentralized way: the model manager proposes an initial global model; then each owner of private data computes an update of the proposed model after running it on her private data set; finally, the model manager improves the global model by modifying it based on the aggregation of the received updates. This process is iterated until the global model classifies well enough all private data sets.

Both centralized deep learning and federated deep learning models achieve very accurate results, but they are black boxes in the sense that it is infeasible for humans to understand how decisions are made. This lack of transparency is problematic for both the developers who train a black-box model and the citizens who are affected by its decisions:

- Developers would like to know how decisions are made by the black-box algorithm to make sure that the algorithm takes the relevant features into account during the training phase. It may happen that wrong features are used to make decisions, such as in the well-known example of [9] where the animal is classified as a wolf if the image has snow background and as a husky dog if the image has grass background, because in the training pictures all the wolves were displayed in a snowy landscape and the huskies were not.
- Citizens are affected by the growing number of automated decisions: credit granting, insurance premiums, medical diagnoses, granting of loans, etc. While transparency measures are being implemented by public administrations worldwide, there is a risk of automated decisions becoming a ubiquitous black box. To protect citizens, explainability requirements are starting to appear in legal regulations and ethics guidelines, such as: article 22 of the EU General Data Protection Regulation [14], which states the right of citizens to an explanation on automated decisions; the European Commission's Ethics Guidelines for Trustworthy AI [2], that insist on the organizations that make automated decisions to be prepared to explain them at the request of the affected citizens; the IEEE report on ethically aligned design for intelligent systems [10].

Since it is necessary to explain the black-box behavior, the generation of explanations must be automated. A usual strategy to generate explanations for decisions made by a black-box machine learning model, such as a deep learning model, is to build a surrogate model based on simpler and more expressive machine learning algorithms, like decision rules [9,13], decision trees [1,11], or linear models [12]. As pointed out in [3], those simpler algorithms are intrinsically understandable by humans as long as they are not too large. The surrogate model can be trained either on the same data set as the black-box model to be explained or on new data points classified by that same model. Global surrogate models explain decisions on points in the whole domain, while local surrogate models build explanations that are relevant for a single point or a small region of the domain.

Contribution and Plan of This Paper

We present an approach which assumes that the party who generates the explanations has unrestricted access to the black-box model and a training data set. The training data set of the surrogate model can be smaller than the entire data used to train the black-box model: a sufficiently representative subset may be enough.

In this work, we use random decision forests to build our surrogate model. Random forests [4] consist of a fixed number of decision trees, each of which has a controlled depth and a measure reflecting the feature importance. The surrogate random forest will be trained on the same data that was used to train the black-box model.

Single decision trees have already been employed in the literature as surrogate models [1,11]. The originality of our proposal lies in using a random decision forest rather than a single decision tree. In this forest, all trees have limited depth, but the structure of each tree may differ. The diversity in the random decision forest makes it possible to have trees whose predictions match the prediction of the black-box model even if the majority of the trees do not match it. This allows scanning the forest for trees that agree with the black-box model and using these trees for the explanation of the black-box decision.

In our work we concentrate on the cases where the black-box model made wrong predictions and our aim is to identify the features involved in those predictions. For example, if some features of the training data set were altered as a result of an attack, we can identify which of the altered features were more influential in the wrong predictions.

Section 2 identifies the desirable properties and the possible risks of explanations via surrogate models. Section 3 deals with outliers and possible attacks on machine learning data sets. Section 4 describes our surrogate model based on a random decision forest. Experimental results are provided in Sect. 5. Finally, in Sect. 6 we gather conclusions and sketch future research lines.

2 Requirements and Risks of Surrogate Models for Explanation

As discussed in [6], explanations of the decisions or predictions of a black-box model should satisfy the following properties:

- **Accuracy:** This property refers to how well an explanatory surrogate model predicts unseen data.
- **Fidelity:** The decisions of the explanatory surrogate model should be close to the decisions of the black-box model on unseen data. If the black-box model has high accuracy and the explanation has high fidelity, then the explanatory surrogate model also has high accuracy.
- **Consistency:** The explanations should apply equally well to any machine learning algorithm trained on the same data set.

- **Stability:** Decisions by the black-box model on similar instances should yield similar explanations.
- **Representativeness:** If the surrogate model can be applied to several decisions on several instances we can call it a highly representative explanation.
- **Certainty:** If the black-box model at study provides a measure of assurance in its predictions, an explanation of these predictions should reflect this measure.
- **Novelty:** It indicates the ability of the explanations of the surrogate model to cover cases far from the training data.
- **Degree of importance:** The explanation should highlight the important features.
- **Comprehensibility:** The explanations provided by the surrogate model should be understandable to humans. Depending on the target users, more or less complex explanations can be acceptable, but in general short explanations are more comprehensible.

No single explanation model in the current literature is able to satisfy all the above properties. In our approach we will focus on accuracy, fidelity, degree of importance, comprehensibility, and representativeness.

It must be realized that a complex surrogate model, even if based on intrinsically understandable algorithms, may fail to be comprehensible to humans. For example, we trained a decision tree on one of the studied data sets (Synthetic data set) that contains 10 feature attributes and a class attribute with two classes. The explanation tree contained 13,404 nodes and had depth 15, which does not make this tree very useful as an explanation to humans: it is very difficult to comprehend it.

3   Outliers and Attacks on Data Sets

Machine learning models rely on the credibility of the data set used for training. However, data sets may contain some abnormal instances, also known as outliers. Although outliers can be legitimate, they can also be the outcome of attacks. On the one hand, cheating respondents may deliberately contribute bad answers when the data set is gathered. On the other hand, third-party attackers may seek to modify the training data set with a view to disrupt the training of the model.

Random attacks are the most common type of attack. In this case, the attacker intends to sabotage the machine learning model by replacing the actual values of one or more of the attributes of the training data with random values in the same range.

4   Random Forest-Based Surrogate Model

To satisfy the properties listed in Sect. 2, we need a method to construct surrogate models that keeps complexity at bay while being able to capture as much as possible of the performance of black-box deep learning models. For that purpose, we use a random forest model. When trained on a fraction or the whole training data set, the random forest model outputs a vector of numerical values assessing the importance of the different features on classification decisions.

More specifically, feature importances are computed as follows [7]. First, let us define the notion of impurity of a data set, which is a measure of the homogeneity of the values in it. Impurity can be measured in several values, including Gini impurity and Shannon entropy. In particular, the Gini impurity of a data set is

$$C = \sum_{i=1}^{L} f_i(1 - f_i),$$

where L is the number of possible different labels and f_i is the relative frequency of values with the i-th label. Clearly, if all the values in the data set correspond to the same label, then $C = 0$; the more diverse the values, the higher C. Now, given a decision tree, the importance of each node j in it is

$$ni_j = w_j C_j - \sum_{k \in \text{Children}_j} w_k C_k,$$

where w_j is the weighted number of samples reaching node j, C_j is the Gini impurity of the samples reaching node j, and Children$_j$ are the children nodes of node j. Thus, the higher the homogeneity gain of a node, the more important it is, where the homogeneity gain is the reduction of impurity between the input set of the node and its output subsets (those that go to its children nodes). In other words, an important node is one that "neatly" classifies the samples that reach it. Then, the raw importance of each feature i is

$$fi_i = \frac{\sum_{j:\text{node } j \text{ splits on feature } i} ni_j}{\sum_{k \in \text{all nodes}} ni_k}.$$

Finally, the normalized feature importance of each feature i is a number between 0 and 1 computed as

$$norm\, fi_i = \frac{fi_i}{\sum_{j \in \text{all features}} fi_j}.$$

From now on, when we mention feature importances we will refer to normalized feature importances.

Algorithm 1 attempts to determine the causes of wrong prediction by the black-box model. First, the algorithm uses the training data set to train both the black-box model and the random forest model. Second, it evaluates both models to make sure that they can be compared with each other. The accuracy of the random forest should not be too inferior to that of the black-box model; otherwise, the explanations obtained from the random forest would be useless. Third, for each wrong prediction of the black-box model on the test data set, the algorithm scans the forest and stores the vector of feature importances of each tree in the forest whose prediction coincides with the wrong black-box prediction. Finally, the algorithm averages all stored feature importance vectors, in order to obtain a vector containing the average importance of every feature in causing wrong decisions.

Algorithm 1: Determine the importance of features in wrong predictions

---

Input: Dataset X
1 Train_X, Test_X ← Split_Train_Test(X);
2 Black_Box ← Train_Black_Box(Train_X);
3 Forest ← Build_Random_Forest(Train_X);
   /* The accuracy of the black-box model and the random forest are
      evaluated   */
4 Score_Black_Box ← Evaluate_Black_Box(Test_X);
5 Score_Forest ← Evaluate_Forest(Test_X);
6 Feature_Importances_List ← {} ; // This list will contain the vectors of
   feature importances for selected trees in the forest
7 **for** *Each Sample in Test_X* **do**
8   **if** *Predict (Black_Box,Sample) not correct* **then**
9   **for** *Each Tree in Forest* **do**
10   **if** *Predict (Black_Box, Sample) == Predict (Tree, Sample)* **then**
                   /* The vector with the feature importances of each tree
                      agreeing with the black-box model is appended to the
                      list   */
11   Append (Feature_Importances_List, Tree.Feature_Importances); `
12   **end**
13   **end**
14   **end**
15 **end**
   /* The vectors in the list are averaged to obtain the average
      feature importance vector   */
16 Feature_Importances←Average(Feature_Importances_List);
17 **Return** Black_Box, Score_Black_Box, Forest, Feature_Importances;

Algorithm 2 attempts to discover whether a black-box model was trained on an attacked data set and which are the training features that are most likely to have been attacked. The algorithm takes as input the outputs of Algorithm 1, that is, a trained black-box model, its reported accuracy score, a random forest explaining the black-box model and the vector of reported feature importances associated with the random forest. The algorithm also takes as input a *reliable* test data set Test_X that will be used to test whether the black-box model was trained on attacked data.

First, the algorithm evaluates the accuracy on the black-box model on the reliable test data Test_X. If the accuracy Score_Black_Box2 on Test_X is much lower than the reported accuracy Score_Black_Box, this suggests that the black-box model was trained on attacked data. In this case:

1. For each wrong prediction of the black-box model on the reliable test data Test_X, the algorithm scans the forest and stores the vector of feature importances of each tree in the forest whose prediction coincides with the wrong black-box prediction.
2. The algorithm averages all stored feature importance vectors, in order to obtain a vector Feature_Importances2 containing the average importance of every feature in causing wrong decisions.
3. The likelihood of each feature to have been attacked in the training data is proportional to the difference of the importance of that feature in Feature_Importances2 and the reported Feature_Importances. In particular, the feature with the largest difference is the most likely to have been attacked.

Algorithm 2: Discover the feature under attack

Input: Black_Box, Score_Black_Box, Forest, Feature_Importances, Test_X

1 Score_Black_Box2 ← Evaluate_Black_Box(Test_X);
2 **if** *Score_Black_Box − Score_Black_Box2 > threshold* **then**
3   Feature_Importance_List2 ← {};
4   **for** *Each Sample in Test_X* **do**
5   **if** *Predict (Black_Box,Sample) not correct* **then**
6   **for** *Each Tree in Forest* **do**
7   **if** *Predict (Black_Box, Sample) == Predict (Tree, Sample)* **then**
8   Append (Feature_Importance_List2,Tree.Feature_Importances);
9   **end**
10   **end**
11   **end**
12   **end**
13   Feature_Importance2←Average(Feature_Importance_List2);
14   Feature_Attack_Likelihoods = |Feature_Importance − Feature_Importance2|;
15   **Return** Feature_Attack_Likelihoods;
16 **end**

---

5   Experimental Results

We have applied the above-described methodology to three data sets: a synthetic numerical data set, a real numerical data set and a real data set with a mix of categorical and numerical attributes.

To keep computation simple, we have made a small change when testing Algorithm 2. Instead of attacking the training data (which would require training both the black-box model and the random forest first with the original training data to get the real importance of features and then again with the attacked training data), we have attacked the test data Test_X used by Algorithm 2.

This avoids re-training but has the same effect: the test data Test_X used in Algorithm 2 depart from the data used to train the model. We can take Test_X as being the good data and the training data as having been attacked.

5.1 Experiments on Synthetic Numerical Data

We generated a data set consisting of $1,000,000$ records, each with 10 numeric continuous attributes and a single binary class labeled using the `make_classif-ication` method from *Scikit-learn*.[1] We reserved 2/3 of the records to train the models, and the remaining 1/3 to test them. We took as a black-box model a neural network denoted by ANN with three hidden layers of 100 neurons each, which achieved 96.55% classification accuracy. We also trained a random forest with 1000 trees, that had maximum depth 5 and an average size of 62 nodes. The forest classification accuracy was 90.8%; this is less than the accuracy of the black-box model, but still high enough for the explanations obtained from the random forest to be useful. Table 1 shows the importances of features on the wrong decisions of the black-box model, as computed by Algorithm 1; features are sorted in descending order of importance. Feature number 5 turns out to be the one that has most influence on wrong decisions.

Table 1. Importance of the features of the synthetic data set on wrong predictions by the black-box model

Feature name	Feature importance
X[5]	0.2916
X[1]	0.2564
X[4]	0.0924
X[6]	0.0769
X[3]	0.0646
X[9]	0.0532
X[7]	0.0462
X[0]	0.0395
X[2]	0.0394
X[8]	0.0392

Also, we ran Algorithm 2 after attacking each feature individually with a threshold for the drop in the accuracy set to 4%. Our attacks were sufficient to exceed the threshold for all features. Algorithm 2 correctly detected the attacked feature in 60% of these single-feature attacks. Table 2 shows the result of attacking the three features with the highest importance (according to Table 1). It can be seen that the algorithm correctly detected attacks on the two most important

---

[1] https://scikit-learn.org/stable/index.html.

features $X[5]$ and $X[1]$ (the only two whose importance is above 10%). However, the algorithm failed when the third most important feature $X[4]$ was attacked (it mistook it for the second most important feature $X[1]$). Table 3 shows the performance of Algorithm 2 when the three features with least importance were attacked. The algorithm detected well the attacks on the least important feature $X[8]$ and the third least important feature $X[0]$, but failed for the second least important feature $X[2]$ (which was mistaken for the second most important feature $X[1]$).

A comparison between Table 2 and Table 3 also suffices to see that the drop in the accuracy of the black-box model was greater when the high-importance features were under attack. Indeed, attacks on low-importance features entailed a milder degradation of accuracy. This also confirms that the feature importances identified by Algorithm 1 are coherent with the impact of those features on accuracy.

Table 2. Attacking the three most important features in the synthetic data set

Attacked feature	Black-box accuracy	Feature detected by Algorithm 2
$X[5]$	64.81%	$X[5]$
$X[1]$	64.9%	$X[1]$
$X[4]$	61.97%	$X[1]$

Table 3. Attacking the three least important features in the synthetic data set

Attacked feature	Black-box accuracy	Feature detected by Algorithm 2
$X[8]$	76.45%	$X[8]$
$X[2]$	73.38%	$X[1]$
$X[0]$	75.99%	$X[0]$

5.2   Experiments on Real Numerical Data

To experiment with large real numerical data we used the "PAMAP2 Physical Activity Monitoring" data set from the UCI Machine Learning Repository.[2] This data set contains continuous measurements of 3 inertial body sensors (placed on the arm, chest, and ankle) and a heart-rate monitor worn by 9 subjects who performed 18 different activities such as walking, cycling, watching TV, etc. First, as recommended by the releasers of the activity data set [8], we created our data set by discarding the transient activity (e.g. going to the next activity location). Second, for simplicity, we mapped the various types of activity into two categories indicating whether the activity involved displacement or not (e.g.

---
[2] https://archive.ics.uci.edu/ml/datasets/PAMAP2+Physical+Activity+Monitoring.

walking and cycling were mapped to "displacement" and watching TV to "not displacement"). As a result, we obtained a data set containing $1,942,872$ records of which $1,136,540$ records were labeled as "displacement" and $806,332$ as "not displacement".

Each record contained 54 numerical attributes corresponding to timestamp, label, heart rate, and 17 sensor data feeds for each of the 3 inertial sensors. Given an unlabeled record, the classification task in the experiment consisted in deciding whether the subject was performing at that instant an activity involving physical displacement. We used the same black-box model and random forest model as in the synthetic data set. In the classification of this data set the black-box achieved 99.9% accuracy and the forest 94.6% accuracy. Algorithm 1 computed the importances of the various features on wrong decisions of the black-box model: these importances ranged from 0.1616 down to 0.00017802.

We then applied Algorithm 2 with a threshold of 5% drop in the accuracy. Only 58.97% of the attacks on single features exceeded this threshold. Algorithm 2 correctly detected the attacked feature in 78.26% of these single-feature attacks. Table 4 shows that the attacks on the three most important features were detected correctly. Table 5 shows three examples of attacked features with low importance: two of those attacked features were correctly detected, but the algorithm was not able to detect the attack on acceleration_6_x_chest.

A comparison between Table 4 and Table 5 shows that accuracy drop was greater when attacking a feature with high importance than a feature with low importance. The exception was acceleration_6_x_chest which, in spite of being low-importance, caused a substantial accuracy drop and was in fact misdetected by Algorithm 2 as the high-importance feature acceleration_6_x_ankle. However, other than that our proposed method showed very promising results in detecting the importance of the features causing the black-box wrong predictions on a real numerical data and on detecting attacks on specific features.

Table 4. Attacking the three most important features in the physical activity data set

Attacked feature	Black-box accuracy	Feature detected by Algorithm 2
magnetometer_z_chest	82.66%	magnetometer_z_chest
gyroscope_z_ankle	76.23%	gyroscope_z_ankle
acceleration_6_x_ankle	86.25%	acceleration_6_x_ankle

Table 5. Attacking the three least important features in the physical activity data set

Attacked feature	Black-box accuracy	Feature detected by Algorithm 2
acceleration_6_x_hand	92.47%	acceleration_6_x_hand
acceleration_6_x_chest	86.73%	acceleration_16_x_ankle
gyroscope_x_ankle	90.68%	gyroscope_x_ankle

5.3   Experiments on Real Categorical Data

To experiment on a real data set containing categorical data we used Adult, which is a standard data set hosted in the UCI Machine Learning Repository.[3] Adult contains $48,842$ records of census income information and has 14 attributes reporting both numerical and categorical values. For each categorical attribute, we recoded categories as numbers to obtain a numerical version of the attribute. We reserved $2/3$ of records to train the models, and the remaining $1/3$ to validate them. We used the same black-box and forest as in the synthetic and physical activity data sets. The black-box achieved 84.53% classification accuracy and the forest achieved 84.42%. Table 6 lists the importance of features on the wrong decisions of the black-box model, as computed by Algorithm 1; features are sorted in descending order of importance.

Table 6. Importance of the features of the Adult data set on wrong predictions by the black-box model

Feature name	Feature importance
marital-status	0.2459
capital-gain	0.1967
relationsip	0.1902
educational-num	0.1645
age	0.012
hours-per-week	0.0451
capital-loss	0.0384
occupation	0.021
gender	0.0177
workclass	0.0035
native-country	0.00172
fnlwgt	0.00171
race	0.00094

We applied Algorithm 2 with a threshold of 3% in the accuracy drop. Only 53.84% of the attacks on single attributes exceeded this threshold. Among this 53.84%, Algorithm 2 correctly detected the attacked attribute in 85.71% of the cases. Table 7 shows the results for the three most important features: attacks on them were all correctly detected. Table 8 reports on three features with low importance: two were well detected but capital-loss was misdetected. Yet capital-loss was the only attribute in the data set whose attack exceeded the accuracy drop threshold but was not correctly detected by Algorithm 2.

---

[3] https://archive.ics.uci.edu/ml/datasets/adult.

A comparison between Table 7 and Table 8 also shows that the drop in the accuracy was greater when attacking a feature with high importance (according to Algorithm 1) than a feature with lower importance.

Our approach holds promise, because: (i) in this real categorical data set only one single-feature attack was misdetected by Algorithm 2; (ii) the feature importances computed by Algorithm 1 anticipate the black-box model accuracy drop when each respective feature is attacked.

Table 7. Attacking the three most important features in the Adult data set

Attacked feature	Black-box accuracy	Feature detected by Algorithm 2
martial-status	80.96%	martial-status
capital-gain	63.62%	capital-gain
relationship	77.51%	relationship

Table 8. Attacking the three least important features in the Adult data set

Attacked feature	Black-box accuracy	Feature detected by Algorithm 2
hours-per-week	81.30%	hours-per-week
capital-loss	79.18%	capital-gain
educational-num	76.48%	educational-num

6   Conclusions and Future Research

We have presented an approach based on random decision forests with small tree depth that provide explanations on the decisions made by black-box machine learning models. Specifically, we have focused on investigating and explaining wrong decisions. Algorithm 1 computes the importance of the various features on the wrong black-box model decisions. Additionally, the visualization of the random forest trees affords further understanding of the decision making process. Algorithm 2 introduces a new way to protect against attacks altering the training data, because it allows detecting which features have been attacked.

As future work we intend to test the performance of our approach on other types of attacks such multi-feature random attacks and poisoning attacks. Also, we plan to satisfy more properties among those described in Sect. 2, in order to improve the quality of the explanations of the black-box model decisions.

Acknowledgments. Partial support to this work has been received from the European Commission (project H2020-871042 "SoBigData++"), the Government of Catalonia (ICREA Acadèmia Prize to J. Domingo-Ferrer and grant 2017 SGR 705), and from the Spanish Government (project RTI2018-095094-B-C21 "Consent" and TIN2016-80250-R "Sec-MCloud"). The authors are with the UNESCO Chair in Data Privacy, but the views in this paper are their own and are not necessarily shared by UNESCO.

References

1. Blanco-Justicia, A., Domingo-Ferrer, J., Martínez, S., Sánchez, D.: Machine learning explainability via microaggregation and shallow decision trees. Knowl. Based Syst. **194**, 105532 (2020)
2. European Commission High-Level Expert Group on Artificial Intelligence: Ethics Guidelines for Trustworthy AI (2019). https://ec.europa.eu/futurium/en/ai-alliance-consultation
3. Guidotti, R., Monreale, A., Ruggieri, S., Turini, F., Giannotti, F., Pedreschi, D.: A survey of methods for explaining black box models. ACM Comput. Surv. **51**(5), 1–42 (2018)
4. Ho, T.K.: Random decision forests. In: Proceedings of the 3rd International Conference on Document Analysis and Recognition. IEEE (1995)
5. Konečný, J., McMahan, H.B., Yu, F.X., Richtárik, P., Suresh, A.T., Bacon, D.: Federated learning: strategies for improving communication efficiency (v2) (2017). arXiv:1610.05492
6. Molnar, C.: Interpretable Machine Learning (2019). https://christophm.github.io/interpretable-ml-book/
7. Pedregosa, F., et al.: Scikit-learn: machine learning in Python. J. Mach. Learn. Res. **12**, 2825–2830 (2011)
8. Reiss, A., Stricker, D.: Introducing a new benchmarked data set for activity monitoring. In: 16th International Symposium on Wearable Computers, pp. 108–109. IEEE (2012)
9. Ribeiro, M.T., Singh, S., Guestrin, C.: "Why should I trust you?" Explaining the predictions of any classifier. In: Proceedings of the 22nd ACM SIGKDD International Conference on Knowledge Discovery and Data Mining, pp. 1135–1144 (2016)
10. Shahriari, K., Shahriari, M.: IEEE standard review – ethically aligned design: a vision for prioritizing human wellbeing with artificial intelligence and autonomous systems. In: 2017 IEEE Canada International Humanitarian Technology Conference (IHTC), pp. 197–201. IEEE (2017)
11. Singh, S., Ribeiro, M.T., Guestrin, C.: Programs as black-box explanations (2016). arXiv:1611.07579
12. Strumbelj, E., Kononenko, I.: An efficient explanation of individual classifications using game theory. J. Mach. Learn. Res. **11**, 1–18 (2010)
13. Turner, R.: A model explanation system. In: IEEE 26th International Workshop on Machine Learning for Signal Processing (MLSP), pp. 1–6. IEEE (2016)
14. Voigt, P., Von dem Bussche, A.: The EU General Data Protection Regulation (GDPR): A Practical Guide, 1st edn. Springer, Cham (2017)

Fair-MDAV: An Algorithm for Fair Privacy by Microaggregation

Julián Salas[1,2]([⊠]) and Vladimiro González-Zelaya[3,4]

[1] Internet Interdisciplinary Institute (IN3), Universitat Oberta de Catalunya,
Barcelona, Spain
jsalaspi@uoc.edu
[2] Center for Cybersecurity Research of Catalonia, Barcelona, Spain
[3] Newcastle University, Newcastle upon Tyne, UK
[4] Escuela de Ciencias Económicas y Empresariales, Universidad Panamericana,
Mexico City, Mexico
cvgonzalez@up.edu.mx

Abstract. Automated decision systems are being integrated to several institutions. The General Data Protection Regulation from the European Union, considers the right to explanation on such decisions, but many systems may require a group-level or community-wide analysis. However, the data on which the algorithms are trained is frequently personal data. Hence, the privacy of individuals should be protected, at the same time, ensuring the fairness of the algorithmic decisions made. In this paper we present the algorithm Fair-MDAV for privacy protection in terms of t-closeness. We show that its microaggregation procedure for privacy protection improves fairness through relabelling, while the improvement on fairness obtained equalises privacy guarantees for different groups. We perform an empirical test on Adult Dataset, carrying out the classification task of predicting whether an individual earns $50,000$ per year, after applying Fair-MDAV with different parameters on the training set. We observe that the accuracy of the results on the test set is well preserved, with additional guarantees of privacy and fairness.

Keywords: Fair classification · t-closeness · Fair privacy

1 Introduction

Machine learning models are trained with large amounts of individual data and automated decisions are made on these predictive models. Privacy and fairness arise as important issues that we must consider to protect people from possibly adverse effects of such systems. We must consider the privacy implications of collecting and using personal information together with the biases embedded in datasets and algorithms, and the consequences of the resulting classifications and

This work was supported by the Spanish Government, in part under Grant RTI2018-095094-B-C22 "CONSENT".

segmentation [9]. Privacy and fairness should be considered by design, as they are key principles of the General Data Protection Regulation (GDPR) from EU. Hence, we must solve some relevant questions, such as how privacy protection technologies may improve or impede the fairness of the systems they affect, and many other questions in the research agenda suggested by [10].

In this paper we show that privacy protection and fairness improvement mechanisms can be applied at the same time, having positive interactions between them. We devise an algorithm based on the well-known MDAV [6] algorithm for providing k-anonymity by microaggregation with fairness guarantees, to answer affirmatively the following two questions for this privacy protection method, as stated in [10].

- Does the system provide comparable privacy protections to different groups of subjects?
- Can privacy protection technologies or policies be used or adapted to enhance the fairness of a system?

From the fair classification point of view, our method provides a natural connection between two families of fairness definitions, namely those based on groups and those based on individuals.

1.1   Roadmap

This paper is organised as follows. In Sect. 2, we review the related work and the state of the art on integrated privacy and fairness, then we present definitions and related work on each separately. In Sect. 3, we define our algorithm to provide fair privacy. Our experimental framework is provided in Sect. 4. Finally, we discuss the conclusions of this work and future work in Sect. 5.

2   Related Work on Integrated Privacy and Fairness

There is plenty of work on privacy literature and on fairness but few studies relate both. In this section we present recent works that carry out research on the intersection of both, then we present some related work and definitions for fairness and for privacy. In [14] k-anonymity was used to protect frequent patterns with fairness. While [8] discussed when fairness in classification implies privacy, and how differential privacy may be related to fairness. New models of differential private and fair logistic regression are provided in [30]. A study on how differential privacy may have disparate impact is performed in [22], they carry out an experiment with the two real-world decisions made using U.S. Census data of allocating federal funds and assignment of voting rights benefits. The position paper [10] argues in general for integrating research on fairness and non-discrimination to socio-technical systems that provide privacy protection.

2.1 Fairness Related Work and Definitions

Automated classifiers decide on a variety of tasks ranging from whether bank loans are approved to deciding on the early release of prisoners based on their likelihood to recidivate. These decisions may be made by classifiers learnt from biased data, causing unfair decisions to be made and even causing these biases to be perpetuated. Many different definitions of fairness have been proposed [19], which are sometimes in contrast with one another. A decision rule satisfying one of the definitions may well prove to be very unfair for a different one [5]. For example, determining university admissions through gender quotas may achieve similar rates of acceptance across this variable, but make the acceptance rates for good students of different genders disparate. We suggest that a classifier's unfair behaviour may be corrected through relabelling for the unfavoured group to match the favoured group's positive ratio at the fairlet level. According to [2], one of the main problems with data relabelling is the loss of prediction accuracy caused by such interventions. In Sect. 4, we show that even with the most aggressive correction level, the accuracy-loss due to our method is relatively low. Our method was inspired by fair clustering [1,4]: we evaluate a dataset's *fairlets*—subsets of datapoints located close to each other with respect to a distance function, with a demographic distribution similar to the overall dataset's—which in the clustering context help in obtaining *fair* clusters. We use fairlets with two distinct purposes: to anonymise data and to enhance fairness on classification tasks meant to be learned over the original data. Fair classification requires defining several concepts in order to properly measure a classifier's level of fairness. These definitions, necessary to understand *group fairness*—the family of fairness metrics we have analysed—follow.

Definition 1 (Positive and Negative Label). *A binary classifier's labels may usually take* positive *or* negative *values, referring to how desirable an outcome in predictions may be. These outcomes could be, for instance, whether an application for college admission is successful or not. In this case, the positive label would refer to getting accepted, and the negative one to being rejected.*

Definition 2 (Protected Attribute). *A* protected attribute *(PA) of a dataset, refers to a feature prone to discrimination, due to many possible factors. In our case we will be dealing with a single binary PA, meaning there will only be two PA groups, with every datapoint belonging to one of these groups.*

Definition 3 (Positive Ratio). *We will call the ratio of the number of positive instances divided by the total number of instances in a specific group the* positive ratio *(PR) of the group.*

Definition 4 (Favoured and Unfavoured Groups). *Among the two PA groups, the one having the highest PR will be referred to as the* favoured *(F) group, while the other one will be referred to as the* unfavoured *(U) group. So we will denote by FPR the favoured positive ratio and by UPR the unfavoured positive ratio.*

Fairness Metrics. There are many different fairness definitions [17,19], on this paper we will focus on two of them.

Definition 5 (Demographic Parity). *A classifier satisfies* demographic parity *if the probability of being classified as positive is the same across PA subgroups:*

$$P\left(\hat{Y} = 1 \mid PA = U\right) = P\left(\hat{Y} = 1 \mid PA = F\right). \tag{1}$$

Definition 6 (Equality of Opportunity). *A classifier satisfies* equality of opportunity *if the probability of being classified as positive for true positives is the same across PA subgroups:*

$$P\left(\hat{Y} = 1 \mid Y = 1, \; PA = U\right) = P\left(\hat{Y} = 1 \mid Y = 1, \; PA = F\right). \tag{2}$$

Demographic parity and equality of opportunity may be expressed as the following ratios, respectively:

$$\text{DPR} := \frac{P\left(\hat{Y} = 1 \mid PA = U\right)}{P\left(\hat{Y} = 1 \mid PA = F\right)}, \quad \text{EOR} := \frac{P\left(\hat{Y} = 1 \mid Y = 1, PA = U\right)}{P\left(\hat{Y} = 1 \mid Y = 1, PA = F\right)}.$$

These ratios will measure how fair a classifier's predictions are with respect to a fairness definition: the closer the ratio is to 1, the fairer a classifier will be.

Correcting for Fair Classification. Fairness-aware machine learning is defined by [11] as a set of preprocessing techniques that modify input data so that any classifier trained on such data will be fair. According to [16], there are four main ways in which to make appropriate adjustments to data to enforce fairness: suppressing certain features, also known as *fairness through unawareness* [12], reweighing features [18], resampling data instances [13,15,24,27] and *massaging* variable values [3]; our method belongs to this last category.

2.2   Privacy Related Work and Definitions

There is a vast and growing literature on privacy protection techniques. Regarding data publishing we may consider that the two main privacy models are k-anonymity (and its enhancements) and differential privacy. There are some differences and interactions between them in different contexts, some of them are discussed in [25]. We will focus on t-closeness, which is an improvement of k-anonymity.

References [28,29] showed that removing all the *identifiers* (IDs), such as social security number or name-surname, does not prevent the individual's re-identification. Unique combinations of attribute values, called *quasi-identifiers* (QIs) may be used instead of the IDs for re-identification. After this, individuals' *sensitive attributes* (SAs) are revealed, e.g. salary, medical conditions, etc. To prevent re-identification, they defined k-anonymity as follows.

Definition 7 (k-Anonymity). *A dataset is k-anonymous if each record is indistinguishable from at least other $k-1$ records within the dataset, when considering the values of its QIs.*

There are several techniques for obtaining k-anonymous datasets, e.g. generalisation, suppression and microaggregation.

Definition 8 (Microaggregation). *To obtain microaggregates in dataset with n records, these are combined to form groups g of size at least k (k-groups). For each attribute, the average value over each group is computed and is used to replace each of the original averaged values. Groups are formed using a criterion of maximal similarity.*

The optimal k-partition (from the information loss point of view) is defined to be the one that maximises within-group homogeneity. The higher the within-group homogeneity, the lower the information loss, since microaggregation replaces values in a group by the group centroid. The sum of squares criterion is common to measure homogeneity in clustering. The within-groups sum of squares SSE is defined as:

$$SSE = \sum_{j=1}^{n_g} \sum_{i=1}^{n_j} d(x_{ij}, \overline{x}_j)^2$$

Where x_{ij} denotes the $i-$th record in the $j-$th group, $\overline{x}_j$ is the average record in the group j, n_g is the number of groups and n_j is the number of elements in the $j-$th group.

MDAV (Maximum Distance to AVerage) [6] is an algorithm for k-anonymisation based on microaggregation it has been used extensively and modified in several different ways. Two recent examples are [26], in which it is used for k-anonymisation of dynamic data and other is [23], in which MDAV's efficiency is largely improved without losing precision. By design, k-anonymity guarantees that a user's record is protected from linkage with probability higher than $1/k$. However, if the SAs are not modified, a low variability of SAs in a group will help an adversary to infer the SAs of a user, even without being able to link her. Hence, some modifications to k-anonymity have been done, such as ℓ-diversity [21] and t-closeness [20].

Definition 9 (t-Closeness). *An equivalence class satisfies t-closeness if the distance between the distribution of the SAs of the individuals in the group to the distribution of the SAs in the whole table is not greater than a threshold t. A dataset D (usually a k-anonymous dataset) satisfies t-closeness if all its equivalence classes (k-groups) satisfy t-closeness.*

3 Fair-MDAV

In this section, we present the Fair-MDAV algorithm, designed to improve fairness while giving some guarantees of user's privacy. From now on, we consider that the PA and the label in sense of fairness are also SAs privacy-wise, i.e. neither the PA nor the label should be inferred easily from the QIs. As a concrete example, in a gender study of income, neither the gender nor the income of an individual should be inferred from the data, to prevent from discrimination and from invasion of privacy. On each iteration Fair-MDAV finds the element that is furthest from the average record, generates a fairlet, removes these elements, and iterates. After obtaining all the fairlets, calculates the center of the fairlet (average record) and replaces all the original records by their corresponding center (Algorithm 1). Finally, Fair-MDAV relabels the labels of the records in the unfavored class in each fairlet while the positive ratio for them is less than τ times the positive ratio for the favored class, see Algorithm 2.

The main properties of Fair-MDAV are that it is able to generate fairlets for fair clustering. It provides $t-$closeness regarding the PA by setting the proportion n_U/n_F to approximate the proportion $|U|/|F|$ in the full set, in the more restrictive case, setting $n_U = |U|$ and $n_F = |F|$. Fair-MDAV also provides $(n_U + n_F)$-anonymity by design and it equalises the privacy guarantees for both classes of PA with respect to the SA label, this is obtained by improving fairness with the Tau function, as the UPR is increased by relabelling while it is less than $\tau * FPR$.

3.1 Toy Example

Table 1 illustrates the full process over a simple dataset, consisting of five individuals with one unprotected and one protected attributes, as well as a binary label. The following list summarises the process:

- Algorithm 1 (Fairlets and Centers):
    1. Assume $n_U = n_F = 1$ and $\tau = 1$.
    2. Find $c = mean(\text{Feat})$: $c = 0.424$.
    3. Use farthest point from c w.r.t. Feat to initialise C_1 (id $= \{4\}$).
    4. Add to C_1 the closest points such that n_U and n_F hold (id $= \{4, 1\}$).
    5. Repeat and find $C_2 = \{2, 5\}$.
    6. Since no more clusters are possible, discard $\{3\}$.
    7. Replace Feat with corresponding centroid values.
- Algorithm 2 (Fairness):
    1. Since $FPR(C_1) = 1$ and $UPR(C_1) = 0$, relabel $\{4\}$.
    2. Since $FPR(C_2) = 0$ and $UPR(C_2) = 0$, no relabelling is performed.

4 Experimental Framework

In this section, we describe the experimental framework used to analyse and compare the information-loss induced and the data-utility retained after performing Fair-MDAV.

Algorithm 1: Microaggregation operations of Fair-MDAV

---

1 **Function** `fairlets`(df, n_U, n_F)
  Input: dataframe, size of U, size of F
2   **while** $|\{x \in df : label(x) = U\}| \geq n_U$ *and* $|\{x \in df : label(x) = F\}| \geq n_F$ **do**
3     $x_r = argmax\{d(x, \overline{x})\}$ where $\overline{x} = avg\{x \in df\}$
4     $U = \{x_U\}$ the n_U closest records to x_r such that $labels(x_U) = U$
5     $F = \{x_F\}$ the n_F closest records to x_r such that $labels(x_F) = F$
6     $g = U \cup F$ (must have $n_U + n_F$ records including x_r)
7   Remove g from df
8   Add g to $\mathcal{G}$ list of fairlets.
9   **end**
10   **return** $\mathcal{G}$ list of fairlets
11 **end**
12 **Function** `centers`$(df, \mathcal{G})$
13   **for** $g \in \mathcal{G}$ **do**
14     $\overline{x} = avg\{x \in g\}$ $x = \overline{x}$ for $x \in g \subset df$ (replace record by group centroid)
15   **end**
16   **return** $\overline{df}$
17 **end**

Algorithm 2: Tau function for relabelling

---

1 **Function** `Tau`$(df, \mathcal{G}, \tau)$
  Input: dataframe, clusters, parameter
2   **for** $C \in \mathcal{G}$ **do**
3     $FPR(C) = \left| \frac{\{x \in C | pa(x) = F \text{ and } label(x) = 1\}}{\{x \in C | pa(x) = F\}} \right|$
4     $UPR(C) = \left| \frac{\{x \in C | pa(x) = U \text{ and } label(x) = 1\}}{\{x \in C | pa(x) = U\}} \right|$
5     $U_0 = \{x \in C \mid pa(x) = U \text{ and } label(x) = 0\}$
6   **while** $UPR(C) < \tau * FPR(C)$ *and* $|U_0| > 0$ **do**
7       $label(x) = 1$ for some $x \in U_0$
8   **end**
9   **end**
10   **return** $\widetilde{df}$
11 **end**

Algorithm 3: Pseudo code of Fair-MDAV

---

1 **Function** `Fair-MDAV`(df, n_U, n_F, τ)
  Input: dataframe, size of U, size of F, parameter
2   $\mathcal{G} = $ `fairlets`(df, n_U, n_F)
3   $\overline{df} = $ `centers`$(\mathcal{G}, df)$
4   $\widetilde{df} = $ `tau`$(\overline{df}, \mathcal{G}, \tau)$
  Output: $\widetilde{df}$
5 **end**

Table 1. Toy example. During the first phase of the method, two clusters are detected, imputing the cluster's centroid values of the unprotected attributes (*Feat*) into each cluster's members; the resulting clusters have been coloured green (C_1) and red (C_2), while the blue datapoint is discarded. On the second phase, fairness is corrected by making $UPR \geq FPR$ ($\tau = 1$) for each cluster.

	Original Data				Microaggregation				Fairness Correction		
id	Feat	PA	Label		Feat	PA	Label		Feat	PA	Label
1	0.95	F	1		0.99	F	1		0.99	F	1
2	0.05	F	0		0.04	F	0		0.04	F	0
3	0.06	F	1	$\rightarrow$	*Dropped*			$\rightarrow$	*Dropped*		
4	1.03	U	0		0.99	U	0		0.99	U	1
5	0.03	U	0		0.04	U	0		0.04	U	0

4.1 Data

Our experiments were performed on the *Adult Dataset* [7]; this dataset was chosen since it is commonly used as benchmark in both privacy and fairness literature. It consists of 14 categorical and integer attributes, with 48842 instances. The usual classification task for this dataset is to predict its `label`: whether an individual earns more than $50,000$ per year. For our experiments, we consider the the `sex` attribute to be the PA. Adult is PA-imbalanced, with one third of the instances being *female* and two thirds being *male*. It also presents PA-bias with respect to the `label`, since only 10% of the females are labelled as positive, while the positive proportion for males is 30%. Prior to our experiments, the following preprocessing was applied to data:

– The attribute `fnlwgt` was dropped, since it is a demographics parameter not usually considered for class prediction.
– Numerical attributes (`age`, `education-num`, `capital-gain`, `capital-loss` and `hours-per-week`) were binned into five categories each.
– All variables were one-hot encoded for better classifier compatibility.

4.2 Experiments

There are two important parameters in Fair-MDAV: The size of each fairlet, represented by k and the amount of fairness correction to be introduced, represented by τ. For each of these, we performed an experiment, fixing the remaining parameters.

In both cases, we set the PA unfavoured/favoured proportion in fairlets to $3/7$, since this is very close to the dataset's PA proportion. The dataset was train/test split to an 80/20 proportion. Only the train sets were transformed using Fair-MDAV, and logistic regression classifiers were learnt from the transformed train sets. We generated 10 train sets with $\tau = 1$ and $k = \{10, 20, \ldots, 100\}$ and 11 train sets with $k = 10$ and $\tau = \{0, 0.1, \ldots, 1.0\}$.

Figure 1 shows the resulting performance in predicting both the test and the original train sets. Even though predicting the train set's label is unusual, we believe that in our setting the performance differences between predicting the train and the test sets are an indicator of how well data is anonymised: better performance over the train set would indicate information leakage into the model, while similar performance is indicative of an adequate generalisation. Interestingly, in our experiments the learnt classifiers performed consistently better over the test set than over the train set.

Classifiers learnt from the transformed sets never outperformed the classifier learnt from the unmodified train set (baseline in Fig. 1), as is to be expected. However, the loss in accuracy was always around 2%, with AUC and F1 scores very close to the baseline at parameter values $k = 10, \tau = 1$. As a general trend, we may say that increasing the τ value seems to improve both AUC and F1, while larger k values tend to underperform (as this implies larger groups for privacy protection).

Regarding fairness, we considered metrics associated to demographic parity and equality of opportunity. As may be seen in Fig. 2, classification fairness drops substantially if no fairness-correction is performed, e.g. when $\tau = 0$. However, large τ values, e.g. $\tau \approx 1$ more than make up for this and yield classifiers outperforming the baseline and in the case of EOR getting close to an optimal EOR of 1. Increasing the value of k also seems to have a negative impact over the resulting classifier's fairness, yet fixing $\tau = 1$ consistently produced fairer classifiers than the baseline one. In a trend similar to the one followed by performance metrics, fairness performance on train set predictions was consistently similar to performance on test predictions.

Finally, to measure the information loss associated to increasing k on the first experiment, we evaluated the average distance of datapoints to their fairlet representative. The resulting mean square errors (MSEs) may be observed on Fig. 3. As expected, larger k-values produced a higher MSE.

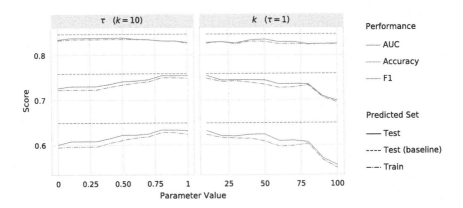

Fig. 1. Accuracy, AUC and F1 scores for different τ and k values, for *test* (as a measure of performance) and *train* (as a potential measure of privacy).

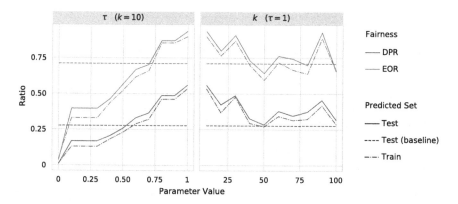

Fig. 2. DPR and EOR for different τ and k values, for *test* (as a measure of performance) and *train* (as a potential measure of privacy).

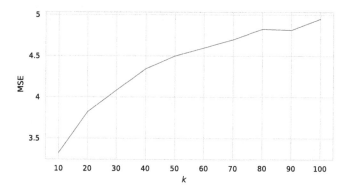

Fig. 3. MSE across k-values for k in $\{10, 20, \ldots, 100\}$.

5 Conclusions and Future Work

In this paper, we have defined the algorithm Fair-MDAV for fair privacy. It provides t–closeness by considering the SAs when performing microaggregation. It integrates a method of relabelling on the microaggregates that improves fairness controlled by a parameter τ. We conducted an empirical evaluation on Adult dataset from US census, and demonstrated that our algorithm is able to reduce information loss, while simultaneously retain data utility and improving fairness in terms of demographic parity and equality of opportunity.

Our microaggregation method generates fairlets that may be used for fair clustering, nonetheless, we leave as future work to compare the quality of the fairlets generated by Fair-MDAV to state-of-the-art fair clustering algorithms. We considered fairness measures of demographic parity and equality of opportunity, still, we would like to extend our results to other measures and apply them on more datasets including synthetically generated data.

References

1. Backurs, A., Indyk, P., Onak, K., Schieber, B., Vakilian, A., Wagner, T.: Scalable fair clustering (2019). arXiv preprint arXiv:1902.03519
2. Berk, R., Heidari, H., Jabbari, S., Kearns, M., Roth, A.: Fairness in criminal justice risk assessments: the state of the art. Sociol. Meth. Res., 0049124118782533 (2018)
3. Chiappa, S., Gillam, T.P.: Path-specific counterfactual fairness (2018). arXiv preprint arXiv:1802.08139
4. Chierichetti, F., Kumar, R., Lattanzi, S., Vassilvitskii, S.: Fair clustering through fairlets. In: Proceedings of the 31st International Conference on Neural Information Processing Systems, NIPS'17, pp. 5036–5044. Curran Associates Inc., Red Hook, NY, USA (2017)
5. Chouldechova, A.: Fair prediction with disparate impact: a study of bias in recidivism prediction instruments. Big Data 5(2), 153–163 (2017)
6. Domingo-Ferrer, J., Torra, V.: Ordinal, continuous and heterogeneous k-anonymity through microaggregation. Data Min. Knowl. Discov. 11(2), 195–212 (2005)
7. Dua, D., Graff, C.: UCI Machine Learning Repository (2017). http://archive.ics.uci.edu/ml
8. Dwork, C., Hardt, M., Pitassi, T., Reingold, O., Zemel, R.: Fairness through awareness. In: Proceedings of the 3rd Innovations in Theoretical Computer Science Conference, ITCS '12, pp. 214–226. Association for Computing Machinery, New York, NY, USA (2012)
9. Dwork, C., Mulligan, D.K.: It's not privacy, and it's not fair. Stan. L. Rev. Online 66, 35 (2013)
10. Ekstrand, M.D., Joshaghani, R., Mehrpouyan, H.: Privacy for all: ensuring fair and equitable privacy protections. In: Friedler, S.A., Wilson, C. (eds.) Proceedings of the 1st Conference on Fairness, Accountability and Transparency. Proceedings of Machine Learning Research, 23–24 February 2018, vol. 81, pp. 35–47. PMLR, New York, NY, USA (2018)
11. Friedler, S.A., Scheidegger, C., Venkatasubramanian, S., Choudhary, S., Hamilton, E.P., Roth, D.: A comparative study of fairness-enhancing interventions in machine learning. In: Proceedings of the Conference on Fairness, Accountability, and Transparency, pp. 329–338. ACM (2019)
12. Gajane, P., Pechenizkiy, M.: On formalizing fairness in prediction with machine learning (2017). arXiv preprint arXiv:1710.03184
13. González Zelaya, C.V., Missier, P., Prangle, D.: Parametrised data sampling for fairness optimisation. In: 2019 XAI Workshop at SIGKDD, Anchorage, AK, USA (2019). http://homepages.cs.ncl.ac.uk/paolo.missier/doc/kddSubmission.pdf
14. Hajian, S., Domingo-Ferrer, J., Monreale, A., Pedreschi, D., Giannotti, F.: Discrimination- and privacy-aware patterns. Data Min. Knowl. Disc. 29(6), 1733–1782 (2014). https://doi.org/10.1007/s10618-014-0393-7
15. Kamiran, F., Calders, T.: Classification with no discrimination by preferential sampling. In: Proceedings of the 19th Machine Learning Conference on Belgium and the Netherlands, pp. 1–6. Citeseer (2010)
16. Kamiran, F., Calders, T.: Data preprocessing techniques for classification without discrimination. Knowl. Inf. Syst. 33(1), 1–33 (2012)
17. Kilbertus, N., Carulla, M.R., Parascandolo, G., Hardt, M., Janzing, D., Schölkopf, B.: Avoiding discrimination through causal reasoning. In: Advances in Neural Information Processing Systems, pp. 656–666 (2017)

18. Krasanakis, E., Spyromitros-Xioufis, E., Papadopoulos, S., Kompatsiaris, Y.: Adaptive sensitive reweighting to mitigate bias in fairness-aware classification. In: Proceedings of the 2018 World Wide Web Conference on World Wide Web, pp. 853–862. International World Wide Web Conferences Steering Committee (2018)
19. Kusner, M.J., Loftus, J., Russell, C., Silva, R.: Counterfactual fairness. In: Advances in Neural Information Processing Systems, pp. 4066–4076 (2017)
20. Li, N., Li, T., Venkatasubramanian, S.: t-closeness: privacy beyond k-anonymity and l-diversity. In: 2007 IEEE 23rd International Conference on Data Engineering, pp. 106–115 (April 2007)
21. Machanavajjhala, A., Kifer, D., Gehrke, J., Venkitasubramaniam, M.: L-diversity: privacy beyond k-anonymity. ACM Trans. Knowl. Discov. Data **1**(1) (2007)
22. Pujol, D., McKenna, R., Kuppam, S., Hay, M., Machanavajjhala, A., Miklau, G.: Fair decision making using privacy-protected data. In: Proceedings of the 2020 Conference on Fairness, Accountability, and Transparency, FAT* '20, pp. 189–199. Association for Computing Machinery, New York, NY, USA (2020)
23. Rodríguez-Hoyos, A., Estrada-Jiménez, J., Rebollo-Monedero, D., Mezher, A.M., Parra-Arnau, J., Forné, J.: The fast maximum distance to average vector (F-MDAV): an algorithm for K-anonymous microaggregation in big data. Eng. Appl. Artif. Intell. **90**, 103531 (2020)
24. Rubin, D.B.: The use of matched sampling and regression adjustment to remove bias in observational studies. Biometrics **29**, 185–203 (1973)
25. Salas, J., Domingo-Ferrer, J.: Some basics on privacy techniques, anonymization and their big data challenges. Math. Comput. Sci. **12**(3), 263–274 (2018)
26. Salas, J., Torra, V.: A general algorithm for k-anonymity on dynamic databases. In: Garcia-Alfaro, J., Herrera-Joancomartí, J., Livraga, G., Rios, R. (eds.) DPM/CBT -2018. LNCS, vol. 11025, pp. 407–414. Springer, Cham (2018). https://doi.org/10.1007/978-3-030-00305-0_28
27. Salimi, B., Rodriguez, L., Howe, B., Suciu, D.: Capuchin: Causal database repair for algorithmic fairness (2019). arXiv preprint arXiv:1902.08283
28. Samarati, P.: Protecting respondents identities in microdata release. IEEE Trans. Knowl. Data Eng. **13**(6), 1010–1027 (2001)
29. Sweeney, L.: k-anonymity: a model for protecting privacy. Int. J. Uncertainty Fuzziness Knowl. Based Syst. **10**(05), 557–570 (2002)
30. Xu, D., Yuan, S., Wu, X.: Achieving differential privacy and fairness in logistic regression. In: Companion Proceedings of the 2019 World Wide Web Conference, WWW '19, pp. 594–599. Association for Computing Machinery, New York, NY, USA (2019)

Author Index

Printed in the United States
by Bookmasters